STUDENT SOLUTIONS MANUAL & STUDY GUIDE

to accompany

Volume 1

PHYSICS

For Scientists and Engineers

Fifth Edition

Serway • Beichner

John R. Gordon
James Madison University

Ralph McGrew
Broome Community College

Raymond A. Serway

with contributions by
Duane Deardorff
North Carolina State University

Saunders College Publishing
A Division of Harcourt College Publishers

Fort Worth   Philadelphia   San Diego   New York   Orlando   Austin
San Antonio   Toronto   Montreal   London   Sydney   Tokyo

Printed in the United States of America

ISBN 0-03-020972-2

012 202 76543

Preface

This <u>Student Solution Manual and Study Guide</u>, has been written to accompany the textbook **Physics for Scientists and Engineers**, Fifth Edition, by Raymond A. Serway and Robert J. Beichner. The purpose of this Student Solution Manual and Study Guide is to provide the students with a convenient review of the basic concepts and applications presented in the textbook, together with solutions to selected end-of-chapter problems from the textbook. This is not an attempt to rewrite the textbook in a condensed fashion. Rather, emphasis is placed upon clarifying typical troublesome points, and providing further drill for methods of problem solving.

Each chapter is divided into several parts, and every textbook chapter has a matching chapter in this book. Very often, reference is made to specific equations or figures in the textbook. Every feature of this Study Guide has been included to insure that it serves as a useful supplement to the textbook. Most chapters contain the following components:

- **Equations and Concepts:** This represents a review of the chapter, with emphasis on highlighting important concepts, and describing important equations and formalisms.

- **Suggestions, Skills, and Strategies:** This offers hints and strategies for solving typical problems that the student will often encounter in the course. In some sections, suggestions are made concerning mathematical skills that are necessary in the analysis of problems.

- **Review Checklist:** This is a list of topics and techniques the student should master after reading the chapter and working the assigned problems.

- **Answers to Selected Conceptual Questions:** Suggested answers are provided for approximately a quarter of the conceptual questions.

- **Solutions to Selected End-of-Chapter Problems:** Solutions are shown for approximately half of the odd-numbered problems from the text which illustrate the important concepts of the chapter.

An important note concerning significant figures: The answers to most of the end-of-chapter problems are stated to three significant figures. For uniformity of style, this practice has been followed even in those cases where the data in the corresponding problem statement is given to only two significant figures. In the latter case, your instructor may request that you "round off" your calculations to two significant figures.

We sincerely hope that this Student Solution Manual and Study Guide will be useful to you in reviewing the material presented in the text, and in improving your ability to solve problems and score well on exams. We welcome any comments or suggestions which could help improve the content of this study guide in future editions; and we wish you success in your study.

John R. Gordon
Harrisonburg, Virginia

Ralph McGrew
Binghamton, New York

Raymond A. Serway
Chapel Hill, North Carolina

Acknowledgments

It is a pleasure to acknowledge the many excellent contributions of Michael Rudmin, of DSC—Publishing, LLC. to this Fifth Edition of the Student Solutions Manual and Study Guide. His graphics skills and technical expertise have combined to produce illustrations which enhance the descriptions and solutions. The attractive layout is a result of his attention to detail; and the overall appearance of the Solution Manual reflects his work on the final camera-ready copy.

Special thanks go to the professional staff at Saunders College Publishing, especially Susan Pashos, Suzanne Hakanen, and Ericka Yeoman for managing all phases of the production of the manual.

Finally, we express our appreciation to our families for their inspiration, patience, and encouragement.

Suggestions for Study

Very often we are asked "How should I study this subject, and prepare for examinations?" There is no simple answer to this question, however, we would like to offer some suggestions which may be useful to you.

1.   It is essential that you understand the basic concepts and principles before attempting to solve assigned problems. This is best accomplished through a careful reading of the textbook before attending your lecture on that material, jotting down certain points which are not clear to you, taking careful notes in class, and asking questions. You should reduce memorization of material to a minimum. Memorizing sections of a text, equations, and derivations does not necessarily mean you understand the material. Perhaps the best test of your understanding of the material will be your ability to solve the problems in the text, or those given on exams.

2.   Try to solve as many problems at the end of the chapter as possible. You will be able to check the accuracy of your calculations to the odd-numbered problems, since the answers to these are given at the back of the text. Furthermore, detailed solutions to approximately half of the odd-numbered problems are provided in this study guide. Many of the worked examples in the text will serve as a basis for your study.

3.   The method of solving problems should be carefully planned. First, read the problem several times until you are confident you understand what is being asked. Look for key words which will help simplify the problem, and perhaps allow you to make certain assumptions. You should also pay special attention to the information provided in the problem. In many cases a simple diagram is a good starting point; and it is always a good idea to write down the given information before proceeding with a solution. After you have decided on the method you feel is appropriate for the problem, proceed with your solution. If you are having difficulty in working problems, we suggest that you again read the text and your lecture notes. It may take several readings before you are ready to solve certain problems, though the solved problems in this Study Guide should be of value to you in this regard. However, your solution to a problem does not have to look just like the one presented here. A problem can sometimes be solved in different ways, starting from different principles. If you wonder about the validity of an alternative approach, ask your instructor.

4.   After reading a chapter, you should be able to define any new quantities that were introduced, and discuss the first principles that were used to derive fundamental formulas. A review is provided in each chapter of the Study Guide for this purpose, and the marginal notes in the textbook (or the index) will help you locate these topics. You should be able to correctly associate with each physical quantity the symbol used to represent that quantity (including vector notation if appropriate) and the SI unit in which the quantity is specified. Furthermore, you should be able to express each important formula or equation in a concise and accurate prose statement.

5.   We suggest that you use this Study Guide to review the material covered in the text, and as a guide in preparing for exams. You should also use the Equations and Concepts to focus in on any points which require further study. Remember that the main purpose of this Study Guide is to improve upon the efficiency and effectiveness of your study hours and your overall understanding of physical concepts. However, it should not be regarded as a substitute for your textbook or individual study and practice in problem solving.

Problem Solving

Besides what you might expect to learn about physics concepts, another very valuable skill you should hope to take away from your physics course is the ability to solve complicated problems. The way physicists approach complex situations and break them down into manageable pieces is extremely useful. Below is a memory aid to help you easily recall the steps required for successful problem solving. When working on problems, the secret is to keep your **GOAL** in mind.

GOAL Problem Solving Steps

Gather information

The first thing to do when approaching a problem is to understand the situation. Carefully read the problem statement, looking for key phrases like "at rest," or "freely falls." What information is given? Exactly what is the question asking? Don't forget to gather information from your own experiences and common sense. What should a reasonable answer look like? You wouldn't expect to calculate the speed of an automobile to be 5×10^6 m / s. Do you know what units to expect? Are there any limiting cases you can consider? What happens when an angle approaches 0° or 90° or a mass gets huge or goes to zero? Also make sure you carefully study any drawings that accompany the problem.

Organize your approach

Once you have a really good idea of what the problem is about, you need to think about what to do next. Have you seen this type of question before? Being able to classify a problem can make it much easier to lay out a plan to solve it. You should almost always make a quick drawing of the situation. Label important events with circled letters. Indicate any known values, perhaps in a table or directly on your sketch. Some kinds of problems require specific drawings, like a free body diagram when analyzing forces. Once you've done this and have a plan of attack, it is time for the next step.

Analyze the problem

Because you have already categorized the problem, it should not be too difficult to select relevant equations that apply to this type of situation. Use algebra (and calculus if necessary) to solve for the unknown variable in terms of what is given. Substitute in the appropriate numbers, calculate the result, and round it to the proper number of significant figures.

Learn from your efforts

This is actually the most important part. Examine your numerical answer. Does it meet your expectations from the first step? What about the algebraic form of the result before you plugged in numbers? Does it make sense? (Try looking at the variables in it to see if the answer would change in a physically meaningful way if they were drastically increased or decreased or even reduced to zero.) Think about how this problem compares to others you have done. How was it similar? In what critical ways did it differ? Why was this problem even assigned? You should have learned something by doing it. Can you figure out what?

For complex problems, you may need to apply these four steps of the GOAL process recursively to subproblems. For very simple problems, you probably don't need this protocol. But when you are looking at a problem and you don't know what to do next, remember what the letters in GOAL stand for and use that as a guide.

Selected problem solutions in this solutions manual follow the GOAL problem solving strategy. These problems should serve as examples of how the GOAL strategy can be applied to nearly any physics problem.

Table of Contents

Chapter	Title	Page
1	Physics and Measurement	2
2	Motion in One Dimension	18
3	Vectors	41
4	Motion in Two Dimensions	59
5	The Laws of Motion	85
6	Circular Motion and Other Applications of Newton's Laws	109
7	Work and Kinetic Energy	129
8	Potential Energy and Conservation of Energy	152
9	Linear Momentum and Collisions	175
10	Rotation of a Rigid Object About a Fixed Axis	202
11	Rolling Motion and Angular Momentum	227
12	Static Equilibrium and Elasticity	253
13	Oscillatory Motion	276
14	The Law of Gravity	301
15	Fluid Mechanics	323
16	Wave Motion	343
17	Sound Waves	361
18	Superposition and Standing Waves	383
19	Temperature	404
20	Heat and the First Law of Thermodynamics	422
21	The Kinetic Theory of Gases	441
22	Heat Engines, Entropy, and the Second Law of Thermodynamics	462

Physics and
Measurement

PHYSICS AND MEASUREMENT

INTRODUCTION

Since the following chapters will be concerned with the laws of physics, we must begin by clearly defining the basic quantities involved in these laws. For example, such physical quantities as force, velocity, volume, and acceleration can be described in terms of more fundamental quantities. In the next several chapters we shall encounter three basic quantities: length (L), mass (M), and time (T). In later chapters we will need to add two other standard units to our list, for temperature (the Kelvin) and for electric current (the ampere). In our study of mechanics, however, we shall be concerned only with the units of length, mass and time.

EQUATIONS AND CONCEPTS

The density of any substance is defined as the ratio of mass to volume. The SI units of density are kg / m^3.

$$\rho \equiv \frac{m}{V} \tag{1.1}$$

The mass per atom for a given element is found by using Avogadro's number, N_A.

$$m_{atom} = \frac{\text{atomic mass of element}}{N_A} \tag{1.2}$$

$$N_A = 6.02 \times 10^{23} \ \frac{\text{atoms}}{\text{mole}}$$

SUGGESTIONS, SKILLS, AND STRATEGIES

GENERAL PROBLEM-SOLVING STRATEGY

Six basic steps are commonly used to develop a problem-solving strategy:

- Read the problem carefully at least twice. Be sure you understand the nature of the problem before proceeding further.

- Draw a suitable diagram with appropriate labels and coordinate axes if needed.

- As you examine what is being asked in the problem, identify the basic physical principle or principles that are involved, listing the knowns and unknowns.

- Select a basic relationship or derive an equation that can be used to find the unknown, and symbolically solve the equation for the unknown.

- Substitute the given values, along with the appropriate units, into the equation.

- Obtain a numerical value for the unknown. The problem is verified if the following questions can be properly answered: Do the units match? Is the answer reasonable? Is the plus or minus sign proper or meaningful?

You should be familiar with some important mathematical techniques:

- Using powers of ten in expressing such numbers as $0.00058 = 5.8 \times 10^{-4}$. Appendix B.1 gives a brief review of such notation, and the algebraic operations of numbers using powers of ten.

- Basic algebraic operations such as factoring, handling fractions, solving quadratic equations, and solving linear equations. For your convenience, some of these techniques are reviewed in Appendix B.2.

- The fundamentals of plane and solid geometry—including the ability to graph functions, calculate the areas and volumes of standard geometric figures and recognize the equations and graphs of a straight line, a circle, an ellipse, a parabola and a hyperbola.

- The basic ideas of trigonometry—definitions and properties of the sine, cosine, and tangent functions; the Pythagorean Theorem, the law of cosines, the law of sines, and some of the basic trigonometric identities.

  For your convenience, reviews of geometry and trigonometry are given in Appendix D of the text.

REVIEW CHECKLIST

▷ Discuss the units of length, mass and time and the standards for these quantities in SI units; and perform a **dimensional analysis** of an equation containing physical quantities whose individual units are known.

▷ **Convert units** from one system to another.

▷ Carry out **order-of-magnitude calculations** or guesstimates.

▷ Become familiar with the meaning of various mathematical symbols and Greek letters. Identify and properly use mathematical notations such as the following: $\propto$ (is proportional to), $<$ (is less than), $\approx$ (is approximately equal to), Δ (change in value), etc.

ANSWERS TO SELECTED CONCEPTUAL QUESTIONS

2. Suppose that the three fundamental standards of the metric system were length, density, and time rather than length, mass, and time. The standard of density in this system is to be defined as that of water. What considerations about water would you need to address to make sure that the standard of density is as accurate as possible?

Answer A number of considerations would be necessary. There are the environmental details related to the water—a standard temperature would have to be defined, as well as a standard pressure. Another consideration is the quality of the water, in terms of defining a lower limit of impurities. Another problem with this scheme is that density cannot be measured directly with a single measurement, as can length, mass and time. As a combination of two measurements (mass, and volume, which itself involves **three** measurements!), it has higher inherent uncertainty than a single measurement.

□   □   □   □

4.   Express the following quantities using the prefixes given in Table 1.4: (a) 3×10^{-4} m (b) 5×10^{-5} s (c) 72×10^2 g.

Answer   (a) 0.300 mm (b) 50.0 μs (c) 7.20 kg.

□   □   □   □

5.   Suppose that two quantities A and B have different dimensions. Determine which of the following arithmetic operations **could** be physically meaningful: (a) $A + B$ (b) A / B (c) $B - A$ (d) AB.

Answer   One cannot add or subtract different dimensions, because the resultant dimension would be unknown. However, different dimensions can be meaningfully multiplied and divided. Therefore, (b) and (d) **could** be physically meaningful.

□   □   □   □

6.   What level of accuracy is implied in an order-of-magnitude calculation?

Answer   An order-of-magnitude calculation implies that even if the exact number is not correct, the answer is within a factor of ten of the "correct" answer.   Thus, if an order-of-magnitude calculation yielded an answer of $(2 \times 10^{11}$ m$)$, one would expect the "correct" answer to be between $(2 \times 10^{10}$ m$)$ and $(2 \times 10^{12}$ m$)$.

□   □   □   □

Chapter 1

SOLUTIONS TO SELECTED END-OF-CHAPTER PROBLEMS

7.   Calculate the mass of an atom of (a) helium, (b) iron, and (c) lead. Give your answers in atomic mass units and in grams. The atomic masses are 4.00, 55.9, and 207 g/mol, respectively, for the atoms given.

Solution

G: Gather information: The mass of an atom of any element is essentially the mass of the protons and neutrons that make up its nucleus since the mass of the electrons is negligible (less than a 0.05% contribution). Since most atoms have about the same number of neutrons as protons, the atomic mass is approximately double the atomic number (the number of protons). We should also expect that the mass of a single atom is a very small fraction of a gram ($\sim 10^{-23}$ g) since one mole (6.02×10^{23}) of atoms has a mass on the order of several grams.

O: Organize: An atomic mass unit is defined as 1/12 of the mass of a carbon-12 atom (which has a molar mass of 12.0 g/mol), so the mass of any atom in atomic mass units is simply the numerical value of the molar mass. The mass in grams can be found by multiplying the molar mass by the mass of one atomic mass unit (u): $1\,u = 1.66 \times 10^{-24}$ g

A: Analyze:

For He, $m = 4.00\,u = (4.00\,u)(1.66 \times 10^{-24}\,g/u) = 6.64 \times 10^{-24}$ g   ◊

For Fe, $m = 55.9\,u = (55.9\,u)(1.66 \times 10^{-24}\,g/u) = 9.28 \times 10^{-23}$ g   ◊

For Pb, $m = 207\,u = (207\,u)(1.66 \times 10^{-24}\,g/u) = 3.44 \times 10^{-22}$ g   ◊

L: Learn: As expected, the mass of the atoms is larger for bigger atomic numbers. If we did not know the conversion factor for atomic mass units, we could use the mass of a proton as a close approximation: $1\,u \approx m_p = 1.67 \times 10^{-24}$ g.

6

13. The displacement of a particle moving under uniform acceleration is some function of the elapsed time and the acceleration. Suppose we write this displacement as $s = ka^m t^n$, where k is a dimensionless constant. Show by dimensional analysis that this expression is satisfied if $m = 1$ and $n = 2$. Can this analysis give the value of k?

Solution

For the equation to be valid, we must choose values of m and n to make it dimensionally consistent. Since s is a displacement, its dimensions are those of length (L). The acceleration, a, is a length divided by the square of a time (L/T²). The variable t has dimensions of time (T), and the constant k has no dimensions. Substituting these dimensions into the equation yields:

$$(L) = \left(\frac{L}{T^2}\right)^m (T)^n = (L)^m (T)^{-2m} (T)^n$$

or

$$(L)^1 (T)^0 = (L)^m (T)^{n-2m}$$

Note that the factor (T)⁰ introduced on the left side of the second equation is equal to 1. This equation can be true only if the powers of length (L) are the same on the two sides of the equation and, simultaneously, the powers of time (T) are the same on both sides. Indeed, if another basic unit such as mass (M) were present, we would also require that its powers be identical on the two sides. Thus, we obtain a set of two simultaneous equations: $1 = m$ and $0 = n - 2m$. The solutions are therefore seen to be: $m = 1$ and $n = 2$ ◊

This technique gives no information about the possible values of the dimensionless constant k. ◊

15. Which of the equations below is dimensionally correct?

(a) $v = v_0 + ax$ (b) $y = (2\text{ m}) \cos(kx)$ where $k = 2\text{ m}^{-1}$

Solution

(a) Write out dimensions for each quantity in the equation $v = v_0 + ax$.

The variables v and v_0 are expressed in units of m/s; therefore, $[v] = [v_0] = LT^{-1}$. The variable a is expressed in units of m/s^2 and $[a] = LT^{-2}$, while the variable x is expressed in meters. Therefore $[ax] = L^2T^{-2}$.

Consider the right-hand member (RHM) of equation (a):

$$[\text{RHM}] = LT^{-1} + L^2T^{-2}$$

Quantities in a sum must be dimensionally the same.
Therefore, equation (a) **is not** dimensionally correct.   ◊

(b) Write out dimensions for each quantity in the equation $y = (2\text{ m})\cos(kx)$

For y, $[y] = L$;   for 2 m, $[2\text{ m}] = L$;   and for (kx), $[kx] = \left[\left(2\text{ m}^{-1}\right)x\right] = L^{-1}L$.

For the left-hand member (LHM) and the right-hand member (RHM) of equation,

$$[\text{LHM}] = [y] = L \qquad\qquad [\text{RHM}] = [2\text{ m}]\left[\cos(kx)\right] = L$$

These are the same, so equation (b) **is** dimensionally correct.   ◊

8

17. The consumption of natural gas by a company satisfies the empirical equation $V = 1.50t + 0.00800t^2$, where V is the volume in millions of cubic feet and t the time in months. **Express this equation in units of cubic feet and seconds. Put the proper units on the coefficients.** Assume that a month is 30.0 days.

Solution Write "millions of cubic feet" as 10^6 ft^3, and use the given units of time and volume to assign units to the equation:

$$V = \left(1.50 \times 10^6 \ \text{ft}^3/\text{mo}\right)t + \left(0.00800 \times 10^6 \ \text{ft}^3/\text{mo}^2\right)t^2$$

To convert the units to seconds, use 1 month = 2.59 × 10^6 s, to obtain:

$$V = \left(1.50 \times 10^6 \ \frac{\text{ft}^3}{\text{mo}}\right)\left(\frac{1 \ \text{mo}}{2.59 \times 10^6 \ \text{s}}\right)t + \left(0.00800 \times 10^6 \ \frac{\text{ft}^3}{\text{mo}^2}\right)\left(\frac{1 \ \text{mo}}{2.59 \times 10^6 \ \text{s}}\right)^2 t^2$$

$$V = \left(0.579 \ \text{ft}^3/\text{s}\right)t + \left(1.19 \times 10^{-9} \ \text{ft}^3/\text{s}^2\right)t^2 \qquad \Diamond$$

---

19. A rectangular building lot is 100 ft by 150 ft. Determine the area of this lot in m^2.

Solution

G: We must calculate the area and convert units. Since a meter is about 3 feet, we should expect the area to be about $A \approx (30 \ \text{m})(50 \ \text{m}) = 1500 \ \text{m}^2$.

O: Area = Length × Width. Use the conversion: 1 m = 3.281 ft.

A: $A = L \times w = (100 \ \text{ft})\left(\dfrac{1 \ \text{m}}{3.281 \ \text{ft}}\right)(150 \ \text{ft})\left(\dfrac{1 \ \text{m}}{3.281 \ \text{ft}}\right) = 1390 \ \text{m}^2 \qquad \Diamond$

L: Our calculated result agrees reasonably well with our initial estimate and has the proper units of m^2. Unit conversion is a common technique that is applied to many problems.

---

25. A solid piece of lead has a mass of 23.94 g and a volume of 2.10 cm³. From these data, calculate the density of lead in SI units (kg/m³).

Solution

G: From Table 1.5, the density of lead is 1.13×10^4 kg / m³, so we should expect our calculated value to be close to this number. This density value tells us that lead is about 11 times denser than water, which agrees with our experience that lead sinks.

O: Density is defined as $\rho = m/V$. We must convert to SI units in the calculation.

A: $\rho = \left(\dfrac{23.94\ \text{g}}{2.10\ \text{cm}^3} \right) \left(\dfrac{1\ \text{kg}}{1000\ \text{g}} \right) \left(\dfrac{100\ \text{cm}}{1\ \text{m}} \right)^3 = 1.14 \times 10^4$ kg / m³   ◊

L: At one step in the calculation, we note that **one million** cubic centimeters make one cubic meter. Our result is indeed close to the expected value. Since the last reported significant digit is not certain, the difference in the two values is probably due to measurement error and should not be a concern. One important common sense check on density values is that objects which sink in water must have a density greater than 1 g/cm³, and objects that float must be less dense than water.

---

29. At the time of this book's printing, the US national debt is about $6 trillion. (a) If payments were made at the rate of $1000/s, how many years would it take to pay off a $6-trillion debt, assuming no interest were charged? (b) A dollar bill is about 15.5 cm long. If 6 trillion dollar bills were laid end to end around the Earth's equator, how many times would they encircle the Earth? Take the radius of the Earth at the equator to be 6378 km. (**Note:** Before doing any of these calculations, try to guess at the answers. You may be very surprised.)

Solution   Part (a):

G: $6 trillion is certainly a large amount of money, so even at a rate of $1000/second, we might guess that it will take a lifetime (~ 100 years) to pay off the debt.

O: The time to repay the debt will be calculated by dividing the total debt by the rate at which it is repaid.

A: (a) $T = \dfrac{\$6 \text{ trillion}}{\$1000 / s} = \dfrac{\$6 \times 10^{12}}{(\$1000 / s)(3.16 \times 10^7 \text{ s} / y)} = 199 \text{ y}$ ◊

L: Our guess was a bit low. $6 trillion really is a lot of money!

Part (b):

G: We might guess that 6 trillion bills would encircle the Earth at least a few hundred times, maybe more since our first estimate was low.

O: The number of bills can be found from the total length of the bills placed end to end divided by the circumference of the Earth.

A: (b) $N = \dfrac{L}{C} = \dfrac{L}{2\pi r} = \dfrac{(\$6 \times 10^{12})(15.5 \text{ cm} / \$)(1 \text{ m} / 100 \text{ cm})}{2\pi(6.378 \times 10^6 \text{ m})} = 2.3 \times 10^4 \text{ times}$ ◊

L: OK, so again our guess was low. Knowing that the bills could encircle the earth more than 20 000 times, it might be reasonable to think that 6 trillion bills could cover the entire surface of the earth, but the calculated result shows that a surprisingly small fraction of the earth's surface area (a band around the equator a mile wide) is covered !

11

31. One gallon of paint (volume = 3.78×10^{-3} m³) covers an area of 25.0 m². What is the thickness of the paint on the wall?

Solution

We assume the paint keeps the same volume in the can and on the wall. The volume of the film on the wall is its surface area multiplied by its thickness t: $V = At$

Therefore,
$$t = \frac{V}{A} = \frac{3.78 \times 10^{-3} \text{ m}^3}{25.0 \text{ m}^2} = 1.51 \times 10^{-4} \text{ m} \qquad \lozenge$$

37. The diameter of our disk-shaped galaxy, the Milky Way, is about 1.0×10^5 light years. The distance to Messier 31 which is in Andromeda, the spiral galaxy nearest to the Milky Way, is about 2.0 million light years. If a scale model represents the Milky Way and Andromeda galaxies as dinner plates 25 cm in diameter, determine the distance between the two plates.

Solution

The scale used in the "dinner plate" model is

$$S = \frac{1.0 \times 10^5 \text{ lightyears}}{25 \text{ cm}} = 4.00 \times 10^3 \frac{\text{lightyears}}{\text{cm}}$$

The distance to Andromeda in the dinner plate model will be

$$D = \frac{2.00 \times 10^6 \text{ lightyears}}{4.00 \times 10^3 \text{ lightyears / cm}} = 5.00 \times 10^2 \text{ cm} = 5.00 \text{ m} \qquad \lozenge$$

39. One cubic meter (1.00 m^3) of aluminum has a mass of 2.70×10^3 kg, and 1.00 m^3 of iron has a mass of 7.86×10^3 kg. Find the radius of a solid aluminum sphere that will balance a solid iron sphere of radius 2.00 cm on an equal-arm balance.

Solution

The requirement of equal masses, $m_{Al} = m_{Fe}$, can be written in terms of density and volume, and the volume can then be written in terms of radius:

$$\rho_{Al} V_{Al} = \rho_{Fe} V_{Fe}$$

Substituting, $\quad \rho_{Al}\left(\frac{4}{3}\pi r_{Al}^3\right) = \rho_{Fe}\left(\frac{4}{3}\pi(2.00 \text{ cm})^3\right)$

Solving, $\quad r_{Al}^3 = \left(\frac{\rho_{Fe}}{\rho_{Al}}\right)(2.00 \text{ cm})^3 = \left(\frac{7.86 \text{ kg} / \text{m}^3}{2.70 \text{ kg} / \text{m}^3}\right)(2.00 \text{ cm})^3 = 23.3 \text{ cm}^3$

Thus, the radius of the aluminum sphere is $\quad\quad\quad\quad r_{Al} = 2.86 \text{ cm} \quad \Diamond$

---

41. Estimate the number of Ping-Pong balls that would fit into an average-size room (without being crushed). In your solution state the quantities you measure or estimate and the values you take for them.

Solution

G: Since the volume of a typical room is much larger than a Ping-Pong ball, we should expect that a very large number of balls (maybe a million) could fit in a room.

O: Since we are only asked to find an estimate, we do not need to be too concerned about how the balls are arranged. Therefore, to find the number of balls we can simply divide the volume of an average-size living room (perhaps $15 \text{ ft} \times 20 \text{ ft} \times 8 \text{ ft}$) by the volume of an individual Ping-Pong ball.

A: We decided a typical room might have dimensions of $15\text{ ft} \times 20\text{ ft} \times 8\text{ ft}$. Using the approximate conversion $1\text{ ft} = 30\text{ cm}$, we find

$$V_{\text{Room}} = (15\text{ ft})(20\text{ ft})(8\text{ ft})(30\text{ cm} / \text{ft})^3 \approx 7 \times 10^7 \text{ cm}^3$$

A Ping-Pong ball has a diameter of about 3 cm, so we can estimate its volume as a cube:

$$V_{\text{ball}} = (3\text{ cm})(3\text{ cm})(3\text{ cm}) \approx 30 \text{ cm}^3$$

The number of Ping-Pong balls that can fill the room is

$$N \approx \frac{V_{\text{Room}}}{V_{\text{ball}}} \approx 2 \times 10^6 \text{ balls} \approx 10^6 \text{ balls} \qquad \lozenge$$

L: So a typical room can hold about a million Ping-Pong balls. This problem gives us a sense of how large a quantity "a million" really is.

---

49. To an order of magnitude, how many piano tuners are there in New York City? The physicist Enrico Fermi was famous for asking questions like this on oral Ph.D. qualifying examinations and for his own facility in making order-of-magnitude calculations.

Solution Assume a total population of 10^7 people. Also, let us estimate that one person in one hundred owns a piano. In addition, assume that in one year a single piano tuner can service about 1000 pianos (about 4 per day for 250 weekdays) assuming each piano is tuned once per year. Therefore,

$$\text{The number of tuners} = \left(\frac{1\text{ tuner}}{1000\text{ pianos}}\right)\left(\frac{1\text{ piano}}{100\text{ people}}\right)(10^7\text{ people}) \approx 100\text{ tuners} \quad \lozenge$$

---

61. A high fountain of water is located at the center of a circular pool as in Figure P1.61. Not wishing to get his feet wet, a student walks around the pool and measures its circumference to be 15.0 m. Next, the student stands at the edge of the pool and uses a protractor to gauge the angle of elevation of the top of the fountain to be 55.0°. How high is the fountain?

Solution Define a right triangle whose legs represent the height and radius of the fountain. From the dimensions of the triangle and fountain, the circumference

$$C = 2\pi r \quad \text{and} \quad \tan\theta = h/r$$

Therefore, $h = r\tan\theta = \left(\dfrac{C}{2\pi}\right)\tan\theta$

or $\quad h = \dfrac{15.0 \text{ m}}{2\pi} \tan 55.0° = 3.41 \text{ m} \quad \lozenge$

Figure P1.61

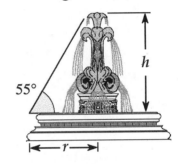

---

63. A useful fact is that there are about $\pi \times 10^7$ s in one year. Find the percentage error in this approximation, where "percentage error" is defined as

$$\text{Percentage error} = \frac{|\text{Assumed value} - \text{true value}|}{\text{True value}} \times 100\%$$

Solution First evaluate the "true value." Remember that every fourth year is a leap year; therefore there are 365.25 days in an average year.

$$1 \text{ y} = (1 \text{ y})\left(\frac{365.25 \text{ d}}{1 \text{ y}}\right)\left(\frac{24 \text{ h}}{1 \text{ d}}\right)\left(\frac{3600 \text{ s}}{1 \text{ h}}\right) = 3.1558 \times 10^7 \text{ s}$$

Now substitute this value, and the assumed value of $\pi \times 10^7 \text{ s} = 3.1416 \times 10^7 \text{ s}$ into the equation given in the problem statement:

$$\text{Percentage error} = \frac{(3.1558 - 3.1416) \times 10^7}{3.1558 \times 10^7} \times 100\% = 0.449\% \text{ too high} \quad \lozenge$$

67. Assume that there are 100 million passenger cars in the United States and that the average fuel consumption rate is 20 mi/gal of gasoline. If the average distance traveled by each car is 10000 mi/yr, how much gasoline would be saved per year if average fuel consumption could be increased to 25 mi/gal?

Solution   We define an average fuel consumption rate based upon the total miles driven, and suppose that this rate is currently at 20 mi/gallon.

$$\text{fuel consumed} = \frac{\text{total miles driven}}{\text{average fuel consumption rate}} \qquad \text{or} \qquad f = \frac{s}{c}$$

Since the total number of miles driven is the same in each case,

At 20 miles/gal, $\qquad \Delta f = \dfrac{\left(100 \times 10^6 \text{ cars}\right)\left(10^4 \text{ (mi/yr)/car}\right)}{20 \text{ mi/gal}} = 5 \times 10^{10} \text{ gal / yr}$

At 25 miles/gal, $\qquad \Delta f = \dfrac{\left(100 \times 10^6 \text{ cars}\right)\left(10^4 \text{ (mi/yr)/car}\right)}{25 \text{ mi/gal}} = 4 \times 10^{10} \text{ gal / yr}$

Thus, we estimate $\quad \Delta f = 1 \times 10^{10} \text{ gal/yr}$ $\hfill \lozenge$

Motion in One Dimension

MOTION IN ONE DIMENSION

INTRODUCTION

As a first step in studying mechanics, it is convenient to describe motion in terms of space and time, ignoring for the present the agents that caused that motion. This portion of mechanics is called **kinematics**. In this chapter we consider motion along a straight line, that is, one-dimensional motion. Starting with the concept of displacement, discussed in the previous chapter, we first define velocity and acceleration. Then, using these concepts, we study the motion of objects traveling in one dimension under a constant acceleration. In many situations, we can treat the moving object as a particle, which in mathematics is defined as a point having no size. In Chapter 4 we shall discuss the motions of objects in two dimensions.

EQUATIONS AND CONCEPTS

The **displacement** Δx of a particle moving from position x_i to position x_f equals the final coordinate minus the initial coordinate.

$$\Delta x \equiv x_f - x_i \qquad (2.1)$$

Displacement should not be confused with distance traveled. When $x_f = x_i$, the displacement is zero; however, if the particle leaves x_i, travels along a path, and returns to x_f, the distance traveled will not be zero.

The average velocity of an object during a time interval is the ratio of the total displacement to the time interval during which the displacement occurred. Note that the instantaneous velocity of the object during the time interval might have different values from instant to instant.

$$\bar{v}_x = \frac{\Delta x}{\Delta t} \qquad (2.2)$$

The average speed of a particle is defined as the ratio of the total distance traveled to the time required to travel that distance.

The **instantaneous velocity** v is defined as the limit of the ratio $\Delta x / \Delta t$ as Δt approaches zero. The instantaneous velocity can be positive, negative, or zero.

$$v_x \equiv \lim_{\Delta t \to 0} \frac{\Delta x}{\Delta t} = \frac{dx}{dt} \qquad (2.4)$$

The average acceleration of an object during a time interval is the ratio of the change in velocity to the time interval during which the change in velocity occurs.

$$\bar{a}_x = \frac{\Delta v_x}{\Delta t} = \frac{v_{xf} - v_{xi}}{t_f - t_i} \qquad (2.5)$$

The **instantaneous acceleration** a is defined as the limit of the ratio $\Delta v_x / \Delta t$ as Δt approaches zero.

$$a_x \equiv \lim_{\Delta t \to 0} \frac{\Delta v_x}{\Delta t} = \frac{dv_x}{dt} \qquad (2.6)$$

Equations 2.8 - 2.12 are called the equations of kinematics, and can be used to describe one-dimensional motion along the x axis with constant acceleration. Note that each equation shows a different relationship among physical quantities: initial velocity, final velocity, acceleration, time, and position.

$$v_{xf} = v_{xi} + a_x t \tag{2.8}$$

$$\bar{v}_x = \frac{v_{xi} + v_{xf}}{2} \tag{2.9}$$

$$x_f - x_i = \bar{v}_x t = \tfrac{1}{2}(v_{xi} + v_{xf})t \tag{2.10}$$

$$x_f - x_i = v_{xi} t + \tfrac{1}{2}a_x t^2 \tag{2.11}$$

$$v_{xf}^2 = v_{xi}^2 + 2a_x \left(x_f - x_i\right) \tag{2.12}$$

We can apply Equations 2.8 - 2.12 to an object in free fall by noting that such an object has an acceleration $(\mathbf{a} = -g\mathbf{j})$ along the y axis, with the $+y$ axis predefined to point upwards.

$$v_{yf} = v_{yi} - gt$$

$$y_f - y_i = \tfrac{1}{2}(v_{yf} + v_{yi})t$$

$$y_f - y_i = v_y t - \tfrac{1}{2}gt^2$$

$$v_{yf}^2 = v_{yi}^2 - 2g(y_f - y_i)$$

SUGGESTIONS, SKILLS, AND STRATEGIES

The following procedure is recommended for solving problems involving accelerated motion:

1. Make sure all the units in the problem are consistent. That is, if distances are measured in meters, be sure that velocities have units of m/s and accelerations have units of m/s^2.

2. Choose a coordinate system.

3. Make a list of all the quantities given in the problem and a separate list of those to be determined.

4. Select from the list of kinematic equations the one or ones that will enable you to determine the unknowns.

5. Construct an appropriate motion diagram and check to see if your answers are consistent with the diagram.

REVIEW CHECKLIST

▷   Define the displacement and average velocity of a particle in motion. (Section 2.1)

▷   Define the instantaneous velocity and understand how this quantity differs from average velocity. (Section 2.2)

▷   Define average acceleration and instantaneous acceleration. (Section 2.3)

▷   Construct a graph of position versus time (given a function such as $x = 5 + 3t - 2t^2$) for a particle in motion along a straight line. From this graph, you should be able to determine both average and instantaneous values of velocity by calculating the slope of the tangent to the graph. (Section 2.2)

▷   Describe what is meant by a body in **free fall** (one moving under the influence of gravity--where air resistance is neglected). Recognize that the equations of kinematics apply directly to a freely falling object and that the acceleration is then given by $a = -g$ (where $g = 9.80 \, \text{m} / \text{s}^2$). (Section 2.6)

▷   Apply the equations of kinematics to any situation where the motion occurs under constant acceleration.

ANSWERS TO SELECTED CONCEPTUAL QUESTIONS

9. A student at the top of a building of height h throws one ball upward at the top of a building of height h throws one ball upward with an initial speed v_{yi} and then throws a second ball downward with the same initial speed. How do the final speeds of the balls compare when they reach the ground?

Answer They are the same, because after the first ball reaches its apex and falls back downwards past the student, it will have a downwards velocity of v_{yi}. This velocity is the same velocity as the second ball, so we can expect that the final speed of each ball will also be the same.

□ □ □ □

11. If the average velocity of an object is zero in some time interval, what can you say about the displacement of the object for that interval?

Answer The displacement is **zero**, since the displacement is proportional to average velocity.

□ □ □ □

13. Two cars are moving in the same direction in parallel lanes along a highway. At some instant , the velocity of car A exceeds the velocity of car B. Does this mean that the acceleration of car A is greater than that of car B? Explain.

Answer No. If Car A has been traveling with cruise control, its velocity will be high (60 mph), but its acceleration will be close to zero. If Car B is pulling onto the highway, it's velocity is likely to be low (30 mph), but its acceleration will be high.

□ □ □ □

15. Consider the following combinations of signs and values for velocity and acceleration of a particle with respect to a one-dimensional x axis:

	Velocity	Acceleration
(a)	Positive	Positive
(b)	Positive	Negative
(c)	Positive	Zero
(d)	Negative	Positive
(e)	Negative	Negative
(f)	Negative	Zero
(g)	Zero	Positive
(h)	Zero	Negative

Describe what the particle is doing in each case, and give a real-life example for an automobile on an east-west one-dimensional axis, with east considered to be the positive direction.

Answer

(a) The particle is moving to the right (in the $+x$ direction) since the velocity is positive, and its speed is increasing, because the acceleration is in the same direction as the velocity. This would be the case in an automobile that is moving toward the east, starting up after waiting for a red light.

(b) The particle is moving to the right, since the velocity is positive, and its speed is decreasing, because the acceleration is in the opposite direction to the velocity. The example for this would be an automobile is moving toward the east, slowing down in preparation for a red light.

(c) The particle is moving to the right, since the velocity is positive, and its speed is constant, because the acceleration is zero. For example, an automobile with its cruise control on is moving eastward along a freeway.

(d) The particle is moving to the left (in the $-x$ direction), since the velocity is negative, and its speed is decreasing, because the acceleration is in the opposite direction to the velocity. An example of this is an automobile that is moving toward the west, slowing down in preparation for a red light.

(e) The particle is moving to the left (in the $-x$ direction), since the velocity is negative, and its speed is increasing, because the acceleration is in the same direction as the velocity. This case might be an automobile moving toward the west, starting up after waiting for a red light.

(f) The particle is moving to the left (in the $-x$ direction), since the velocity is negative, and its speed is constant, because the acceleration is zero. In this case, the example automobile is moving toward the west, moving at constant speed on a straight freeway.

(g) The particle is momentarily at rest, and, in the next instant, will be moving to the right. This situation can only exist for an instant, since as soon as the particle begins moving, the velocity will no longer be zero. This is the situation for an automobile just before it begins to move to the east from a standstill at a red light.

(h) The particle is momentarily at rest, and, in the next instant, will be moving to the left. This situation can only exist for an instant, since as soon as the particle begins moving, the velocity will no longer be zero. This is the situation for an automobile just before it begins to move to the west from a standstill at a red light.

SOLUTIONS TO SELECTED END-OF-CHAPTER PROBLEMS

3. The displacement versus time for a certain particle moving along the x axis is shown in Figure P2.3. Find the average velocity in the time intervals (a) 0 to 2 s, (b) 0 to 4 s, (c) 2 s to 4 s, (d) 4 s to 7 s, (e) 0 to 8 s.

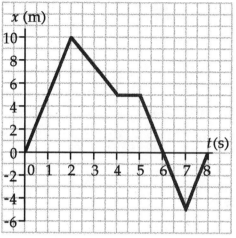

Figure P2.3

Solution On this graph, we can tell positions to two significant figures:

(a) $x = 0$ at $t = 0$ and $x = 10.0$ m at $t = 2$ s

$$\bar{v}_x = \frac{\Delta x}{\Delta t} = \frac{10.0 \text{ m} - 0}{2 \text{ s} - 0} = 5.00 \text{ m/s} \quad \lozenge$$

(b) $x = 5.0$ m at $t = 4$ s

$$\bar{v}_x = \frac{\Delta x}{\Delta t} = \frac{5.0 \text{ m} - 0}{4 \text{ s} - 0} = 1.25 \text{ m/s} \quad \lozenge$$

(c) $\bar{v}_x = \frac{\Delta x}{\Delta t} = \frac{5.0 \text{ m} - 10.0 \text{ m}}{4 \text{ s} - 2 \text{ s}} = -2.50 \text{ m/s} \quad \lozenge$

(d) $\bar{v}_x = \frac{\Delta x}{\Delta t} = \frac{-5.0 \text{ m} - 5.0 \text{ m}}{7 \text{ s} - 4 \text{ s}} = -3.33 \text{ m/s} \quad \lozenge$

(e) $\bar{v}_x = \frac{\Delta x}{\Delta t} = \frac{0.0 \text{ m} - 0.0 \text{ m}}{8 \text{ s} - 0 \text{ s}} = 0 \text{ m/s} \quad \lozenge$

5. A person walks first at a constant speed of 5.00 m/s along a straight line from point A to point B and then back along the line from B to A at a constant speed of 3.00 m/s. What are (a) her average speed over the entire trip and (b) her average velocity over the entire trip?

Solution

(a) The average speed during any time interval is equal to the total distance of travel divided by the total time:

$$\bar{v}_x = \frac{\Delta x}{\Delta t} = \frac{d_{AB} + d_{BA}}{t_{AB} + t_{BA}}$$

But $\qquad d_{AB} = d_{BA} = d, \quad t_{AB} = d/v_{AB}, \qquad$ and $\qquad t_{BA} = d/v_{BA}$

so $\qquad \bar{v}_x = \frac{d+d}{(d/v_{AB}) + (d/v_{AB})} = \frac{2(v_{AB})(v_{BA})}{v_{AB} + v_{BA}}$

so $\qquad \bar{v}_x = 2\left[\frac{(5.00 \text{ m}/\text{s})(3.00 \text{ m}/\text{s})}{5.00 \text{ m}/\text{s} + 3.00 \text{ m}/\text{s}}\right] = 3.75 \text{ m}/\text{s}$ ◊

(b) The average velocity during any time interval equals total displacement divided by elapsed time.

$$\bar{\mathbf{v}} = \frac{\Delta \mathbf{x}}{\Delta t}$$

Since the walker returns to the starting point, $\Delta \mathbf{x} = 0$ and $\bar{\mathbf{v}} = 0$. ◊

26

9. A position–time graph for a particle moving along the x axis is shown in Figure P2.9. (a) Find the average velocity in the time interval $t = 1.5$ s to $t = 4.0$ s. (b) Determine the instantaneous velocity at $t = 2.0$ s by measuring the slope of the tangent line shown in the graph. (c) At what value of t is the velocity zero?

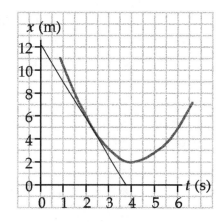

Figure P2.9

Solution   (a)   From the graph:

At $t_i = 1.5$ s, $x_i = 8.0$ m

At $t_f = 4.0$ s, $x_f = 2.0$ m

Thus, $\overline{v} = \dfrac{\Delta x}{\Delta t} = \dfrac{2.0 \text{ m} - 8.0 \text{ m}}{4.0 \text{ s} - 1.5 \text{ s}} = -2.40 \text{ m / s}$ ◊

(b)   Choose two points along line which is tangent to the curve at $t = 2.0$ s. We will use the two points:

$$\left(t_i = 1.0 \text{ s}, \ x_i = 9.0 \text{ m}\right) \quad \text{and} \quad \left(t_f = 3.5 \text{ s}, \ x_f = 1.0 \text{ m}\right)$$

Instantaneous velocity equals the slope of the tangent line, so

$$v_x = \frac{x_f - x_i}{t_f - t_i} = \frac{1.0 \text{ m} - 9.0 \text{ m}}{3.5 \text{ s} - 1.0 \text{ s}} = -3.20 \text{ m / s}$$

The negative sign indicates that the **direction** of **v** is in the negative x direction.   ◊

(c)   The velocity will be zero when the slope of the tangent line is zero. This occurs for the point on the graph where x has its minimum value. Therefore,

$$v_x = 0 \quad \text{at} \quad t \cong 4 \text{ s} \quad \text{(visible to 1 significant figure)} \qquad ◊$$

17. A particle moves along the x axis according to the equation $x = 2.00 + 3.00t - t^2$, where x is in meters and t is in seconds. At $t = 3.00$ s, find (a) the position of the particle, (b) its velocity, and (c) its acceleration.

Solution With the position given by $x = 2.00 + 3.00t - t^2$, we can use the rules for differentiation to write expressions for the velocity and acceleration as functions of time:

$$v_x = \frac{dx}{dt} = 3.00 - 2.00t \qquad \text{and} \qquad a_x = \frac{dv_x}{dt} = -2.00$$

Now we can evaluate x, v_x, and a_x at $t = 3.00$ s.

(a) $x = (2.00 \text{ m}) + (3.00 \text{ m} / \text{s})(3.00 \text{ s}) - (1.00 \text{ m} / \text{s}^2)(3.00 \text{ s})^2 = 2.00 \text{ m}$ ◊

(b) $v_x = (3.00 \text{ m} / \text{s}) - (2.00 \text{ m} / \text{s}^2)(3.00 \text{ s}) = -3.00 \text{ m} / \text{s}$ ◊

(c) $a_x = -2.00 \text{ m} / \text{s}^2$ ◊

---

25. A body moving with uniform acceleration has a velocity of 12.0 cm/s in the positive x direction when its x coordinate is 3.00 cm. If its x coordinate 2.00 s later is –5.00 cm, what is the magnitude of its acceleration?

Solution

G: Since the object must slow down as it moves to the right and then speeds up to the left, the acceleration must be negative and should have units of cm/s^2.

O: First we should sketch the problem to see what is happening (shown at the right). We can see that the object travels along the x-axis, first to the right, slowing down, then speeding up as it travels to

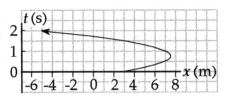

the left in the negative x direction. We can show position and velocity as a function of time with the notation: $x(t)$ and $v(t)$

$$x_i = x(0) = 3.00 \text{ cm} \qquad x_f = x(2.00) = -5.00 \text{ cm} \qquad v_i = v(0) = 12.0 \text{ cm / s}$$

A: Use the kinematic equation $x_f - x_i = v_{xi}t + \frac{1}{2}a_x t^2$, and solve for a_x:

$$a_x = 2\left(x_f - x_i - v_{xi} t\right)\big/t^2$$

$$a_x = \frac{2\left(-5.00 \text{ cm} - 3.00 \text{ cm} - (12.0 \text{ cm / s})(2.00 \text{ s})\right)}{(2.00 \text{ s})^2} = -16.0 \text{ cm / s}^2 \qquad \Diamond$$

L: The acceleration is negative as expected and it has the correct units of cm/s^2. It also makes sense that the magnitude of the acceleration must be greater than 12 cm/s^2 since this is the acceleration that would cause the object to stop after 1 second and just return the object to its starting point after 2 seconds.

---

31. A jet plane lands with a speed of 100 m/s and can accelerate at a maximum rate of -5.00 m/s² as it comes to rest. (a) From the instant the plane touches the runway, what is the minimum time needed before it can come to rest? (b) Can this plane land on a small tropical island airport where the runway is 0.800 km long?

Solution

(a) Assume that the acceleration of the plane is **constant** at the maximum rate, $a_x = -5.00 \text{ m / s}^2$. Given $v_{xi} = 100 \text{ m}$ and $v_{xf} = 0$, use the equation $v_{xf} = v_{xi} + a_x t$ and solve for t:

$$t = \frac{v_{xf} - v_{xi}}{a_x} = \frac{0 - 100 \text{ m / s}}{-5.00 \text{ m / s}^2} = 20.0 \text{ s} \qquad \Diamond$$

(b) Find the required stopping distance and compare this to the length of the runway. Taking x_i to be zero, we get

$$v_{xf}^2 = v_{xi}^2 + 2a_x x \qquad \text{or} \qquad x_f = \frac{v_{xf}^2 - v_{xi}^2}{2a_x} = \frac{0 - (100 \text{ m / s})^2}{2(-5.00 \text{ m / s}^2)} = 1000 \text{ m}$$

The stopping distance is greater than the length of the runway; therefore the plane **cannot land.** ◊

---

37. For many years, the world's land-speed record was held by Colonel John P. Stapp, USAF. On March 19, 1954, he rode a rocket-propelled sled that moved down a track at 632 mi/h. He and the sled were safely brought to rest in 1.40 s. Determine (a) the acceleration he experienced and (b) the distance he traveled during this acceleration.

Solution We assume the acceleration remains constant during the 1.40 s period of negative acceleration.

$$v_{xi} = 632 \; \frac{\text{mi}}{\text{h}} \; = \; 632 \; \frac{\text{mi}}{\text{h}} \left(\frac{1609 \text{ m}}{1 \text{ mi}} \right) \left(\frac{1 \text{ h}}{3600 \text{ s}} \right) = 282 \text{ m / s}$$

(a) Taking $v_{xf} = v_{xi} + a_x t$ with $v_{xf} = 0$,

$$a_x = \frac{v_{xf} - v_{xi}}{t} = \frac{0 - 282 \text{ m / s}}{1.40 \text{ s}} = -202 \text{ m / s}^2 \qquad \qquad ◊$$

This is approximately 20*g* !

(b) $$x_{xf} - x_{xi} = \tfrac{1}{2}(v_{xi} + v_{xf})t = \tfrac{1}{2}(282 \text{ m / s} + 0)(1.40 \text{ s}) = 198 \text{ m} \qquad ◊$$

---

43. A student throws a set of keys vertically upward to her sorority sister, who is in a window 4.00 m above. The keys are caught 1.50 s later by the sister's outstretched hand. (a) With what initial velocity were the keys thrown? (b) What was the velocity of the keys just before they were caught?

Solution

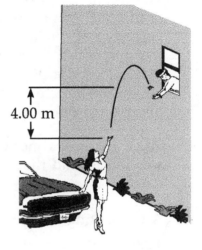

4.00 m

(a) Taking $y_i = 0$ (at the position of the thrower) and given that $y_f = 4.00$ m at $t = 1.50$ s, we find (with $a_y = -9.80 \text{ m / s}^2$):

For a falling object, $y_f = y_i + v_{yi}t + \frac{1}{2}a_y t^2$

Or
$$v_{yi} = \left(y_f - \frac{1}{2}a_y t^2\right)/t$$

$$v_{yi} = \frac{4.00 \text{ m} - \frac{1}{2}(-9.80 \text{ m / s}^2)(1.50 \text{ s})^2}{1.50 \text{ s}}$$

Thus, $v_{yi} = 10.0$ m / s   ◊

(b) The velocity at any time $t > 0$ is given by $v_{xf} = v_{xi} + a_x t$. Therefore, at $t = 1.50$ s,

$$v_{xf} = 10.0 \text{ m / s} - \left(9.80 \text{ m / s}^2\right)(1.50 \text{ s}) = -4.68 \text{ m / s}$$ ◊

The negative sign means that the keys are moving **downward** just before they are caught.

47. A baseball is hit so that it travels straight upward after being struck by the bat. A fan observes that it requires 3.00 s for the ball to reach its maximum height. Find (a) its initial velocity and (b) the maximum height it reaches.

Solution

G: The initial speed of the ball is probably somewhat greater than the speed of the pitch, which might be about 60 mph (~30 m/s), so an initial upward velocity off the bat of somewhere between 20 and 100 m/s would be reasonable. We also know that the length of a ball field is about 300 ft. (~100m), and a pop-fly usually does not go higher than this distance, so a maximum height of 10 to 100 m would be reasonable for the situation described in this problem.

O: Since the ball's motion is entirely vertical, we can use the equation for free fall to find the initial velocity and maximum height from the elapsed time.

A: After leaving the bat, the ball is in free fall for $t = 3.00$ s and has constant acceleration, $a = -g = -9.80 \text{ m}/\text{s}^2$. Solve the equation $v_{yf} = v_{yi} + a_y t$ with $a_y = -g$ to obtain v_{yi} when $v_{yf} = 0$ (and the ball reaches its maximum height).

(a) $v_{yi} = v_{yf} + gt = 0 + \left(9.80 \text{ m}/\text{s}^2\right)(3.00 \text{ s}) = 29.4 \text{ m}/\text{s}$ (upward) ◊

(b) The maximum height in the vertical direction is $y_f = v_{yi}t - \frac{1}{2}gt^2$:

$$y_f = (29.4 \text{ m}/\text{s})(3.00 \text{ s}) - \frac{1}{2}\left(9.80 \text{ m}/\text{s}^2\right)(3.00 \text{ s})^2 = 44.1 \text{ m}$$ ◊

L: The calculated answers seem reasonable since they lie within our expected ranges, and they have the correct units and direction. We must remember that it is possible to solve a problem like this correctly, yet the answers may not seem reasonable simply because the conditions stated in the problem may not be physically possible (e.g. a time of 10 seconds for a pop fly would not be realistic).

49. A daring ranch hand sitting on a tree limb wishes to drop vertically onto a horse galloping under the tree. The speed of the horse is 10.0 m/s, and the distance from the limb to the saddle is 3.00 m. (a) What must be the horizontal distance between the saddle and limb when the ranch hand makes his move? (b) How long is he in the air?

Solution

The most convenient way to solve this problem is to first solve part (b) and use the answer from that part in the solution to part (a) as shown below.

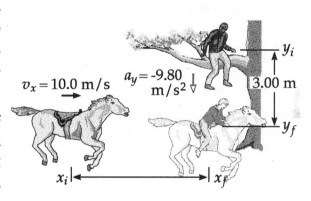

(b) Consider the vertical motion of the man after leaving the limb (with $v_{yi} = 0$ at $y_i = 3.00$ m) until reaching the saddle (at $y_f = 0$). We find his time of fall from

$$y_f - y_i = v_{yi}\,t + \tfrac{1}{2}a_y\,t^2$$

When $v_{yi} = 0$,

$$t = \sqrt{\frac{2(y_f - y_i)}{a_y}} = \sqrt{\frac{2(-3.00\text{ m})}{-9.80\text{ m/s}^2}} = 0.782\text{ s} \qquad \diamond$$

(a)   Since the horse moves with constant velocity ($a_x = 0$), the horizontal distance it travels during the 0.782 s the ranch hand is falling (and therefore the horizontal distance that should exist between him and the saddle when he makes his move) is given by:

$$x_f - x_i = v_x\,t = (10.0\text{ m / s})(0.782\text{ s}) = 7.82\text{ m} \qquad \diamond$$

53. Automotive engineers refer to the time rate of change of acceleration as the "jerk." If an object moves in one dimension such that its jerk J is constant, (a) determine expressions for its acceleration a_x, velocity v_x, and position x, given that its initial acceleration, speed, and position are a_{xi}, v_{xi}, and x_i, respectively. (b) Show that $a_x^2 = a_{xi}^2 + 2J(v_x - v_{xi})$.

Solution   We are given $J = da_x/dt =$ constant, so we know that $da_x = J\,dt$.

(a)   Integrating, $a_x = J \int dt = Jt + c_1$

    When $t = 0$, $a_x = a_{xi}$: $c_1 = a_{xi}$

    Therefore, $a_x = Jt + a_{xi}$ ◊

    From $a_x = dv_x/dt$, $dv_x = a_x dt$

    Integration yields $v_x = \int a_x dt = \int (Jt + a_{xi})dt = \frac{1}{2}Jt^2 + a_{xi}t + c_2$

    When $t = 0$, $v_x = v_{xi}$: $c_2 = v_{xi}$

    and $v_x = \frac{1}{2}Jt^2 + a_{xi}t + v_{xi}$ ◊

    From $v_x = dx/dt$, $dx = v_x dt$: $x = \int v_x dt = \int \left(\frac{1}{2}Jt^2 + a_{xi}t + v_{xi}\right)dt$

    and $x = \frac{1}{6}Jt^3 + \frac{1}{2}a_{xi}t^2 + v_{xi}t + c_3$

    Since $x = x_i$ when $t = 0$, $c_3 = x_i$

    Therefore, $x = \frac{1}{6}Jt^3 + \frac{1}{2}a_{xi}t^2 + v_{xi}t + x_i$ ◊

(b)   Squaring the acceleration. $a_x^2 = (Jt + a_{xi})^2 = J^2t^2 + a_{xi}^2 + 2Ja_{xi}t$

    Solving, $a_x^2 = a_{xi}^2 + 2J(\frac{1}{2}Jt^2 + a_{xi}t)$

    The expression for v was $v_x = \frac{1}{2}Jt^2 + a_{xi}t + v_{xi}$

    So $(v_x - v_{xi}) = \frac{1}{2}Jt^2 + a_{xi}t$

    Therefore, $a_x^2 = a_{xi}^2 + 2J(v_x - v_{xi})$ ◊

63. In a 100-m race, Maggie and Judy cross the finish line in a dead heat, both taking 10.2 s. Accelerating uniformly, Maggie took 2.00 s and Judy 3.00 s to attain maximum speed, which they maintained for the rest of the race. (a) What was the acceleration of each sprinter? (b) What were their respective maximum speeds? (c) Which sprinter was ahead at the 6.00-s mark, and by how much?

Solution

(a)   Maggie moves with constant positive acceleration $a_{x,M}$ for 2.00 s, then with constant speed $v_{xf,M}$ (zero acceleration) for 8.20 s, covering a distance of $\Delta x_{M1}+\Delta x_{M2}=100$ m. The equation for each distance is

$$\Delta x_{M1}=\tfrac{1}{2}a_{x,M}(2.00\text{ s})^2 \quad\text{and}\quad \Delta x_{M2}=v_{xf,M}(8.20\text{ s}),$$

Since
$$v_{xf,M}=0+a_{x,M}(2.00\text{ s})$$

by substitution
$$\tfrac{1}{2}a_{x,M}(2.00\text{ s})^2+a_{x,M}(2.00\text{ s})(8.20\text{ s})=100\text{ m}$$
$$a_{x,M}=5.43\text{ m}/\text{s}^2 \qquad \Diamond$$

For Judy,
$$\Delta x_{J1}=\tfrac{1}{2}a_{x,J}(3.00\text{ s})^2$$
$$\Delta x_{J2}=v_{xf,J}(7.20\text{ s})$$
$$v_{xf,J}=a_{x,J}(3.00\text{ s})$$

In a similar way, we use these values in the equation $x_{J1}+x_{J2}=100$ m

to get
$$\tfrac{1}{2}a_{x,J}(3.00\text{ s})^2+a_{x,J}(3.00\text{ s})(7.20\text{ s})=100\text{ m}$$
$$a_{x,J}=3.83\text{ m}/\text{s}^2 \qquad \Diamond$$

(b)   Maggie's and Judy's speeds after accelerating are

$$v_{xf,M}=a_{x,M}(2.00\text{ s})=\left(5.43\text{ m}/\text{s}^2\right)(2.00\text{ s})=10.9\text{ m}/\text{s} \qquad \Diamond$$
$$\text{and}\qquad v_{xf,J}=a_{x,J}(3.00\text{ s})=\left(3.83\text{ m}/\text{s}^2\right)(3.00\text{ s})=11.5\text{ m}/\text{s} \qquad \Diamond$$

(c)   In the first 6.00 s, Maggie covers a distance

$$x_{f,M} = \frac{1}{2}a_{x,M}(2.00\ s)^2 + v_{xf,M}(4.00\ s)$$

$$x_{f,M} = \frac{1}{2}(5.43\ m\ /\ s^2)(2.00\ s)^2 + (10.9\ m\ /\ s)(4.00\ s) = 54.3\ m$$

and Judy runs a distance

$$x_{f,J} = \frac{1}{2}a_{x,J}(3.00\ s)^2 + v_{xf,J}(3.00\ s)$$

$$x_{f,J} = \frac{1}{2}(3.83\ m\ /\ s^2)(3.00\ s)^2 + (11.5\ m\ /\ s)(3.00\ s) = 51.7\ m$$

So Maggie is ahead by 54.3 m − 51.7 m = 2.62 m   ◊

---

67.   An inquisitive physics student and mountain climber climbs a 50.0-m cliff that overhangs a calm pool of water. He throws two stones vertically downward, 1.00 s apart, and observes that they cause a single splash. The first stone has an initial speed of 2.00 m/s. (a) How long after release of the first stone do the two stones hit the water? (b) What was the initial velocity of the second stone? (c) What is the velocity of each stone at the instant they hit the water?

Solution

Set $y_i = 0$ at the top of the cliff, and find the time required for the first stone to reach the water using the equation

$$y_f - y_i = v_{yi}t + \frac{1}{2}a_y t^2, \quad \text{or in quadratic form,} \quad -\frac{1}{2}a_y t^2 - v_{yi}t + (y_f - y_i) = 0$$

☐

(a) If we take the direction downward to be negative, $y_f = -50.0\ m$,
$v_i = -2.00\ m\ /\ s$, and $a = -9.80\ m\ /\ s^2$. Substituting these values into the equation, we find

$$(4.90\ m\ /\ s^2)t^2 + (2.00\ m\ /\ s)t - 50.0\ m = 0$$

Using the quadratic formula, and noting that only the positive root describes this physical situation,

$$t = \frac{-2.00 \text{ m/s} + \sqrt{(2.00 \text{ m/s})^2 - 4(4.90 \text{ m/s}^2)(-50.0 \text{ m})}}{2(-4.90 \text{ m/s}^2)} = 3.00 \text{ s} \qquad \Diamond$$

(b) For the second stone, the time of travel is $t = 3.00 \text{ s} - 1.00 \text{ s} = 2.00 \text{ s}$

Since $y_f - y_i = v_{yi} t + \frac{1}{2} a_y t^2$,

$$v_{yi} = \frac{(y_f - y_i) - \frac{1}{2} a_y t^2}{t} = \frac{(-50.0 \text{ m}) - \frac{1}{2}(-9.80 \text{ m/s}^2)(2.00 \text{ s})^2}{2.00 \text{ s}} = -15.3 \text{ m/s} \quad \Diamond$$

The negative value indicates the downward direction of the initial velocity of the second stone.

(c) For the first stone,

$$v_{yf} = v_{yi} + a_y t_1 = -2.00 \text{ m/s} + (-9.80 \text{ m/s}^2)(3.00 \text{ s}) = -31.4 \text{ m/s} \qquad \Diamond$$

For the second stone,

$$v_{yf} = v_{yi} + a_y t_2 = -15.2 \text{ m/s} + (-9.80 \text{ m/s}^2)(2.00 \text{ s}) = 34.8 \text{ m/s} \qquad \Diamond$$

---

69. Kathy Kool buys a sports car that can accelerate at the rate of 4.90 m/s². She decides to test the car by racing with another speedster, Stan Speedy. Both start from rest, but experienced Stan leaves the starting line 1.00 s before Kathy. If Stan moves with a constant acceleration of 3.50 m/s² and Kathy maintains an acceleration of 4.90 m/s², find (a) the time it takes Kathy to overtake Stan, (b) the distance she travels before she catches him, and (c) the speeds of both cars at the instant she overtakes him.

Solution

(a) Let the times of travel for Kathy and Stan be t_K and t_S where $t_S = t_K + 1.00$ s. Both start from rest ($v_{xi, K} = v_{xi, S} = 0$), so the expressions for the distances traveled are:

$$x_K = \tfrac{1}{2} a_{x, K} t_K^2 = \tfrac{1}{2}(4.90 \text{ m} / \text{s}^2) t_K^2$$

$$x_S = \tfrac{1}{2} a_{x, S} t_S^2 = \tfrac{1}{2}(3.50 \text{ m} / \text{s}^2)(t_K + 1.00 \text{ s})^2$$

When Kathy overtakes Stan, the two distances will be equal; setting $x_K = x_S$,

$$\tfrac{1}{2}(4.90 \text{ m} / \text{s}^2) t_K^2 = \tfrac{1}{2}(3.50 \text{ m} / \text{s}^2)(t_K + 1.00 \text{ s})^2$$

This can be simplified and written in the standard form of a quadratic as

$$t_K^2 - 5.00 t_K \text{ s} - 2.50 \text{ s}^2 = 0$$

and solved using the quadratic formula to find

$$t_K = 5.46 \text{ s} \qquad \lozenge$$

(b) Use equation from part (a) for distance of travel,

$$x_K = \tfrac{1}{2} a_{x, K} t_K^2 = \tfrac{1}{2}(4.90 \text{ m} / \text{s}^2)(5.46 \text{ s})^2 = 73.0 \text{ m} \qquad \lozenge$$

(c) The final velocities will be (remember $v_{xi, K} = v_{xi, S} = 0$):

$$v_{xf, K} = a_{x, K} t_K = (4.90 \text{ m} / \text{s}^2)(5.46 \text{ s}) = 26.7 \text{ m} / \text{s} \qquad \lozenge$$

$$v_{xf, S} = a_{x, S} t_S = (3.50 \text{ m} / \text{s}^2)(6.46 \text{ s}) = 22.6 \text{ m} / \text{s} \qquad \lozenge$$

73. Two objects, A and B, are connected by a rigid rod that has a length L. The objects slide along perpendicular guide rails, as shown in Figure P2.73. If A slides to the left with a constant speed v, find the velocity of B when $\alpha = 60.0°$.

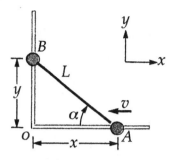

Figure 2.73

Solution

G: The solution to this problem may not seem obvious, but if we consider the range of motion of the two objects, we realize that B will have the same speed as A when $\alpha = 45°$. When $\alpha = 90°$, $v_B = 0$. Therefore when $\alpha = 60°$, we should expect v_B to be between 0 and v.

O: Since we know a distance relationship and we are looking for a velocity, we might try differentiating with respect to time to go from what we know to what we want. We can express the fact that the distance between A and B is always L with the Pythagorean theorem: $x^2 + y^2 = L^2$. By differentiating this equation with respect to time, we can find the value of $v_B = dy/dt$ in terms of $v_A = -v = dx/dt$.

A: Differentiating $x^2 + y^2 = L^2$ gives us
$$2x\frac{dx}{dt} + 2y\frac{dy}{dt} = 0$$

Substituting each velocity,
$$2x(-v) + 2y(v_B) = 0$$

Solving for the speed of B,
$$v_B = (v)(x/y)$$

From the geometry of the figure, $y/x = \tan\alpha$:
$$v_B = \left(\frac{1}{\tan\alpha}\right)v$$

When $\alpha = 60°$, $\quad v_B = \dfrac{v}{\tan\ 60°} = \dfrac{v}{\sqrt{3}} = 0.577v$ (B is moving up) $\qquad \Diamond$

L: Our answer seems reasonable since we have specified both a magnitude and direction for the velocity of B, and the speed is between 0 and v in agreement with our earlier prediction. In this and many other physics problems, we can find it helpful to examine the limiting cases that define boundaries for the answer.

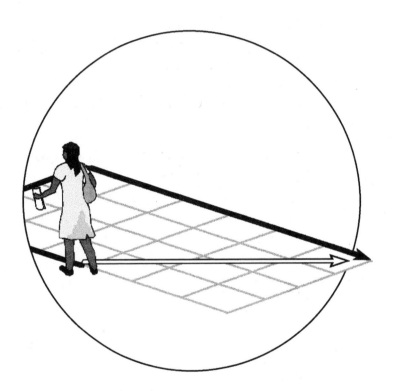

Vectors

VECTORS

INTRODUCTION

Physical quantities that have both numerical and directional properties are represented by vectors. Some examples of vector quantities are force, displacement, velocity, and acceleration. This chapter is primarily concerned with vector algebra and with some general properties of vector quantities. The addition and subtraction of vector quantities is discussed, together with some common applications to physical situations. Discussion of the products of vector quantities shall be delayed until these operations are needed.

EQUATIONS AND CONCEPTS

The location of a point P in a plane can be specified by either cartesian coordinates, x and y, or polar coordinates, r and θ. If one set of coordinates is known, values for the other set can be calculated.

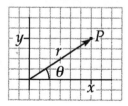

$$x = r \cos\theta \qquad (3.1)$$

$$y = r \sin\theta \qquad (3.2)$$

$$\tan \theta = \frac{y}{x} \qquad (3.3)$$

$$r = \sqrt{x^2 + y^2} \qquad (3.4)$$

Vector quantities obey the **commutative law of addition.** To add vector **A** to vector **B** using the graphical method, first construct **A**, and then draw **B** such that the tail of **B** starts at the head of **A**. The sum of **A** + **B** is the vector that completes the triangle by connecting the tail of **A** to the head of **B**.

$$A + B = B + A \tag{3.5}$$

When three or more vectors are added, the sum is independent of the manner in which the vectors are grouped. This is the **associative law** of addition.

$$A + (B + C) = (A + B) + C \tag{3.6}$$

When more than two vectors are to be added, they are all connected head-to-tail in any order. The resultant or sum is the vector which joins the tail of the first vector to the head of the last vector.

$$R = A + B + C + D$$

When two or more vectors are to be added, all of them must represent the same physical quantity—that is, have the same units. In the graphical or geometrical method of vector addition, the length of each vector is proportional to the magnitude of the vector. Also, each vector must be pointed along a direction which makes the proper angle relative to the others.

The operation of vector subtraction utilizes the definition of the negative of a vector. The negative of vector **A** is the vector which has a magnitude equal to the magnitude of **A**, but acts or points along a direction opposite the direction of **A**.

$$\mathbf{A} - \mathbf{B} = \mathbf{A} + (-\mathbf{B}) \tag{3.7}$$

A vector **A** in a two-dimensional coordinate system can be resolved into its components along the x and y directions. The projection of **A** onto the x axis is the x component of **A**; and the projection of **A** onto the y axis is the y component of **A**.

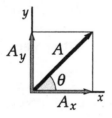

The magnitude of **A** and the angle, θ, which the vector makes with the positive x axis can be determined from the values of the x and y components of **A**.

$$A_x = A\cos\theta \tag{3.8}$$

$$A_y = A\sin\theta \tag{3.9}$$

$$A = \sqrt{A_x^2 + A_y^2} \tag{3.10}$$

$$\theta = \tan^{-1}\left(\frac{A_y}{A_x}\right) \tag{3.11}$$

A vector **A** lying in the x-y plane, having rectangular components A_x and A_y, can be expressed in unit vector notation.

$$\mathbf{A} = A_x\mathbf{i} + A_y\mathbf{j} \qquad (3.12)$$

When two vectors are added, the resultant vector can be expressed in terms of the components of the two vectors.

If $\mathbf{R} = \mathbf{A} + \mathbf{B}$,

$$\mathbf{R} = (A_x + B_x)\mathbf{i} + (A_y + B_y)\mathbf{j} \qquad (3.14)$$

SUGGESTIONS, SKILLS, AND STRATEGIES

When two or more vectors are to be added, the following step-by-step procedure is recommended:

• Select a coordinate system.

• Draw a sketch of the vectors to be added (or subtracted), with a label on each vector.

• Find the x and y components of all vectors.

• Find the resultant components (the algebraic sum of the components) in both the x and y directions.

• Use the Pythagorean theorem to find the magnitude of the resultant vector.

• Use a suitable trigonometric function to find the angle the resultant vector makes with the x axis.

To add vector **A** to vector **B** graphically, first construct **A**, and then draw **B** such that the tail of **B** starts at the head of **A**. The sum **A** + **B** is the vector that completes the triangle as shown in figure (a) below. The procedure for adding more than two vectors (the polygon rule) is illustrated in figure (b).

44

In order to subtract two vectors **graphically,** recognize that **A** – **B** is equivalent to the operation **A** + (**–B**). Since the vector **–B** is a vector whose magnitude is B and is opposite in direction to **B**, the construction in figure (c) below is obtained:

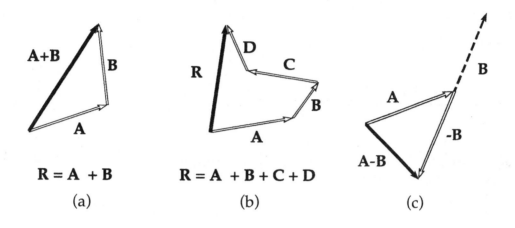

Adding vectors by (a) the triangle rule and (b) the polygon rule. (c) Subtracting two vectors graphically.

REVIEW CHECKLIST

▷ Locate a point in space using both cartesian coordinates and polar coordinates.

▷ Describe the basic properties of vectors such as the rules of vector addition and graphical solutions for addition of two or more vectors.

▷ Resolve a vector into its rectangular components. Determine the magnitude and direction of a vector from its rectangular components.

▷ Understand the use of unit vectors to express any vector in unit vector notation.

ANSWERS TO SELECTED CONCEPTUAL QUESTIONS

4. Vector **A** lies in the *xy* plane. For what orientations of vector **A** will both of its components be negative? For what orientations will its components have opposite signs?

Answer The vector will have negative values for both components when it lies in Quadrant III, and its angle counterclockwise from the positive *x* axis between 180° and 270°.

The vector will have components with opposite signs when it lies either in Quadrant II or Quadrant IV, so that its direction angle is either between 90° and 180°, or else between 270° and 360°.

□   □   □   □

9. Is it possible to add a vector quantity to a scalar quantity? Explain.

Answer No, it is not possible. Addition of a vector to a scalar is not defined. The following is one argument against any definition.

A vector quantity has both a magnitude and a direction, and can be represented as an arrow pointing in a single direction for a specific distance. If you add two vectors that point in the same direction, then you must add the magnitudes. On the other hand, if you add two vectors that point in opposite directions, then you must subtract the magnitudes. Since a scalar quantity does not have a direction, then in adding it to a vector you would not know whether to add to the magnitude of the vector or subtract from it.

□   □   □   □

SOLUTIONS TO SELECTED END-OF-CHAPTER PROBLEMS

1. The polar coordinates of a point are $r = 5.50$ m and $\theta = 240°$. What are the cartesian coordinates of this point?

Solution   When the polar coordinates (r, θ) of a point P are known, the cartesian coordinates can be found:

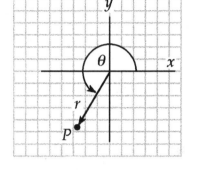

$$x = r \cos \theta \qquad y = r \sin \theta$$

$$x = (5.50 \text{ m})\cos 240° = -2.75 \text{ m} \qquad \lozenge$$

$$y = (5.50 \text{ m})\sin 240° = -4.76 \text{ m} \qquad \lozenge$$

5. A fly lands on one wall of a room. The lower left-hand corner of the wall is selected as the origin of a two-dimensional cartesian coordinate system. If the fly is located at the point having coordinates (2.00, 1.00) m, (a) how far is it from the corner of the room? (b) What is its location in polar coordinates?

Solution

(a) Assume the wall is in the x-y plane so that the coordinates are $x = 2.00$ m and $y = 1.00$ m; and the fly is located at point P. The distance between two points in the x, y plane is

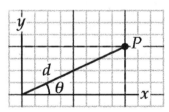

$$d = \sqrt{(x_2 - x_1)^2 + (y_2 - y_1)^2}$$

$$d = \sqrt{(2.00 \text{ m} - 0)^2 + (1.00 \text{ m} - 0)^2} = 2.24 \text{ m} \qquad \lozenge$$

(b) From the figure,

$$\theta = \tan^{-1}\left(\frac{y}{x}\right) = \tan^{-1}\left(\frac{1.00}{2.00}\right) = 26.6° \qquad \text{and} \qquad r = d$$

Therefore, the polar coordinates of the point P are (2.24 m, 26.6°)   ◊

---

9.   A surveyor measures the distance across a river by the following method: Starting directly across from a tree on the opposite bank, she walks 100 m along the riverbank to establish a baseline. Then she sights across to the tree. The angle from her baseline to the tree is 35.0°. How wide is the river?

Solution

Make a sketch of the area as viewed from above. Assume the river-banks are straight and parallel, show the location of the tree, and the original location of the surveyor. The width w between tree and surveyor must be perpendicular to the riverbanks, as the surveyor chooses to start "directly across" from the tree. Now draw the 100-meter baseline, b, and the line showing the "line of sight" to the tree.

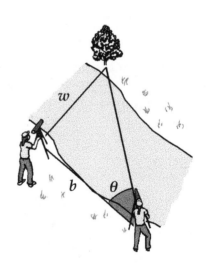

From the figure, $\tan\theta = \dfrac{w}{b}$

and $w = b\tan\theta = (100 \text{ m})\tan 35.0° = 70.0 \text{ m}$ ◊

---

13. A person walks along a circular path of radius 5.00 m. If the person walks around one half of the circle, find (a) the magnitude of the displacement vector and (b) how far the person walked. (c) What is the magnitude of the displacement if the person walks all the way around the circle?

Solution

See the sketch to the right.

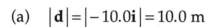

(a) $|\mathbf{d}| = |-10.0\mathbf{i}| = 10.0$ m ◊

since the displacement is a straight line from point *A* to point *C*.

(b) The actual distance walked is not equal to the straight-line displacement. The distance follows the curved path of the semicircle (*ABC*).

$$s = \left(\tfrac{1}{2}\right)(2\pi r) = 5.00\pi \text{ m} = 15.7 \text{ m} \qquad ◊$$

(c) If the circle is complete, **d** begins and ends at point *A*. Hence, $|\mathbf{d}| = 0$. ◊

----

15. Each of the displacement vectors **A** and **B** shown in Figure P3.15 has a magnitude of 3.00 m. Find graphically (a) **A** + **B**, (b) **A** − **B**, (c) **B** − **A**, (d) **A** − 2**B**. Report all angles counterclockwise from the positive *x* axis.

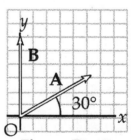

Figure P3.15

Solution   To find these vectors graphically, we draw each set of vectors as indicated by these drawings. Measurements of the results can be taken using a ruler and protractor.

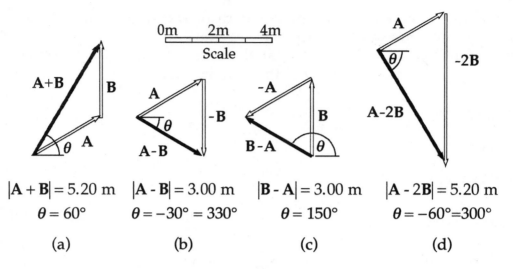

$$|A + B| = 5.20 \text{ m} \quad |A - B| = 3.00 \text{ m} \quad |B - A| = 3.00 \text{ m} \quad |A - 2B| = 5.20 \text{ m}$$
$$\theta = 60° \qquad \theta = -30° = 330° \qquad \theta = 150° \qquad \theta = -60° = 300°$$

(a)   (b)   (c)   (d)

Remember: when adding vectors graphically, they are connected "head-to-tail," represented by arrows whose lengths correspond to their magnitudes. Also, the relative directions of the vectors must be maintained. When subtracting vectors, remember also that $A - B = A + (-B)$.   ◊

---

17. A roller coaster moves 200 ft horizontally and then rises 135 ft at an angle of 30.0° above the horizontal. It then travels 135 ft at an angle of 40.0° downward. What is its displacement from its starting point? Use graphical techniques.

Solution

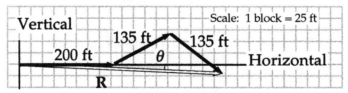

Your sketch when drawn to scale should look somewhat like the one above. (You will probably only be able to obtain a measurement to one or two significant figures). The distance R and the angle θ can be measured to give, upon use of your scale factor, the values of:

$$R = 423 \text{ ft at about } 2.63° \text{ below the horizontal.} \qquad ◊$$

---

25. A vector has an x component of –25.0 units and a y component of 40.0 units. Find the magnitude and direction of this vector.

Solution

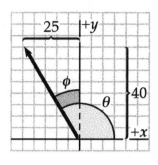

G: First we should visualize the vector either in our mind or with a sketch. Since the hypoteneuse of the right triangle must be greater than either the x or y components that form the legs, we can estimate the magnitude of the vector to be about 50 units. The direction of the vector (θ) appears to be about 120° from the $+x$ axis.

O: The graphical analysis and visual estimates above may suffice for some situations, but we can use trigonometry to obtain a more precise result.

A: The magnitude can be found by the Pythagorean theorem: $r = \sqrt{x^2 + y^2}$

$$r = \sqrt{(-25.0 \text{ units})^2 + (40.0 \text{ units})^2} = 47.2 \text{ units} \qquad \lozenge$$

We observe that $\tan\phi = x/y$ (if we consider x and y to both be positive).

$$\phi = \tan^{-1}\left(\frac{x}{y}\right) = \tan^{-1}\left(\frac{25.0 \text{ units}}{40.0 \text{ units}}\right) = \tan^{-1}(0.625) = 32.0°$$

The angle from the $+x$ axis can be found by adding 90° to ϕ.

$$\theta = \phi + 90° = 122° \qquad \lozenge$$

L: Our calculated results agree with our graphical estimates. We should always remember to check that our answers are reasonable and make sense, especially for problems like this where it is easy to mistakenly calculate the wrong angle by confusing coordinates or overlooking a minus sign.

Quite often the direction angle of a vector can be specified in more than one way, and we must choose a notation that makes the most sense for the given problem. If compass directions were stated in this question, we could have reported the vector angle to be 32.0° west of north or a compass heading of 328°.

---

31. Consider two vectors $\mathbf{A} = 3\mathbf{i} - 2\mathbf{j}$ and $\mathbf{B} = -\mathbf{i} - 4\mathbf{j}$. Calculate (a) $\mathbf{A} + \mathbf{B}$, (b) $\mathbf{A} - \mathbf{B}$, (c) $|\mathbf{A} + \mathbf{B}|$, (d) $|\mathbf{A} - \mathbf{B}|$, (e) the directions of $\mathbf{A} + \mathbf{B}$ and $\mathbf{A} - \mathbf{B}$.

Solution   Use the property of vector addition that states that

$$\text{If} \quad \mathbf{R} = \mathbf{A} + \mathbf{B} \quad \text{then} \quad R_x = A_x + B_x \quad \text{and} \quad R_y = A_y + B_y$$

(a) $\mathbf{A} + \mathbf{B} = (3\mathbf{i} - 2\mathbf{j}) + (-\mathbf{i} - 4\mathbf{j}) = 2\mathbf{i} - 6\mathbf{j}$ ◊

(b) $\mathbf{A} - \mathbf{B} = (3\mathbf{i} - 2\mathbf{j}) - (-\mathbf{i} - 4\mathbf{j}) = 4\mathbf{i} + 2\mathbf{j}$ ◊

For a vector, $\mathbf{R} = R_x\mathbf{i} + R_y\mathbf{j}$ $|\mathbf{R}| = \sqrt{R_x{}^2 + R_y{}^2}$

(c) $|\mathbf{A} + \mathbf{B}| = \sqrt{2^2 + (-6)^2} = 6.32$ ◊

(d) $|\mathbf{A} - \mathbf{B}| = \sqrt{4^2 + 2^2} = 4.47$ ◊

The direction of a vector relative to the positive x axis is $\theta = \tan^{-1}(R_y/R_x)$

(e)   For $\mathbf{A} + \mathbf{B}$, $\theta = \tan^{-1}(-6/2) = -71.6° = 288°$ ◊

For $\mathbf{A} - \mathbf{B}$, $\theta = \tan^{-1}(2/4) = 26.6°$ ◊

---

33. Obtain expressions in component form for the position vectors having polar coordinates (a) 12.8 m, 150°; (b) 3.30 cm, 60.0°; (c) 22.0 in., 215°.

Solution Find the x and y components of each vector using $x = R \cos\theta$ and $y = R\sin\theta$. In unit vector notation, $\mathbf{R} = R_x\mathbf{i} + R_y\mathbf{j}$.

(a) $x = (12.8 \text{ m})\cos 150° = -11.1 \text{ m}$ $\qquad y = (12.8 \text{ m})\sin 150° = 6.40 \text{ m}$

$\mathbf{R} = (-11.1\mathbf{i} + 6.40\mathbf{j}) \text{ m}$ $\qquad\qquad\qquad\qquad\qquad$ ◊

(b) $x = (3.30 \text{ cm})\cos 60.0° = 1.65 \text{ cm}$ $\qquad y = (3.30 \text{ cm})\sin 60.0° = 2.86 \text{ cm}$

$\mathbf{R} = (1.65\mathbf{i} + 2.86\mathbf{j}) \text{ cm}$ $\qquad\qquad\qquad\qquad\qquad$ ◊

(c) $x = (22.0 \text{ in})\cos 215° = -18.0 \text{ in}$ $\qquad y = (22.0 \text{ in})\sin 215° = -12.6 \text{ in}$

$\mathbf{R} = (-18.0\mathbf{i} - 12.6\mathbf{j}) \text{ in}$ $\qquad\qquad\qquad\qquad\qquad$ ◊

---

35. A particle undergoes the following consecutive displacements: 3.50 m south, 8.20 m northeast, and 15.0 m west. What is the resultant displacement?

Solution Take the direction east to be along +**i**. The three displacements can be written as:

$$\mathbf{d}_1 = -3.50\mathbf{j} \text{ m}$$

$$\mathbf{d}_2 = (8.20 \cos 45.0°)\mathbf{i} \text{ m} + (8.20 \sin 45.0°)\mathbf{j} \text{ m} = 5.80\mathbf{i} \text{ m} + 5.80\mathbf{j} \text{ m}$$

$$\mathbf{d}_3 = -15.0\mathbf{i} \text{ m}$$

The resultant, $\mathbf{R} = \mathbf{d}_1 + \mathbf{d}_2 + \mathbf{d}_3$

$$\mathbf{R} = (0 + 5.80 - 15.0)\mathbf{i}\ m + (-3.50 + 5.80 + 0)\mathbf{j}\ m$$

$$\mathbf{R} = (-9.20\mathbf{i} + 2.30\mathbf{j})\ m \qquad\qquad \lozenge$$

The magnitude of the resultant displacement is

$$|\mathbf{R}| = \sqrt{R_x^2 + R_y^2} = \sqrt{(-9.20\ m)^2 + (2.30\ m)^2} = 9.48\ m \qquad \lozenge$$

The direction of the resultant vector is given by

$$\tan^{-1}\left(\frac{R_y}{R_x}\right) = \tan^{-1}\left(\frac{2.30}{-9.20}\right) = -14.0°$$

or relative to the positive x axis, $\theta = 180° - 14.0° = 166° \qquad \lozenge$

---

43. The vector **A** has x, y, and z components of 8.00, 12.0, and –4.00 units, respectively. (a) Write a vector expression for **A** in unit-vector notation. (b) Obtain a unit-vector expression for a vector **B** one-fourth the length of **A** pointing in the same direction as **A**. (c) Obtain a unit-vector expression for a vector **C** three times the length of **A** pointing in the direction opposite the direction of **A**.

Solution

(a) $\mathbf{A} = A_x\mathbf{i} + A_y\mathbf{j} + A_z\mathbf{k}$ $\mathbf{A} = 8.00\ \mathbf{i} + 12.0\ \mathbf{j} - 4.00\ \mathbf{k}$ $\lozenge$

(b) $\mathbf{B} = \mathbf{A}/4$ $\mathbf{B} = 2.00\ \mathbf{i} + 3.00\ \mathbf{j} - 1.00\ \mathbf{k}$ $\lozenge$

(c) $\mathbf{C} = -3\mathbf{A}$ $\mathbf{C} = -24.0\ \mathbf{i} - 36.0\ \mathbf{j} + 12.0\ \mathbf{k}$ $\lozenge$

---

49. Vector **A** has a negative x component 3.00 units in length and a positive y component 2.00 units in length. (a) Determine an expression for **A** in unit-vector notation. (b) Determine the magnitude and direction of **A**. (c) What vector **B**, when added to **A**, gives a resultant vector with no x component and a negative y component 4.00 units in length?

Solution

$A_x = -3.00$ units $\qquad A_y = 2.00$ units

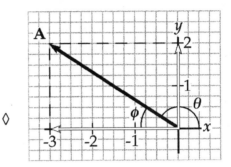

(a) $\quad \mathbf{A} = A_x\mathbf{i} + A_y\mathbf{j} = -3.00\mathbf{i} + 2.00\mathbf{j}$ units $\qquad \Diamond$

(b) $\quad |\mathbf{A}| = \sqrt{A_x^2 + A_y^2} = \sqrt{(-3.00)^2 + (2.00)^2}$

$\quad |\mathbf{A}| = 3.61$ units

$$\tan\phi = \left|\frac{A_y}{A_x}\right| = \left|\frac{2.00}{-3.00}\right| = 0.667 \qquad \text{so} \qquad \phi = 33.7° \text{ (relative to the } -x \text{ axis)}$$

A is in the **second quadrant:** $\qquad \theta = 180° - \phi = 146° \qquad \Diamond$

(c) We are given that $R_x = 0$ and $R_y = -4.00$; since $\mathbf{R} = \mathbf{A} + \mathbf{B}$, $\mathbf{B} = \mathbf{R} - \mathbf{A}$

$$B_x = R_x - A_x = 0 - (-3.00) = 3.00$$

$$B_y = R_y - A_y = -4.00 - 2.00 = -6.00$$

Therefore, $\qquad \mathbf{B} = B_x\mathbf{i} + B_y\mathbf{j} = (3.00\mathbf{i} - 6.00\mathbf{j})$ units $\qquad \Diamond$

51. Three vectors are oriented as shown in Figure P3.51, where $|\mathbf{A}| = 20.0$ units, $|\mathbf{B}| = 40.0$ units, and $|\mathbf{C}| = 30.0$ units. Find (a) the x and y components of the resultant vector (expressed in unit-vector notation) and (b) the magnitude and direction of the resultant vector.

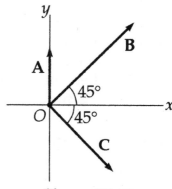

Figure P3.51

Solution   We first identify the cartesian components of **A**, **B**, and **C**, and then we add the components to obtain the resultant components R_x and R_y:

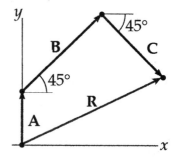

(a) $A_x = (20.0 \text{ units}) \cos 90° = 0$

$A_y = (20.0 \text{ units}) \sin 90° = 20.0 \text{ units}$

$B_x = (40.0 \text{ units}) \cos 45° = 28.3 \text{ units}$

$B_y = (40.0 \text{ units}) \sin 45° = 28.3 \text{ units}$

$C_x = (30.0 \text{ units}) \cos 315° = 21.2 \text{ units}$

$C_y = (30.0 \text{ units}) \sin 315° = -21.2 \text{ units}$

$R_x = A_x + B_x + C_x = (0 + 28.3 + 21.2) \text{ units } = 49.5 \text{ units}$

$R_y = A_y + B_y + C_y = (20 + 28.3 - 21.2) \text{ units} = 27.1 \text{ units}$

Therefore, in unit vector notation, $\mathbf{R} = 49.5\mathbf{i} + 27.1\mathbf{j}$ units   ◊

(b) $|\mathbf{R}| = \sqrt{R_x{}^2 + R_y{}^2} = \sqrt{(49.5 \text{ units})^2 + (27.1 \text{ units})^2} = 56.4 \text{ units}$ ◊

$\theta = \tan^{-1}\left(\dfrac{R_y}{R_x}\right) = \tan^{-1}\left(\dfrac{27.1}{49.5}\right) = 28.7°$ ◊

57. A person going for a walk follows the path shown in Figure P3.57. The total trip consists of four straight-line paths. At the end of the walk, what is the person's resultant displacement measured from the starting point?

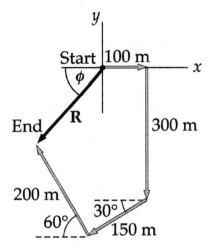

Figure P3.57

Solution   The resultant displacement **R** is equal to the sum of the four individual displacements, $\mathbf{R} = \mathbf{d}_1 + \mathbf{d}_2 + \mathbf{d}_3 + \mathbf{d}_4$

Thus, we first rewrite the individual displacements in unit vector notation:

$\mathbf{d}_1 = 100\mathbf{i}$ m

$\mathbf{d}_2 = -300\mathbf{j}$ m

$\mathbf{d}_3 = (-150 \cos 30°)\mathbf{i}$ m $+ (-150 \sin 30°)\mathbf{j}$ m $= -130\mathbf{i}$ m $- 75\mathbf{j}$ m

$\mathbf{d}_4 = (-200 \cos 60°)\mathbf{i}$ m $+ (200 \sin 60°)\mathbf{j}$ m $= -100\mathbf{i}$ m $+ 173\mathbf{j}$ m

Summing the components together,

$R_x = d_{1x} + d_{2x} + d_{3x} + d_{4x} = (100 + 0 - 130 - 100)$ m $= -130$ m
$R_y = d_{1y} + d_{2y} + d_{3y} + d_{4y} = (0 - 300 - 75 + 173)$ m $= -202$ m

$\mathbf{R} = -130\mathbf{i}$ m $- 202\mathbf{j}$ m   ◊

$|\mathbf{R}| = \sqrt{R_x{}^2 + R_y{}^2} = \sqrt{(-130 \text{ m})^2 + (-202 \text{ m})^2} = 240$ m   ◊

$\phi = \tan^{-1}\left(\dfrac{R_y}{R_x}\right) = \tan^{-1}\left(\dfrac{-202}{-130}\right) = 57.2°$

or $\theta = 57.2° + 180° = 237°$ relative to +x axis   ◊

Motion in Two Dimensions

Chapter 4

MOTION IN TWO DIMENSIONS

INTRODUCTION

In this chapter we use the kinematic equations to describe two dimensional motion. We treat motion in a plane with constant acceleration (projectile motion) and uniform circular motion as special cases of motion in two dimensions, .

EQUATIONS AND CONCEPTS

A particle whose position vector changes from $\mathbf{r}_i$ to $\mathbf{r}_f$ undergoes a **displacement** $\Delta\mathbf{r} \equiv \mathbf{r}_f - \mathbf{r}_i$.

$$\Delta\mathbf{r} \equiv \mathbf{r}_f - \mathbf{r}_i \tag{4.1}$$

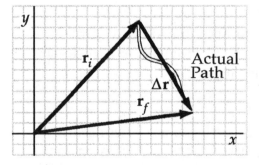

The **average velocity** of a particle which undergoes a displacement $\Delta\mathbf{r}$ in a time interval Δt equals the ratio $\Delta\mathbf{r} / \Delta t$.

$$\overline{\mathbf{v}} \equiv \frac{\Delta\mathbf{r}}{\Delta t} \tag{4.2}$$

The **instantaneous velocity** of a particle equals the limit of the average velocity as $\Delta t \to 0$.

$$\mathbf{v} \equiv \lim_{\Delta t \to 0} \frac{\Delta\mathbf{r}}{\Delta t} = \frac{d\mathbf{r}}{dt} \tag{4.3}$$

59

The **average acceleration** of a particle which undergoes a change in velocity $\Delta\mathbf{v}$ in a time interval Δt equals the ratio $\Delta\mathbf{v}\,/\,\Delta t$.

$$\bar{\mathbf{a}} \equiv \frac{\mathbf{v}_f - \mathbf{v}_i}{t_f - t_i} = \frac{\Delta\mathbf{v}}{\Delta t} \qquad (4.4)$$

The **instantaneous acceleration** is defined as the limit of the average velocity as $\Delta t \to 0$.

$$\mathbf{a} \equiv \lim_{\Delta t \to 0} \frac{\Delta\mathbf{v}}{\Delta t} = \frac{d\mathbf{v}}{dt} \qquad (4.5)$$

The **velocity** of a particle as a function of time moving with **constant acceleration** : (at $t = 0$, the velocity is $\mathbf{v}_i$).

$$\mathbf{v}_f = \mathbf{v}_i + \mathbf{a}t \qquad (4.8)$$

The **position vector** as a function of time for a particle moving with **constant acceleration** : (at $t = 0$, the position vector is $\mathbf{r}_i$ and the velocity is $\mathbf{v}_i$).

$$\mathbf{r}_f = \mathbf{r}_i + \mathbf{v}_i t + \frac{1}{2}\mathbf{a}t^2 \qquad (4.9)$$

A projectile has a constant horizontal velocity, as shown in the diagram at the right:

$v_x = v_i \cos \theta_i = \text{constant}$

$v_y = v_i \sin \theta_i - gt$

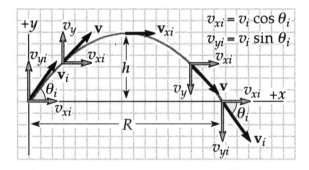

Equations 4.10 and 4.11 give the x and y position coordinates for a projectile, as a function of time.

$$x_f = v_{xi}t = (v_i \cos \theta_i)t \qquad (4.10)$$

$$y_f = v_{yi}t + \frac{1}{2}a_y t^2 = (v_i \sin \theta_i)t - \frac{1}{2}gt^2 \qquad (4.11)$$

The **maximum height** of a projectile can be written in terms of v_i and θ_i.

$$h = \frac{v_i^2 \sin^2 \theta_i}{2g} \qquad (4.13)$$

Likewise, the **horizontal range** of a projectile can also be stated in terms of v_i and θ_i.

$$R = \frac{v_i^2 \sin 2\theta_i}{g} \qquad (4.14)$$

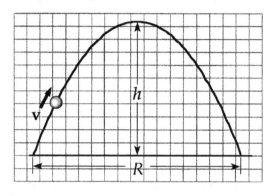

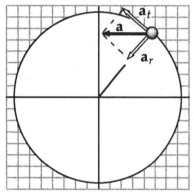

A particle moving in a circle of radius r with speed v undergoes a **centripetal acceleration** a_r equal in magnitude to v^2/r.

$$a_r = \frac{v^2}{r} \qquad (4.15)$$

If a particle which moves with constant speed v in a circular path, the period is equal to the total distance traveled in one revolution, divided by the speed.

$$T = \frac{2\pi r}{v}$$

In general, a particle moving on a curved path can have both a centripetal and a tangential component of acceleration. $\mathbf{a}_r$ is directed towards the center of curvature; $\mathbf{a}_t$ is tangent to the path.

$$\mathbf{a} = \mathbf{a}_r + \mathbf{a}_t \qquad (4.16)$$

If the speed of a particle moving on a curved path changes with time, the particle has a **tangential** component of acceleration equal in magnitude to dv / dt.

$$a_t = \frac{d|\mathbf{v}|}{dt} \qquad (4.17)$$

Galilean transformation equations relate the position and velocity of a particle measured by an observer in a moving frame of reference to those values measured by an observer in a fixed frame (moving with the object).

$$\mathbf{r}' = \mathbf{r} - \mathbf{v}_0 t \qquad (4.20)$$

$$\mathbf{v}' = \mathbf{v} - \mathbf{v}_0 \qquad (4.21)$$

SUGGESTIONS, SKILLS, AND STRATEGIES

- You should be familiar with the mathematical expression for a **parabola.** In particular, the equation which describes the trajectory of a projectile moving under the influence of gravity is given by

$$y = Ax - Bx^2$$

  where $\qquad A = \tan \theta_i \qquad$ and $\qquad B = \dfrac{g}{2v_i^2 \cos^2 \theta_i}$

Note that this expression for y assumes that the particle leaves the origin at $t = 0$, with a velocity $\mathbf{v}_i$. A sketch of y versus x for this situation is shown at the right.

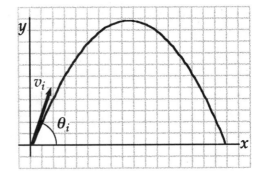

- If you are given v_i and θ_i, you should be able to make a point-by-point plot of the trajectory using the expressions for $x(t)$ and $y(t)$. Furthermore, you should know how to calculate the velocity component v_y at any time t. (Note that the component $v_{xf} = v_{xi} = v_i \cos\theta_i = $ constant, since $a_x = 0$.)

- Assuming that you have values for x and y at any time $t > 0$, you should be able to write an expression for the position vector $\mathbf{r}$ at that time using the relation $\mathbf{r} = x\mathbf{i} + y\mathbf{j}$. From this you can find the **displacement** r, where $r = \sqrt{x^2 + y^2}$. Likewise, if v_x and v_y are known at any time $t > 0$, you can express the velocity vector $\mathbf{v}$ in the formula $\mathbf{v} = v_x\mathbf{i} + v_y\mathbf{j}$. From this, you can find the **speed** at any time, since $v = \sqrt{v_x^2 + v_y^2}$.

PROBLEM-SOLVING STRATEGY: PROJECTILE MOTION

We suggest that you use the following approach to solving projectile motion problems:

- Select a coordinate system.

- Resolve the initial velocity vector into x and y components.

- Treat the horizontal motion and the vertical motion independently.

- Follow the techniques for solving problems with constant velocity to analyze the horizontal motion of the projectile.

- Follow the techniques for solving problems with constant acceleration to analyze the vertical motion of the projectile.

REVIEW CHECKLIST

▷ Describe the displacement, velocity, and acceleration of a particle moving in the $x - y$ plane.

▷ Recognize that two-dimensional motion in the $x - y$ plane with constant acceleration is equivalent to two independent motions along the x and y directions with constant acceleration components a_x and a_y.

▷ Recognize the fact that if the initial speed v_i and initial angle θ_i of a projectile are known at a given point at $t = 0$, the velocity components and coordinates can be found at any later time t. Furthermore, one can also calculate the horizontal range R and maximum height h if v_i and θ_i are known.

▷ Understand the nature of the acceleration of a particle moving in a circle with constant speed. In this situation, note that although $|\mathbf{v}| = $ constant, the **direction** of $\mathbf{v}$ varies in time, the result of which is the radial, or centripetal acceleration.

▷ Describe the components of acceleration for a particle moving on a curved path, where both the magnitude and direction of $\mathbf{v}$ are changing with time. In this case, the particle has a tangential component of acceleration and a radial component of acceleration.

▷ Realize that the outcome of a measurement of the motion of a particle (its position, velocity, and acceleration) depends on the frame of reference of the observer.

ANSWERS TO SELECTED CONCEPTUAL QUESTIONS

3. If you know the position vectors of a particle at two points along its path and also know the time it took to get from one point to the other, can you determine the particle's instantaneous velocity? Its average velocity? Explain.

Answer Its instantaneous velocity cannot be determined from this information, since it could have accelerated and decelerated several times between the two points. If that were the case, then there might be a different instantaneous velocity at every moment and position.

However, the average velocity can be determined from its definition and the given information, since to find the average velocity you simply subtract the first position vector from the second, and divide by the total elapsed time.

□   □   □   □

5. Explain whether or not the following particles have an acceleration: (a) a particle moving in a straight line with constant speed and (b) a particle moving around a curve with constant speed.

Answer (a) The acceleration is zero, since $|\mathbf{v}|$ and its direction remain constant. (b) The particle has an acceleration since the direction of $\mathbf{v}$ changes.

□   □   □   □

9. A spacecraft drifts through space at a constant velocity. Suddenly, a gas leak in the side of the spacecraft causes a constant acceleration perpendicular to the initial velocity. The orientation of the spacecraft does not change, and so the acceleration remains perpendicular to the original direction of the velocity. What is the shape of the path followed by the spacecraft in this situation?

Answer The spacecraft will follow a parabolic path. This is equivalent to a projectile thrown off a cliff with a horizontal velocity. For the projectile, gravity provides an acceleration which is always perpendicular to the initial velocity, resulting in a parabolic path. For the spacecraft, the initial velocity plays the role of the horizontal velocity of the projectile. The acceleration provided by the leaking gas provides an acceleration which plays the role of gravity for the projectile.

If the orientation of the spacecraft were to change in response to the gas leak (which is by far the more likely result), then the acceleration would change direction and the motion could become quite complicated.

□   □   □   □

19. A projectile is fired at some angle to the horizontal with some initial speed v_i, and air resistance is neglected. Is the projectile a freely falling body? What is its acceleration in the vertical direction? What is its acceleration in the horizontal direction?

Answer Yes. The projectile is a freely falling body, because nothing counteracts the force of gravity. The vertical acceleration will be the local gravitational acceleration, g; the horizontal acceleration will be zero.

□   □   □   □

SOLUTIONS TO SELECTED END-OF-CHAPTER PROBLEMS

1. A motorist drives south at 20.0 m/s for 3.00 min, then turns west and travels at 25.0 m/s for 2.00 min, and finally travels northwest at 30.0 m/s for 1.00 min. For this 6.00-min trip, find (a) the total vector displacement, (b) the average speed, and (c) the average velocity. Use a coordinate system in which east is the positive x axis.

Solution

(a) Her displacements are

S: (20.0 m / s)(180 s)   W: (25.0 m / s)(120 s)   NW: (30.0 m / s)(60.0 s)

Choosing $\mathbf{i}$ = east and $\mathbf{j}$ = north, we have

$$\Delta\mathbf{r} = (3.60 \text{ km})(-\mathbf{j}) + (3.00 \text{ km})(-\mathbf{i}) + (1.80 \text{ km cos } 45°)(-\mathbf{i}) + (1.80 \text{ km sin } 45°)(\mathbf{j})$$

$$\Delta\mathbf{r} = (3.00 + 1.27) \text{ km } (-\mathbf{i}) + (1.27 - 3.60) \text{ km } \mathbf{j} = (-4.27\mathbf{i} - 2.33\mathbf{j}) \text{ km} \qquad ◊$$

(In polar form, this would be $\Delta r = 4.87$ km at 209° from east) $\qquad ◊$

(b) The total path-length traveled is (3.60 + 3.00 + 1.80) km = 8.40 km

So   average speed $= \dfrac{8.40 \text{ km}}{6.00 \text{ min}}\left(\dfrac{1.00 \text{ min}}{60.0 \text{ s}}\right)\left(\dfrac{1000 \text{ m}}{\text{km}}\right) = 23.3 \text{ m / s}$ $\qquad ◊$

(c) $\mathbf{v}_{avg} = \dfrac{\Delta\mathbf{r}}{t} = \dfrac{(-4.27 \text{ } \mathbf{i} - 2.33 \text{ } \mathbf{j}) \text{ km}}{360 \text{ s}}$

$\mathbf{v}_{avg} = (-11.9 \text{ } \mathbf{i} - 6.47 \text{ } \mathbf{j}) \text{ m / s}$ $\qquad ◊$

Or in polar form, $\mathbf{v}_{avg} = \dfrac{\Delta\mathbf{r}}{t} = \dfrac{4.87 \text{ km}}{360 \text{ s}} = 13.5 \text{ m / s at } 209°$ $\qquad ◊$

7. A fish swimming in a horizontal plane has velocity $v_i = (4.00i + 1.00j)$ m/s at a point in the ocean whose displacement from a certain rock is $r_i = (10.0i - 4.00j)$ m. After the fish swims with constant acceleration for 20.0 s, its velocity is $v = (20.0i - 5.00j)$ m/s. (a) What are the components of the acceleration? (b) What is the direction of the acceleration with respect to the unit vector **i**? (c) Where is the fish at $t = 25.0$ s and in what direction is it moving?

Solution At $t = 0,$ $v_i = (4.00i + 1.00j)$ m/s and $r_i = (10.0i - 4.00j)$ m

At $t = 20.0$ s, $v = (20.0i - 5.00j)$ m/s

(a) $a_x = \dfrac{\Delta v_x}{\Delta t} = \dfrac{20.0 \text{ m / s} - 4.00 \text{ m / s}}{20.0 \text{ s}} = 0.800 \text{ m / s}^2$ ◊

$a_y = \dfrac{\Delta v_y}{\Delta t} = \dfrac{-5.00 \text{ m / s} - 1.00 \text{ m / s}}{20.0 \text{ s}} = -0.300 \text{ m / s}^2$ ◊

(b) $\theta = \tan^{-1}\left(\dfrac{a_y}{a_x}\right) = \tan^{-1}\left(\dfrac{-0.300 \text{ m / s}^2}{0.800 \text{ m / s}^2}\right) = -20.6° = 339°$ from the +x axis   ◊

(c) The fish's coordinates are $x_f = x_i + v_{xi}t + \frac{1}{2}a_xt^2$ and $y_f = y_i + v_{yi}t + \frac{1}{2}a_yt^2$

while its velocity is $v_{yf} = v_{yi} + a_yt$ and $v_{xf} = v_{xi} + a_xt$

At $t = 25.0$ s,

$x = 10.0 \text{ m} + (4.00 \text{ m / s})(25.0 \text{ s}) + \frac{1}{2}(0.800 \text{ m / s}^2)(25.0 \text{ s})^2 = 360 \text{ m}$

$y = -4.00 \text{ m} + (1.00 \text{ m / s})(25.0 \text{ s}) + \dfrac{(-0.300 \text{ m / s}^2)(25.0 \text{ s})^2}{2} = -72.8 \text{ m}$

Thus, at $t = 25.0$ s, $r = 360i - 72.8j$ m   ◊

$v_{yf} = v_{yi} + a_yt = (1.00 \text{ m / s}) - (0.300 \text{ m / s}^2)(25.0 \text{ s}) = -6.50 \text{ m / s}$

$v_{xf} = v_{xi} + a_xt = (4.00 \text{ m / s}) + (0.800 \text{ m / s}^2)(25.0 \text{ s}) = 24.0 \text{ m / s}$

$\theta = \tan^{-1}(v_y/v_x) = \tan^{-1}\left(\dfrac{-6.50 \text{ m / s}}{24.0 \text{ m / s}}\right) = -15.2° = 345°$ from the +x axis. ◊

9. In a local bar, a customer slides an empty beer mug down the counter for a refill. The bartender is momentarily distracted and does not see the mug, which slides off the counter and strikes the floor 1.40 m from the base of the counter. If the height of the counter is 0.860 m, (a) with what velocity did the mug leave the counter, and (b) what was the direction of the mug's velocity just before it hit the floor?

Solution

G: Based on our everyday experiences and the description of the problem, a reasonable speed of the mug would be a few m/s and it will hit the floor at some angle between 0° and 90°, probably about 45°.

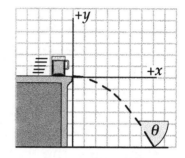

O: We are looking for two different velocities, but we are only given two distances. Our approach will be to separate the vertical and horizontal motions. By using the height that the mug falls, we can find the time of the fall. Once we know the time, we can find the horizontal and vertical components of the velocity. For convenience, we will set the origin to be the point where the mug leaves the counter.

A: Vertical motion: $y = -0.860 \text{ m} \quad v_{yi} = 0 \quad v_y = ? \quad a_y = -9.80 \text{ m} / \text{s}^2$

Horizontal motion: $x = 1.40 \text{ m} \quad v_x = ? = \text{constant} \quad a_x = 0$

(a) To find the time of fall, we use the free fall equation: $y = v_{yi} t + \frac{1}{2} a_y t^2$

Solving, $-0.860 \text{ m} = 0 - \frac{1}{2}\left(9.80 \text{ m} / \text{s}^2\right)t^2$ so $t = 0.419 \text{ s}$

Then $v_x = \dfrac{x}{t} = \dfrac{1.40 \text{ m}}{0.419 \text{ s}} = 3.34 \text{ m} / \text{s}$ ◊

(b) The mug hits the floor with a vertical velocity of $v_{yf} = v_{yi} + a_y t$ and an impact angle of $\theta = \tan^{-1}(v_y/v_x)$ below the horizontal.

Solving for v_y, $\qquad v_y = 0 - (9.80 \text{ m} / \text{s}^2)(0.419 \text{ s}) = -4.11 \text{ m} / \text{s}$

Thus, $\qquad \theta = \tan^{-1}\left[(-4.11 \text{ m} / \text{s})/(3.34 \text{ m} / \text{s})\right] = 50.9° \; \lozenge$

L: This was a multi-step problem that required several physics equations to solve; yet our answers do agree with our initial expectations. Since the problem did not ask for the time, we could have eliminated this variable by substitution, but then we would have had to substitute the algebraic expression $t = \sqrt{2y / g}$ into two other equations. So in this case it was easier to find a numerical value for the time as an intermediate step. Sometimes the most efficient method is not realized until each alternative solution is attempted.

---

15. A projectile is fired in such a way that its horizontal range is equal to three times its maximum height. What is the angle of projection? Give your answer to three significant figures.

Solution   In this problem, we want to find θ_i such that $R = 3h$. We can use the equations for the range and height of a projectile path, $R = v_i^2 \sin(2\theta_i)/g$ and $h = v_i^2 \sin^2 \theta_i/2g$. Since we require that $R = 3h$,

$$\frac{v_i^2 \sin(2\theta_i)}{g} = \frac{3v_i^2 \sin^2 \theta_i}{2g} \qquad \text{or} \qquad \frac{2}{3} = \frac{\sin^2 \theta_i}{\sin(2\theta_i)}$$

But $\sin(2\theta_i) = 2(\sin \theta_i)(\cos \theta_i)$ so $\dfrac{\sin^2 \theta_i}{\sin(2\theta_i)} = \dfrac{\sin^2 \theta_i}{2 \sin \theta_i \cos \theta_i} = \dfrac{\tan \theta_i}{2}$

Substituting and solving for θ_i, $\qquad \theta_i = \tan^{-1}\left(\dfrac{4}{3}\right) = 53.1° \qquad \lozenge$

---

19. A placekicker must kick a football from a point 36.0 m (about 40 yards) from the goal, and the crowd hopes the ball will clear the crossbar, which is 3.05 m high. When kicked, the ball leaves the ground with a speed of 20.0 m/s at an angle of 53.0° to the horizontal. (a) By how much does the ball clear or fall short of clearing the crossbar? (b) Does the ball approach the crossbar while still rising or while falling?

Solution

(a) To find the actual height of the football when it reaches the goal line, (where $x = 36.0$ m, $v_i = 20.0$ m/s, and $\theta_i = 53.0°$) use Equation 4.12:

$$y = x \tan \theta_i - \frac{gx^2}{2v_i^2 \cos^2 \theta_i}$$

$$y = (36.0 \text{ m})(\tan 53.0°) - \frac{(9.80 \text{ m/s}^2)(36.0 \text{ m})^2}{(2)(20.0 \text{ m/s})^2 \cos^2 53.0°} = 3.989 \text{ m}$$

The ball clears the bar by $(3.939 - 3.050)$ m $= 0.889$ m. ◊

(b) The time the ball takes to reach the maximum height ($v_y = 0$) is

$$t_1 = \frac{(v_i \sin \theta_i) - v_y}{g} = \frac{(20.0 \text{ m/s})(\sin 53.0°) - 0}{9.80 \text{ m/s}^2} = 1.63 \text{ s}$$

The time to travel 36.0 m horizontally is $t_2 = x / v_{ix}$

$$t_2 = \frac{36.0 \text{ m}}{(20.0 \text{ m/s})(\cos 53.0°)} = 2.99 \text{ s}$$

Since $t_2 > t_1$, the ball clears the goal on its way down. ◊

25. The athlete shown in Figure P4.25 of the text rotates a 1.00-kg discus along a circular path of radius 1.06 m. The maximum speed of the discus is 20.0 m/s. Determine the magnitude of the maximum radial acceleration of the discus.

Solution   The maximum radial acceleration occurs when maximum tangential speed is attained.

Here, $$a_r = \frac{v^2}{r} = \frac{(20.0 \text{ m/s})^2}{(1.06 \text{ m})} = 377 \text{ m/s}^2 \qquad \lozenge$$

---

31. A train slows down as it rounds a sharp, horizontal curve, slowing from 90.0 km/h to 50.0 km/h in the 15.0 s that it takes to round the curve. The radius of the curve is 150 m. Compute the acceleration at the moment the train speed reaches 50.0 km/h. Assume that the magnitude of the tangential acceleration is constant over the 15.0-s interval.

Solution

G: If the train is taking this turn at a safe speed, then its acceleration should be significantly less than g, perhaps a few m/s^2 (otherwise it might jump the tracks!), and it should be directed toward the center of the curve and backward since the train is slowing.

O: Since the train is changing both its speed and direction, the acceleration vector will be the vector sum of the tangential and radial acceleration components. The tangential acceleration can be found from the changing speed and elapsed time, while the radial acceleration can be found from the radius of curvature and the train's speed.

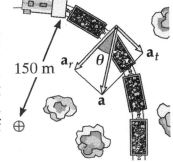

A: First, let's convert the speed units from km/h to m/s:

$$v_i = 90.0 \text{ km / h} = (90.0 \text{ km / h})(10^3 \text{ m / km})(1 \text{ h / 3600 s}) = 25.0 \text{ m / s}$$

$$v_f = 50.0 \text{ km / h} = (50.0 \text{ km / h})(10^3 \text{ m / km})(1 \text{ h / 3600 s}) = 13.9 \text{ m / s}$$

The tangential acceleration and radial acceleration are:

$$a_t = \frac{\Delta v}{\Delta t} = \frac{13.9 \text{ m / s} - 25.0 \text{ m / s}}{15.0 \text{ s}} = -0.741 \text{ m / s}^2 \quad \text{(backward)}$$

$$a_r = \frac{v^2}{r} = \frac{(13.9 \text{ m / s})^2}{150 \text{ m}} = 1.29 \text{ m / s}^2 \quad \text{(inward)}$$

$$a = \sqrt{a_r^2 + a_t^2} = \sqrt{(1.29 \text{ m / s}^2)^2 + (-0.741 \text{ m / s}^2)^2} = 1.48 \text{ m / s}^2 \qquad ◊$$

$$\theta = \tan^{-1}\left(\frac{a_r}{a_t}\right) = \tan^{-1}\left(\frac{0.741 \text{ m / s}^2}{1.48 \text{ m / s}^2}\right) = 26.6° \qquad ◊$$

(This angle points behind the center of the curve)

L: The acceleration is clearly less than g, and it appears that most of the acceleration comes from the radial component, so it makes sense that the acceleration vector should point mostly toward the center of the curve and slightly backwards due to the negative tangential acceleration.

33. Figure P4.33 represents the total acceleration of a particle moving clockwise in a circle of radius 2.50 m at a given instant of time. At this instant, find (a) the centripetal acceleration, (b) the speed of the particle, and (c) its tangential acceleration.

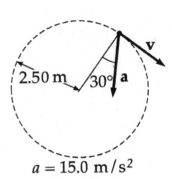

$a = 15.0 \ \text{m}/\text{s}^2$

Solution

(a) The acceleration has an inward radial component:

Figure P4.33

$$a_r = a\cos 30.0° = \left(15.0 \ \text{m}/\text{s}^2\right)\cos 30.0°$$

$$a_r = 13.0 \ \text{m}/\text{s}^2$$

◊

(b) The speed at the instant shown can be found by using

$$a_r = \frac{v^2}{r} \qquad \text{or} \qquad v = \sqrt{a_r r} = \sqrt{(13.0 \ \text{m}/\text{s}^2)(2.50 \ \text{m})} = 5.70 \ \text{m}/\text{s}$$

◊

(c) The acceleration also has a tangential component:

$$a_t = a \sin 30.0° = (15.0 \ \text{m}/\text{s}^2) \sin 30.0° = 7.50 \ \text{m}/\text{s}^2$$

◊

37. A river has a steady speed of 0.500 m/s. A student swims upstream a distance of 1.00 km and swims back to the starting point. If the student can swim at a speed of 1.20 m/s in still water, how long does the trip take? Compare this with the time the trip would take if the water were still.

Solution

G: If we think about the time for the trip as a function of the stream's speed, we realize that if the stream is flowing at the same rate or faster than the student can swim, he will never reach the 1.00 km mark even after an infinite amount of time. Since the student can swim 1.20 km in 1000 s, we should expect that the trip will definitely take longer than in still water, maybe about 2000 s (~30 minutes).

O: The total time in the river is the longer time upstream (against the current) plus the shorter time downstream (with the current). For each part, we will use the basic equation $t = d / v$, where v is the speed of the student relative to the shore.

A:
$$t_{up} = \frac{d}{v_{student} - v_{stream}} = \frac{1000 \text{ m}}{1.20 \text{ m}/\text{s} - 0.500 \text{ m}/\text{s}} = 1430 \text{ s}$$

$$t_{dn} = \frac{d}{v_{student} + v_{stream}} = \frac{1000 \text{ m}}{1.20 \text{ m}/\text{s} + 0.500 \text{ m}/\text{s}} = 588 \text{ s}$$

Total time in river, $t_{river} = t_{up} + t_{dn} = 2.02 \times 10^3 \text{ s}$ ◊

In still water, $t_{still} = \dfrac{d}{v} = \dfrac{2000 \text{ m}}{1.20 \text{ m}/\text{s}} = 1.67 \times 10^3 \text{ s}$ or $t_{still} = 0.827 t_{river}$ ◊

L: As we predicted, it does take the student longer to swim up and back in the moving stream than in still water (21% longer in this case), and the amount of time agrees with our estimation.

43. A science student is riding on a flatcar of a train traveling along a straight horizontal track at a constant speed of 10.0 m/s. The student throws a ball into the air along a path that he judges to make an initial angle of 60.0° with the horizontal and to be in line with the track. The student's professor, who is standing on the ground nearby, observes the ball to rise vertically. How high does she see the ball rise?

Solution   Shown on the right, $\mathbf{v}_{be} = \mathbf{v}_{bc} + \mathbf{v}_{ce}$

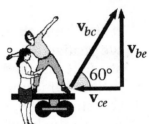

with $\mathbf{v}_{bc}$ = the velocity of the ball relative to the car

$\mathbf{v}_{be}$ = the velocity of the ball relative to the Earth

and $\mathbf{v}_{ce}$ = car velocity relative to the Earth = 10.0 m/s

From the figure, we have $v_{ce} = v_{bc} \cos 60.0°$

So
$$v_{bc} = \frac{10.0 \text{ m/s}}{\cos 60.0°} = 20.0 \text{ m/s}$$

Again from the figure, $v_{be} = v_{bc} \sin 60.0° + 0 = (20.0 \text{ m/s})(0.866) = 17.3 \text{ m/s}$

This is the initial velocity of the ball relative to the Earth. Now from $v_y{}^2 = v_{iy}{}^2 + 2ah$, we can calculate the maximum height that the ball rises:

$$0 = (17.3 \text{ m/s})^2 + 2(-9.80 \text{ m/s}^2)h \qquad \text{and} \qquad h = 15.3 \text{ m} \qquad \lozenge$$

---

49. A home run is hit in such a way that the baseball just clears a wall 21.0 m high, located 130 m from home plate. The ball is hit at an angle of 35.0° to the horizontal, and air resistance is negligible. Find (a) the initial speed of the ball, (b) the time it takes the ball to reach the wall, and (c) the velocity components and the speed of the ball when it reaches the wall. (Assume the ball is hit at a height of 1.00 m above the ground.)

Solution   Let the initial speed of the ball be v_i, and the initial angle be $\theta_i = 35.0°$. Set the starting point at

$$(x_i, y_i) = (0 \text{ m}, 1 \text{ m})$$

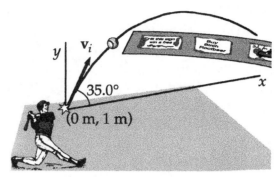

When $x = 130$ m, $y = 21.0$ m

$$x_f = x_i + v_{xi} t = v_i t \, \cos 35.0°$$

$$v_i t \, \cos 35.0° = 130 \text{ m}$$

$$v_{xi} = v_i \cos 35.0°$$
$$v_{yi} = v_i \sin 35.0°$$

and $v_i t = \dfrac{130 \text{ m}}{\cos 35.0°} = 158.7$ m

Next, $y_f = y_i + v_{yi} t - \frac{1}{2}gt^2 = 1.00 \text{ m} + (v_i \sin 35.0°)t - \left(\frac{1}{2}\right)(9.80 \text{ m / s}^2)t^2$

(b)   We substitute for $v_i t$ in the second term:

$$20.0 \text{ m} = (158.7 \text{ m})(\sin 35.0°) - \left(\tfrac{1}{2}\right)(9.80 \text{ m / s}^2)t^2$$

and $t = \sqrt{\dfrac{71.0 \text{ m}}{4.90 \text{ m / s}^2}} = 3.81$ s   ◊

(a)   Therefore, $v_i = \dfrac{158.7 \text{ m}}{3.81 \text{ s}} = 41.7$ m / s   ◊

(c) $v_{yf} = v_{yi} - gt = v_i \sin 35.0° - gt = (23.9 \text{ m / s}) - \left(9.80 \text{ m / s}^2\right)t$

At $t = 3.81$ s, $v_{yf} = -13.4$ m / s   ◊

$$v_{xf} = v_{xi} = v_i \cos 35.0° = (41.7 \text{ m / s}) \cos 35.0° = 34.1 \text{ m / s} \quad ◊$$

and $|\mathbf{v}| = \sqrt{v_x^{\,2} + v_y^{\,2}} = \sqrt{34.1^2 + (-13.4)^2} = 36.6$ m / s   ◊

77

59. A bomber is flying horizontally over level terrain, with a speed of 275 m/s relative to the ground, at an altitude of 3000 m. Neglect the effects of air resistance. (a) How far will a bomb travel horizontally between its release from the plane and its impact on the ground? (b) If the plane maintains its original course and speed, where will it be when the bomb hits the ground? (c) At what angle from the vertical should the telescopic bombsight be set so the bomb will hit the target seen in the sight at the time of release?

Solution

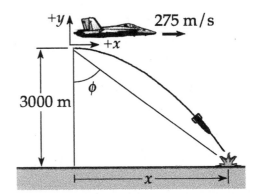

(a) In horizontal flight,

$$v_{yi} = 0 \qquad \text{and} \qquad v_{xi} = 275 \text{ m/s}$$

Therefore, $y_f = y_i - \tfrac{1}{2}gt^2 = -\tfrac{1}{2}gt^2$

and $x = v_{xi}t$

Eliminate t between these two equations to find

$$x_f = v_{xi}\sqrt{\frac{-2y}{g}} = (275 \text{ m / s})\sqrt{\frac{(-2)(-3000 \text{ m})}{9.80 \text{ m / s}^2}} = 6800 \text{ m} \qquad \Diamond$$

(b) The plane and the bomb have the same constant **horizontal** velocity. Therefore, the plane will be 3000 m above the bomb at impact, and 6800 m from the point of release. $\qquad \Diamond$

(c) If θ is the angle between the vertical and the direct line of sight to the target at the time of release,

$$\theta = \tan^{-1}\left(\frac{x}{y}\right) = \tan^{-1}\left(\frac{6800}{3000}\right) = 66.2° \qquad \Diamond$$

61. A hawk is flying horizontally at 10.0 m/s in a straight line, 200 m above the ground. A mouse it has been carrying struggles free from its grasp. The hawk continues on its path at the same speed for 2.00 s before attempting to retrieve its prey. To accomplish the retrieval, it dives in a straight line at constant speed and recaptures the mouse 3.00 m above the ground. (a) Assuming no air resistance, find the diving speed of the hawk. (b) What angle did the hawk make with the horizontal during its descent? (c) For how long did the mouse "enjoy" free fall?

Solution

G: We should first recognize that the hawk cannot instantaneously change from slow horizontal motion to rapid downward motion. The hawk cannot move with infinite acceleration, but we assume that the time required for the hawk to accelerate is short compared to two seconds. Based on our everyday experiences, a reasonable diving speed for the hawk might be about 100 mph (~ 50 m/s) at some angle close to 90° and should last only a few seconds.

O: We know the distance that the mouse and hawk fall, but to find the diving speed of the hawk, we must know the time of descent. If the hawk and mouse both maintain their original horizontal velocity of 10 m/s (as they should without air resistance), then the hawk only needs to think about diving straight down, but to a ground-based observer, the path will appear to be a straight line angled less than 90° below horizontal.

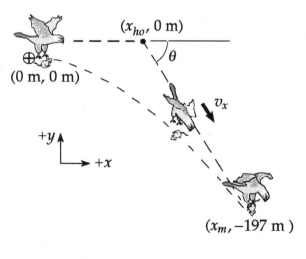

A: The mouse falls a total vertical distance, $y = 200 \text{ m} - 3.00 \text{ m} = 197 \text{ m}$

The time of fall is found from $y = v_{yi}t - \frac{1}{2}gt^2$: $t = \sqrt{\dfrac{2(197 \text{ m})}{9.80 \text{ m}/\text{s}^2}} = 6.34 \text{ s}$ ◊

To find the diving speed of the hawk, we must first calculate the total distance covered from the vertical and horizontal components. We already know the vertical distance, y, so we just need the horizontal distance during the same time (minus the 2.00 s late start).

$$x = v_{xi}(t - 2.00 \text{ s}) = (10.0 \text{ m}/\text{s})(6.34 \text{ s} - 2.00 \text{ s}) = 43.3 \text{ m}$$

The total distance is $\qquad d = \sqrt{x^2 + y^2} = \sqrt{(43.3 \text{ m})^2 + (197 \text{ m})^2} = 202 \text{ m}$

So the hawk's diving speed is $\quad v_{\text{hawk}} = \dfrac{d}{t - 2.00 \text{ s}} = \dfrac{202 \text{ m}}{4.34 \text{ s}} = 46.5 \text{ m}/\text{s}$ ◊

at $\quad \theta = \tan^{-1}\left(\dfrac{y}{x}\right) = \tan^{-1}\left(\dfrac{197 \text{ m}}{43.4 \text{ m}}\right) = 77.6°$ below the horizontal ◊

L: The answers appear to be consistent with our predictions, even thought it is not possible for the hawk to reach its diving speed instantaneously. Sometimes we must make simplifying assumptions to solve complex physics problems, and sometimes these assumptions are not physically possible. Once the idealized problem is understood, we can attempt to analyze the more complex, real-world problem. For this problem, if we considered the realistic effects of air resistance and the maximum diving acceleration attainable by the hawk, we might find that the hawk could not catch the mouse before it hit the ground.

65. A car is parked on a steep incline overlooking the ocean, where the incline makes an angle of 37.0° below the horizontal. The negligent driver leaves the car in neutral, and the parking brakes are defective. The car rolls from rest down the incline with a constant acceleration of 4.00 m/s², traveling 50.0 m to the edge of a vertical cliff. The cliff is 30.0 m above the ocean. Find (a) the speed of the car when it reaches the edge of the cliff and the time it takes to get there, (b) the velocity of the car when it lands in the ocean, (c) the total time the car is in motion, and (d) the position of the car when it lands in the ocean, relative to the base of the cliff.

Solution   (a) From point i to point e, the car undergoes constant acceleration from rest $(v_i = 0)$. Therefore,

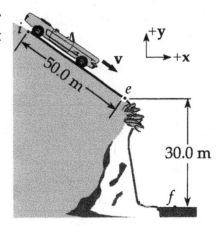

$$v_e^2 = v_i^2 + 2a(\Delta s) = 2a(\Delta s)$$

$$v_e = \sqrt{(2)(4.00 \text{ m} / \text{s}^2)(50.0 \text{ m})}$$

Solving, we find that $v_e = 20.0 \text{ m} / \text{s}$ ◊

We can find the elapsed time from $v_e = v_i + at$:

$$t_{ie} = \frac{v_e - v_i}{a} = \frac{20.0 \text{ m} / \text{s} - 0}{4.00 \text{ m} / \text{s}^2} = 5.00 \text{ s}$$ ◊

(b)   At the edge of the cliff, the components of velocity v_e are:

$$v_{ye} = (-20.0 \text{ m} / \text{s}) \sin 37.0° = -12.0 \text{ m} / \text{s}$$

$$v_{xe} = (20.0 \text{ m} / \text{s}) \cos 37.0° = 16.0 \text{ m} / \text{s}$$

There is no further horizontal acceleration, so $v_{xf} = v_{xe} = 16.0 \text{ m} / \text{s}$

Free fall gives the car a vertical acceleration: $v_{yf} = \pm\sqrt{2a_y(\Delta y) + v_{ye}{}^2}$

$$v_{yf} = \pm\sqrt{2(-9.80 \text{ m / s}^2)(-30.0 \text{ m}) + (-12.0 \text{ m / s})^2} = -27.1 \text{ m / s}$$

$$v_f = \sqrt{v_{xf}{}^2 + v_{yf}{}^2} = \sqrt{(16.0 \text{ m / s})^2 + (-27.1 \text{ m / s})^2} = 31.4 \text{ m / s} \qquad \lozenge$$

(c) From point e to f, $\quad t_{ef} = \dfrac{v_{yf} - v_{ye}}{a_y} = \dfrac{(-27.1 \text{ m / s}) - (-12.0 \text{ m / s})}{(-9.80 \text{ m / s}^2)} = 1.54 \text{ s}$

The total time is $\quad t_{if} = t_{ie} + t_{ef} = 6.54 \text{ s} \qquad \lozenge$

(d) The horizontal distance is $\quad \Delta x = v_{xe} t_{ef} = (16.0 \text{ m / s})(1.54 \text{ s}) = 24.6 \text{ m} \qquad \lozenge$

---

67. A skier leaves the ramp of a ski jump with a velocity of 10.0 m/s, 15.0° above the horizontal, as in Figure P4.67. The slope is inclined at 50.0°, and air resistance is negligible. Find (a) the distance from the ramp to where the jumper lands and (b) the velocity components just before the landing. (How do you think the results might be affected if air resistance were included? Note that jumpers lean forward in the shape of an airfoil, with their hands at their sides, to increase their distance. Why does this work?)

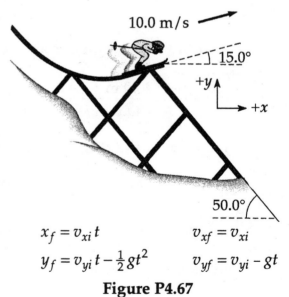

10.0 m/s

15.0°

$+y$

$+x$

50.0°

$x_f = v_{xi}t$

$y_f = v_{yi}t - \frac{1}{2}gt^2$

$v_{xf} = v_{xi}$

$v_{yf} = v_{yi} - gt$

Figure P4.67

Solution

Set point 'i' where the skier takes off, and point 'f' where the skier lands. Define the coordinate system as shown in the figure.

$$v_i = 10.0 \text{ m/s at } 15.0°$$

$$v_{xi} = v_i \cos \theta_i = (10.0 \text{ m / s}) \cos 15.0° = 9.66 \text{ m / s}$$

$$v_{yi} = v_i \sin \theta_i = (10.0 \text{ m / s}) \sin 15.0° = 2.59 \text{ m / s}$$

(a) The skier travels

horizontally $\qquad x_f - x_i = v_{xi}t = (9.66 \text{ m / s})t$ \hfill (1)

and vertically, $\qquad y_f - y_i = v_{yi}t - \frac{1}{2}gt^2 = (2.59 \text{ m / s})t - (4.90 \text{ m / s}^2)t^2$ \hfill (2)

landing when $\qquad \dfrac{y_f - y_i}{x_f - x_i} = \tan(-50.0°) = -1.19$ \hfill (3)

Substituting (1) and (2) into (3), $\qquad \dfrac{(2.59 \text{ m / s})t - (4.90 \text{ m / s}^2)t^2}{(9.66 \text{ m / s})t} = -1.19$

Since we ignore the solution $t = 0$, $\quad -4.90t + 14.1 = 0$

and $t = 2.88$ s. Solving (1), $\qquad x_f - x_i = (9.66 \text{ m/s})t = 27.8 \text{ m}$.

By Figure P4.67, $\qquad d = \dfrac{x_f - x_i}{\cos 50.0°} = \dfrac{27.8 \text{ m}}{\cos 50.0°} = 43.2 \text{ m}$ ◊

(b) $v_{xf} = v_{xi} = 9.66 \text{ m / s}$ $\qquad v_{yf} = v_{yi} - gt = 2.59 \text{ m / s} - (9.80 \text{ m / s}^2)t$

When $t = 2.88$ s, $\qquad v_{xf} = 9.66 \text{ m/s}$ $\qquad\qquad v_{yf} = -25.6 \text{ m/s}$ ◊

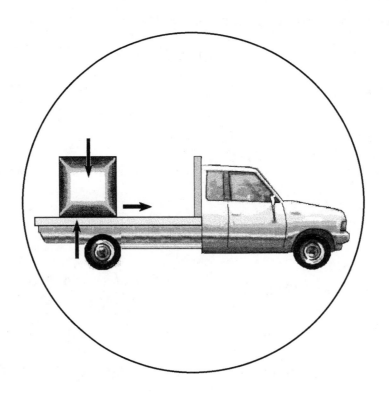

The Laws of Motion

THE LAWS OF MOTION

INTRODUCTION

In Chapters 2 and 4, we described the motion of particles based on the definitions of displacement, velocity, and acceleration. In this chapter, we use the concepts of force and mass to describe the change in motion of particles. We then discuss the three basic laws of motion, which are based on experimental observations and were formulated more than three centuries ago by Newton.

Force laws which describe the quantitative method of calculating the force on an object if its environment is known are discussed in this chapter. These force laws, together with the laws of motion, are the foundations of classical mechanics. We learn how it is possible to describe an object's acceleration in terms of its mass and the resultant force acting on the object. Mass is a measure of the object's inertia, that is, its tendency to resist an acceleration when a force acts on it.

EQUATIONS AND CONCEPTS

A quantitative measurement of mass (the term used to measure inertia) can be made by comparing the accelerations that a given force will produce on different bodies. If a given force acting on a body of mass m_1 produces an acceleration a_1 and the same force acting on a body of mass m_2 produces an acceleration a_2, the ratio of the two masses equals the inverse of the ratio of the two accelerations.

$$\frac{m_1}{m_2} \equiv \frac{a_2}{a_1} \tag{5.1}$$

The acceleration of an object is proportional to the resultant force acting on it and inversely proportional to its mass. This is a statement of Newton's second law.

$$\Sigma \mathbf{F} = m\mathbf{a} \qquad (5.2)$$

When several forces act on an object, it is often convenient to write the equation expressing Newton's second law as component equations. The orientation of the coordinate system can often be chosen so that the object has a nonzero acceleration along only one direction.

$$\Sigma F_x = ma_x \qquad (5.3)$$

$$\Sigma F_y = ma_y$$

$$\Sigma F_z = ma_z$$

Calculations with Equation 5.2 must be made using a consistent set of units for the quantities force, mass, and acceleration. The SI unit of force is the newton (N), defined as the force that, when acting on a 1-kg mass, produces an acceleration of 1 m/s².

$$1 \text{ N} \equiv 1 \text{ kg·m/s}^2 \qquad (5.4)$$

$$1 \text{ lb} \equiv 1 \text{ slug·ft/s}^2 \qquad (5.5)$$

$$1 \text{ lb} \equiv 4.448 \text{ N}$$

Weight (the force of gravity on a mass) is not an inherent property of a body, but depends on the local value of g and varies with location.

$$\mathbf{F}_g = m\mathbf{g} \qquad (5.6)$$

Newton's third law states that the force exerted on body 1 by body 2 is equal in magnitude and opposite in direction to the force exerted on body 2 by body 1.

$$\mathbf{F}_{12} = -\mathbf{F}_{21} \qquad (5.7)$$

The force of static friction between two surfaces in contact, but not in motion relative to each other, cannot be greater than $\mu_s n$, where n is the normal (perpendicular) force between the two surfaces and μ_s (coefficient of static friction) is a dimensionless constant which depends on the nature of the pair of surfaces.

$$f_s \leq \mu_s n \qquad (5.8)$$

When two surfaces are in relative motion, the force of kinetic friction on each body is directed opposite to the direction of motion of the body.

$$f_k = \mu_k n \qquad (5.9)$$

SUGGESTIONS, SKILLS, AND STRATEGIES

The following procedure is recommended for problems involving objects in equilibrium:

- Make a sketch of the object under consideration.

- Draw a free-body diagram and label all external forces acting on the object. Try to guess the correct direction for each force. If you select a direction that leads to a negative sign in your solution for a force, do not be alarmed; this merely means that the direction of the force is the opposite of what you assumed.

- Resolve all forces into x and y components, choosing a convenient coordinate system.

- Use the equations $\Sigma F_x = 0$ and $\Sigma F_y = 0$. Remember to keep track of the signs of the various force components.

- Application of the step above leads to a set of equations with several unknowns. All that is left is to solve the simultaneous equations for the unknowns in terms of the known quantities.

The following procedure is recommended when dealing with problems involving the application of Newton's second law:

- Draw a simple, neat diagram of the system.

- Isolate the object of interest whose motion is being analyzed. Draw a free-body diagram for this object; that is, a diagram showing **all external forces acting on the object.** For systems containing more than one object, draw **separate** diagrams for each object. Do not include forces that the object exerts on its surroundings.

- Establish convenient coordinate axes for each body and find the components of the forces along these axes.

- Apply Newton's second law, $\Sigma \mathbf{F} = m\mathbf{a}$, in the x and y directions for each object under consideration.

- Solve the component equations for the unknowns. Remember that you must have as many independent equations as you have unknowns in order to obtain a complete solution.

- Often in solving such problems, one must also use the equations of kinematics (motion with constant acceleration) to find all the unknowns.

REVIEW CHECKLIST

▷ State in your own words a description of Newton's laws of motion, recall physical examples of each law, and identify the action-reaction force pairs in a multiple-body interaction problem as specified by Newton's third law.

▷ Express the normal force in terms of other forces acting on an object and write out the equation which relates the coefficient of friction, force of friction and normal force between an object and surface on which it rests or moves.

▷ Apply Newton's laws of motion to various mechanical systems using the recommended procedure discussed in Section 5.7. Most important, you should identify all external forces acting on the system, draw the **correct** free-body diagrams which apply to each body of the system, and apply Newton's second law, $\Sigma\mathbf{F} = m\mathbf{a}$, in **component** form.

▷ Apply the equations of kinematics (which involve the quantities displacement, velocity, and acceleration) as described in Chapter 2 along with those methods and equations of Chapter 5 (involving mass, force, and acceleration) to the solutions of problems where **both** the kinematical and dynamic aspects are present.

▷ Be familiar with solving several linear equations simultaneously for the unknown quantities. Recall that you must have as many **independent** equations as you have unknowns.

ANSWERS TO SELECTED CONCEPTUAL QUESTIONS

5.   Identify the action - reaction pairs in the following situations: a man takes a step; a snowball hits a woman in the back; a baseball player catches a ball; a gust of wind strikes a window.

Answer   As a man takes a step, the action is the force his foot exerts on the Earth; the reaction is the force of the Earth on his foot. In the second case, the action is the force exerted on the woman's back by the snowball; the reaction is the force exerted on the snowball by the woman's back. The third action is the force of the glove on the ball; the reaction is the force of the ball on the glove. The fourth action is the force exerted on the window by the air molecules; the reaction is the force on the air molecules exerted by the window.

□   □   □   □

9.   A rubber ball is dropped onto the floor. What force causes the ball to bounce?

Answer   When the ball hits the floor, it is compressed. As the ball returns to its original shape, it exerts a force on the floor, and the reaction to this thrusts it back into the air.

□   □   □   □

18.   A weight lifter stands on a bathroom scale. He pumps a barbell up and down. What happens to the reading on the scale as this is done? Suppose he is strong enough to actually **throw** the barbell upward. How does the reading on the scale vary now?

Answer   If the barbell is not moving, the reading on the bathroom scale is the combined weight of the weightlifter and the barbell. At the beginning of the lift of the barbell, the barbell accelerates upward. By Newton's third law, the barbell pushes downward on the hands of the weightlifter with more force than its weight, in order to accelerate. As a result, he is pushed with more force into the scale, increasing its reading. Near the top of the lift, the weightlifter reduces the upward force, so that the acceleration of the barbell is downward, causing it to come to rest. While the barbell is coming to rest, it pushes with less force on the weightlifter's hands, so the reading on the scale is below the combined stationary weight. If the barbell is held at rest for a moment at the top of the lift, the scale reading is simply the combined weight. As it begins to be brought down, the reading decreases, as the force of the weightlifter on the barbell is reduced. The reading increases as the barbell is slowed down at the bottom.

If we now consider the throwing of the barbell, we have the same behavior as before, except that the variations in scale reading will be larger, since more force must be applied to throw the barbell upward rather than just lift it. Once the barbell leaves the weightlifter's hands, the reading will suddenly drop to just the weight of the weightlifter, and will rise suddenly when the barbell is caught.

□　　□　　□　　□

20. In the motion picture **It Happened One Night** (Columbia Pictures, 1934), Clark Gable is standing inside a stationary bus in front of Claudette Colbert, who is seated. The bus suddenly starts moving forward, and Clark falls into Claudette's lap. Why did this happen?

Answer When the bus starts moving, the mass of Claudette is accelerated by the force of the back of the seat on her body. Clark is standing, however, and the only force on him is the friction between his shoes and the floor of the bus. Thus, when the bus starts moving, his feet start accelerating forward, but the rest of his body experiences almost no accelerating force (only that due to his being attached to his accelerating feet!). As a consequence, his body tends to stay almost at rest, according to Newton's first law, relative to the ground. Relative to Claudette, however, he is moving toward her and falls into her lap.

□　　□　　□　　□

SOLUTIONS TO SELECTED END-OF-CHAPTER PROBLEMS

3. A 3.00-kg mass undergoes an acceleration given by $\mathbf{a} = (2.00\mathbf{i} + 5.00\mathbf{j})$ m/s^2. Find the resultant force $\mathbf{F}$ and its magnitude.

Solution $\quad \sum \mathbf{F} = m\mathbf{a} = (3.00 \text{ kg})(2.00\mathbf{i} + 5.00\mathbf{j}) \text{ m/s}^2 = (6.00\mathbf{i} + 15.0\mathbf{j}) \text{ N}$ ◊

$$|\mathbf{F}| = \sqrt{(F_x)^2 + (F_y)^2} = \sqrt{(6.00 \text{ N})^2 + (15.0 \text{ N})^2} = 16.2 \text{ N}$$ ◊

---

9. A 4.00-kg object has a velocity of $3.00\mathbf{i}$ m/s at one instant. Eight seconds later, its velocity increases to $(8.00\mathbf{i} + 10.0\mathbf{j})$ m/s. Assuming the object was subject to a constant total force, find (a) the components of the force and (b) its magnitude.

Solution We are given the mass, the starting and ending velocities, and the elapsed time:

$$m = 4.00 \text{ kg}, \quad \mathbf{v}_i = 3.00\mathbf{i} \text{ m/s}, \quad \mathbf{v}_f = (8.00\mathbf{i} + 10.00\mathbf{j}) \text{ m/s}, \quad t = 8.00 \text{ s}$$

The object's acceleration is

$$\mathbf{a} = \frac{\Delta \mathbf{v}}{\Delta t} = \frac{\mathbf{v}_f - \mathbf{v}_i}{\Delta t} = \frac{(8.00\mathbf{i} + 10.0\mathbf{j} - 3.00\mathbf{i}) \text{ m/s}}{8.00 \text{ s}} = \frac{(5.00\mathbf{i} + 10.0\mathbf{j})}{8.00} \text{ m/s}^2$$

(a) So the total force on it must be

$$\sum \mathbf{F} = m\mathbf{a} = (4.00 \text{ kg})\frac{(5.00\mathbf{i} + 10.0\mathbf{j})}{8.00} \text{ m/s}^2 = (2.50\mathbf{i} + 5.00\mathbf{j}) \text{ N}$$ ◊

(b) The magnitude of the force is $\left|\sum \mathbf{F}\right| = \sqrt{(2.50 \text{ N})^2 + (5.00 \text{ N})^2} = 5.59 \text{ N}$ ◊

---

11. An electron of mass 9.11×10^{-31} kg has an initial speed of 3.00×10^5 m/s. It travels in a straight line, and its speed increases to 7.00×10^5 m/s in a distance of 5.00 cm. Assuming its acceleration is constant, (a) determine the force on the electron and (b) compare this force with the weight of the electron, which we neglected.

Solution

G: We should expect that only a very small force is required to accelerate an electron because of its small mass, but this force is probably much greater than the weight of the electron if the gravitational force can be neglected.

O: Since this is simply a linear acceleration problem, we can use Newton's second law to find the force as long as the electron does not approach relativistic speeds (much less than 3×10^8 m/s), which is certainly the case for this problem. We know the initial and final velocities, and the distance involved, so from these we can find the acceleration needed to determine the force.

A: From $v_f^2 = v_i^2 + 2ax$ and $\sum F = ma$ we can solve for the acceleration and the force.

$$a = \frac{(v_f^2 - v_i^2)}{2x} \qquad \text{and so} \qquad \sum F = \frac{m(v_f^2 - v_i^2)}{2x}$$

(a) $\sum F = \dfrac{\left(9.11 \times 10^{-31} \text{ kg}\right)\left(\left(7.00 \times 10^5 \text{ m/s}\right)^2 - \left(3.00 \times 10^5 \text{ m/s}\right)^2\right)}{(2)(0.0500 \text{ m})}$

$\sum F = 3.64 \times 10^{-18}$ N ◊

(b) The electron's weight is

$F_g = mg = \left(9.11 \times 10^{-31} \text{ kg}\right)\left(9.80 \text{ m/s}^2\right) = 8.93 \times 10^{-30}$ N ◊

The ratio of the accelerating force to the weight is $F/F_g = 4.08 \times 10^{11}$ ◊

L: The force that causes the electron to accelerate is indeed a small fraction of a newton, but it is much greater than the gravitational force. For this reason, it is quite reasonable to ignore the weight of the electron in electric charge problems.

15. Two forces, F_1 and F_2, act on a 5.00-kg mass. If $F_1 = 20.0$ N and $F_2 = 15.0$ N, find the accelerations in (a) and (b) of Figure P5.15.

(a)   (b)

Figure P5.15

Solution   We are given that $m = 5.00$ kg.

(a) $\Sigma F = F_1 + F_2 = (20.0i + 15.0j)$ N

$$a = \frac{\Sigma F}{m} = (4.00i + 3.00j) \text{ m/s}^2 = 5.00 \text{ m/s}^2 \text{ at } 36.7° \qquad \Diamond$$

(b) $\Sigma F = F_1 + F_2 = [20.0i + (15.0 \cos 60° \, i + 15.0 \sin 60° \, j)]$ N $= (27.5i + 13.0j)$ N

$$a = \frac{\Sigma F}{m} = (5.50i + 2.60j) \text{ m/s}^2 = 6.08 \text{ m/s}^2 \text{ at } 25.3° \qquad \Diamond$$

25. A bag of cement of weight F_g hangs from three wires as shown in Figure P5.24. Two of the wires make angles θ_1 and θ_2 with the horizontal. If the system is in equilibrium, show that the tension in the left-hand wire is

$$T_1 = F_g \cos \theta_2 / \sin (\theta_1 + \theta_2)$$

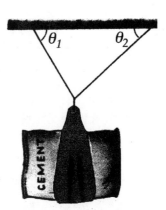

Figure P5.24

Solution

(a) Draw a free-body diagram for the knot where the three ropes are joined. Choose the x axis to be horizontal and apply Newton's second law in component form.

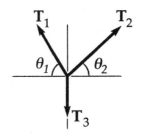

$\Sigma F_x = 0$: $T_2 \cos \theta_2 - T_1 \cos \theta_1 = 0$ (1)

$\Sigma F_y = 0$: $T_2 \sin \theta_2 + T_1 \sin \theta_1 - F_g = 0$ (2)

Solve equation (1) for $T_2 = \dfrac{T_1 \cos \theta_1}{\cos \theta_2}$

Substitute this expression for T_2 into Equation (2):

$$\left(\frac{T_1 \cos \theta_1}{\cos \theta_2} \right) \sin \theta_2 + T_1 \sin \theta_1 = F_g$$

Solve for $T_1 = \dfrac{F_g \cos \theta_2}{\cos \theta_1 \sin \theta_2 + \sin \theta_1 \cos \theta_2}$

Use the trigonometric identity $\sin(\theta_1 + \theta_2) = \cos \theta_1 \sin \theta_2 + \sin \theta_1 \cos \theta_2$ to find

$$T_1 = F_g \cos \theta_2 / \sin (\theta_1 + \theta_2) \qquad \qquad \lozenge$$

29. A 1.00-kg mass is observed to accelerate at 10.0 m/s^2 in a direction 30.0° north of east (Fig. P5.29). The force $\mathbf{F}_2$ acting on the mass has a magnitude of 5.00 N and is directed north. Determine the magnitude and direction of the force $\mathbf{F}_1$ acting on the mass.

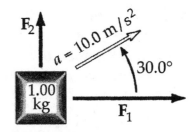

Figure P5.29

Solution

G: The net force acting on the mass is
$$\Sigma F = ma = (1\text{ kg})(10\text{ m}/s^2) = 10\text{ N}.$$ If we sketch a vector diagram of the forces drawn to scale ($\mathbf{F}_2 = \Sigma\mathbf{F} - \mathbf{F}_1$), we see that $F_1 \approx 9\text{ N}$, to the east.

O: We can find a more precise result by examining the forces in terms of vector components. For convenience, we choose directions east and north along $\mathbf{i}$ and $\mathbf{j}$, respectively.

A: $\mathbf{a} = \left[(10.0\cos 30.0°)\mathbf{i} + (10.0\sin 30.0°)\mathbf{j}\right]\text{ m}/s^2 = (8.66\mathbf{i} + 5.00\mathbf{j})\text{ m}/s^2$

From Newton's second law,
$$\Sigma\mathbf{F} = m\mathbf{a} = (1.00\text{ kg})\left(8.66\mathbf{i}\text{ m}/s^2 + 5.00\mathbf{j}\text{ m}/s^2\right) = (8.66\mathbf{i} + 5.00\mathbf{j})\text{ N}$$

and $\quad \Sigma\mathbf{F} = \mathbf{F}_1 + \mathbf{F}_2$

so $\quad \mathbf{F}_1 = \Sigma\mathbf{F} - \mathbf{F}_2 = (8.66\mathbf{i} + 5.00\mathbf{j} - 5.00\mathbf{j})\text{ N} = 8.66\mathbf{i}\text{ N} = 8.66\text{ N east} \qquad ◊$

L: Our calculated answer agrees with the prediction from the force diagram.

---

33. A block is given an initial velocity of 5.00 m/s up a frictionless 20.0° incline (see Fig. P5.32). How far up the incline does the block slide before coming to rest?

Solution Every successful physics student learns to solve inclined-plane problems.

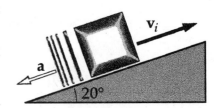

**Figure P5.32
(modified)**

Trick one: Try to set the axes in the direction of motion. In this case, take the x axis along the incline, so that $a_y = 0$.

Trick two: Recognize that the 20.0° angle between the x axis and horizontal implies a 20.0° angle between the weight vector and the y axis. Why? Because "angles are equal if their sides are perpendicular, right side to right side and left side to left side." Either you learned this theorem in geometry class, or you learn it now, since it is the theorem used often in physics.

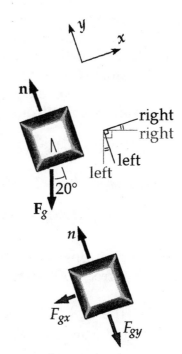

Trick three: The 20.0° angle lies between $\mathbf{F}_g$ and the y axis, so split the weight vector into its x and y components:

$$F_{gx} = -F_g \sin 20.0° \qquad\qquad F_{gy} = -F_g \cos 20.0°$$

Now, Newton's law applies for **each** axis. Applying it to the x axis,

$\Sigma F_x = ma_x$: $\qquad\qquad -F_g \sin 20° = ma_x$

Since $F_g = mg$: $\qquad\qquad a_x = -g \sin 20° = -\left(9.80 \text{ m / s}^2\right)\sin 20° = -3.35 \text{ m / s}^2$

From Eq. 2.11, $\qquad\qquad v_{xf}^2 = v_{xi}^2 + 2a_x\left(x_f - x_i\right)$

$$0 = (5.00 \text{ m / s})^2 + 2\left(-3.35 \text{ m / s}^2\right)\left(x_f - x_i\right)$$

Solving, $\qquad\qquad \left(x_f - x_i\right) = 3.73 \text{ m}$ $\qquad\qquad$ ◊

37. In the system shown in Figure P5.37, a horizontal force F_x acts on the 8.00-kg mass. (a) For what values of F_x does the 2.00-kg mass accelerate upward? (b) For what values of F_x is the tension in the cord zero? (c) Plot the acceleration of the 8.00-kg mass versus F_x. Include values of F_x from −100 N to +100 N.

Figure P5.37

Solution The blocks' weights are:

$$F_{g1} = m_1 g = (8.00 \text{ kg})(9.80 \text{ m}/\text{s}^2) = 78.4 \text{ N}$$
$$F_{g2} = m_2 g = (2.00 \text{ kg})(9.80 \text{ m}/\text{s}^2) = 19.6 \text{ N}$$

Let **T** be the tension in the connecting cord and draw a free-body diagram for each block.

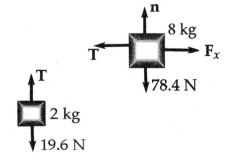

(a) For the 2-kg mass, with the y-axis directed upwards, $\Sigma F_y = ma_y$ yields

$$T - 19.6 \text{ N} = (2.00 \text{ kg})a_y \qquad (1)$$

Thus, we find that $a_y > 0$ when $T > 19.6 \text{ N}$. For acceleration by the system of two blocks, $F_x \geq T$, so $F_x > 19.6 \text{ N}$ whenever the 2-kg mass accelerates upward. ◊

(b) Looking at the free-body diagram for the 8.00-kg mass, and taking the $+x$ direction to be directed to the right, we can apply Newton's law in the horizontal direction:

From $\Sigma F_x = ma_x$, $\qquad\qquad -T + F_x = (8.00 \text{ kg})a_x \qquad (2)$

If $T = 0$, and $a_x \leq -9.80 \text{ m}/\text{s}^2$ $\qquad F_x \leq -78.4 \text{ N}$ ◊

(c)   If $F_x \geq -78.4$ N, then both equations (1) and (2) apply. Substituting the value for T from the (1) into (2),

$$-(2.00 \text{ kg})a_y - 19.6 \text{ N} + F_x = (8.00 \text{ kg})a_x$$

In this case $a_x = a_y$: $F_x = (8.00 \text{ kg} + 2.00 \text{ kg})a_x + 19.6 \text{ N}$

$$a_x = \frac{F_x}{10.0 \text{ kg}} - 1.96 \text{ m/s}^2 \quad (F_x \geq -78.4 \text{ N}) \qquad (3)$$

From part (b), we find that if $F_x \leq -78.4$ N, then $T = 0$ and equation (2) becomes $F_x = (8.00 \text{ kg})a_x$:

$$a_x = F_x/8.00 \text{ kg} \qquad\qquad (F_x \leq -78.4 \text{ N}) \qquad (4)$$

We can now graph equations (3) and (4), and list some sample values, below:

F_x	a_x
–100 N	-12.5 m/s²
–50.0 N	–6.96 m/s²
0	–1.96 m/s²
50.0 N	3.04 m/s²
100 N	8.04 m/s²
150 N	13.04 m/s²

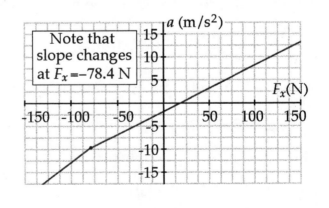

39. A 72.0-kg man stands on a spring scale in an elevator. Starting from rest, the elevator ascends, attaining its maximum speed of 1.20 m/s in 0.800 s. It travels with this constant speed for 5.00 s. The elevator then undergoes a uniform acceleration in the negative y direction for 1.50 s and comes to rest. What does the spring scale register (a) before the elevator starts to move? (b) during the first 0.800 s? (c) while the elevator is traveling at constant speed? (d) during the time it is slowing down?

Solution

G: Based on sensations experienced riding in an elevator, we expect that the man should feel slightly heavier when the elevator first starts to ascend, lighter when it comes to a stop, and his normal weight when the elevator is not accelerating. His apparent weight is registered by the spring scale beneath his feet, so the scale force should correspond to the force he feels through his legs (Newton's third law).

O: We should draw free body diagrams for each part of the elevator trip and apply Newton's second law to find the scale force. The acceleration can be found from the change in speed divided by the elapsed time.

A: Consider the free-body diagram of the man shown below. The force F is the upward force exerted on the man by the scale, and his weight is

$$F_g = mg = (72.0\ \text{kg})(9.80\ \text{m/s}^2) = 706\ \text{N}$$

With $+y$ defined to be up, Newton's 2nd law gives

$$\sum F_y = +F_s - F_g = ma$$

Thus, we calculate the upward scale force to be

$$F_s = 706\ \text{N} + (72.0\ \text{kg})a \tag{1}$$

Where a is the acceleration the man experiences as the elevator changes speed.

(a) Before the elevator starts moving, the elevator's acceleration is zero $(a = 0)$. Therefore, Equation 1 gives the force exerted by the scale on the man as 706 N (upward), and the man exerts a downward force of 706 N on the scale. ◊

(b) During the first 0.800 s of motion, the man accelerates at a rate of $a = \Delta v / \Delta t = (1.20 \text{ m} / \text{s} - 0)/0.800 \text{ s} = 1.50 \text{ m} / \text{s}^2$. Substituting a into equation (1) then gives:

$$F = 706 \text{ N} + (72.0 \text{ kg})(+1.50 \text{ m}/\text{s}^2) = 814 \text{ N} \qquad ◊$$

(c) While the elevator is traveling upward at constant speed, the acceleration is zero and equation (1) again gives a scale force

$$F = 706 \text{ N} \qquad ◊$$

(d) During the last 1.50 s, the elevator starts with an upward velocity of 1.20 m/s, and comes to rest with an acceleration of $a = \Delta v / \Delta t = (0 - 1.20 \text{ m} / \text{s})/1.50 \text{ s} = -0.800 \text{ m} / \text{s}^2$. Thus, the force of the man on the scale is:

$$F = 706 \text{ N} + (72.0 \text{ kg})(-0.800 \text{ m}/\text{s}^2) = 648 \text{ N} \qquad ◊$$

L: The calculated scale forces are consistent with our predictions. This problem could be extended to a couple of extreme cases. If the acceleration of the elevator were $+9.80 \text{ m}/\text{s}^2$, then the man would feel twice as heavy, and if $a = -9.80 \text{ m}/\text{s}^2$ (free fall), then he would feel "weightless", even though his true weight ($F_g = mg$) would remain the same.

45. A 3.00-kg block starts from rest at the top of a 30.0° incline and slides a distance of 2.00 m down the incline in 1.50 s. Find (a) the magnitude of the acceleration of the block, (b) the coefficient of kinetic friction between block and plane, (c) the frictional force acting on the block, and (d) the speed of the block after it has slid 2.00 m.

Solution

(a)   At constant acceleration, $x_f - x_i = v_i t + \frac{1}{2} a t^2$

So, $$a = \frac{2(x_f - x_i - v_i t)}{t^2} = \frac{2(2.00 \text{ m} - 0 - 0)}{(1.50 \text{ s})^2} = 1.78 \text{ m / s}^2 \qquad \lozenge$$

Solving for answer (c) next:

(c)   Choose the x axis parallel to the incline, take the positive direction down the incline (in the direction of the acceleration) and apply the second law.

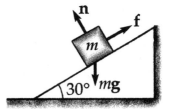

$$\Sigma F_x = mg \sin \theta - f = ma$$

or $$f = m(g \sin \theta - a)$$

$$f = (3.00 \text{ kg})\left[(9.80 \text{ m / s}^2)\sin 30.0° - 1.78 \text{ m / s}^2\right] = 9.36 \text{ N} \qquad \lozenge$$

(b)   By Newton's second law, $\Sigma F_y = n - mg \cos \theta = 0$, so $n = mg \cos \theta$

Also, $f = \mu_k n$ so $$\mu_k = \frac{f}{mg \cos \theta} = \frac{9.36 \text{ N}}{(3.00 \text{ kg})(9.80 \text{ m/s}^2)\cos 30.0°} = 0.368 \quad \lozenge$$

(d) $v_f^2 = v_i^2 + 2a(x_f - x_i)$

$$v_f = \sqrt{0 + (2)(1.78 \text{ m / s}^2)(2.00 \text{ m} - 0)} = 2.67 \text{ m / s} \qquad \lozenge$$

51. Two blocks connected by a rope of negligible mass are being dragged by a horizontal force **F** (Fig. P5.51). Suppose that $F = 68.0$ N, $m_1 = 12.0$ kg, $m_2 = 18.0$ kg, and the coefficient of kinetic friction between each block and the surface is 0.100. (a) Draw a free-body diagram for each block. (b) Determine the tension, T, and the magnitude of the acceleration of the system.

Figure P5.51

Solution   (a)   The free-body diagrams for m_1 and m_2 are:

(b)   Use the free-body diagrams to apply Newton's second law.

For m_1: $\Sigma F_x = T - f_1 = m_1 a$ or $T = m_1 a + f_1$ (1)

          $\Sigma F_y = n_1 - m_1 g = 0$ or $n_1 = m_1 g$

Also, $f_1 = \mu_1 n_1 = (0.100)(12.0 \text{ kg})(9.80 \text{ m / s}^2) = 11.8$ N

For m_2: $\Sigma F_x = F - T - f_2 = m_2 a$ or $T = F - m_2 a - f_2$ (2)

          $\Sigma F_y = n_2 - m_2 g = 0$ or $n_2 = m_2 g$

Also, $f_2 = \mu n_2 = (0.100)(18.0 \text{ kg})(9.80 \text{ m / s}^2) = 17.6$ N

Substituting T from equation (1) into (2), we get $\quad m_1 a + f_1 = F - m_2 a - f_2$

Solving for a. $a = \dfrac{F - f_1 - f_2}{m_1 + m_2} = \dfrac{(68.0 - 11.8 - 17.6) \text{ N}}{(12.0 + 18.0) \text{ kg}} = 1.29 \text{ m / s}^2$ ◊

From Eq. (1), $T = m_1 a + f_1 = (12.0 \text{ kg})(1.29 \text{ m / s}^2) + 11.8 \text{ N} = 27.2 \text{ N}$ ◊

55. An inventive child named Pat wants to reach an apple in a tree without climbing the tree. Sitting in a chair connected to a rope that passes over a frictionless pulley (Fig. P5.55), Pat pulls on the loose end of the rope with such a force that the spring scale reads 250 N. Pat's weight is 320 N, and the chair weighs 160 N. (a) Draw free-body diagrams for Pat and the chair considered as separate systems and draw another diagram for Pat and the chair considered as one system. (b) Show that the acceleration of the system is **upward** and find its magnitude. (c) Find the force Pat exerts on the chair.

Figure P5.55

Solution   (a)

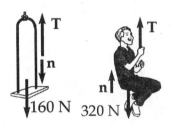 

(b) First consider Pat and the chair as the system. Note that **two** ropes support the system, and $T = 250$ N in each rope.

$\Sigma F = ma$: $\qquad\qquad 2T - (160 + 320)\ \text{N} = ma$

where $\qquad\qquad m = \dfrac{480\ \text{N}}{9.80\ \text{m/s}^2} = 49.0\ \text{kg}$

Solving for a gives $\qquad a = \dfrac{(500 - 480)\ \text{N}}{49.0\ \text{kg}} = 0.408\ \text{m/s}^2$

(c) $\sum F\ (\text{on Pat}) = n + T - 320\ \text{N} = ma$ $\qquad$ where $\qquad m = \dfrac{320\ \text{N}}{9.80\ \text{m/s}^2} = 32.7\ \text{kg}$

$n = ma + 320\ \text{N} - T = 32.7\ \text{kg}\ (0.408\ \text{m/s}^2) + 320\ \text{N} - 250\ \text{N} = 83.3\ \text{N}$

59. A mass M is held in place by an applied force **F** and a pulley system as shown in Figure P5.59. The pulleys are massless and frictionless. Find (a) the tension in each section of rope, T_1, T_2, T_3, T_4, and T_5, and (b) the magnitude of **F**. (**Hint:** draw a free-body diagram.)

Solution Draw free-body diagrams for each element, and apply Newton's 2nd law. (All forces are along the y axis.)

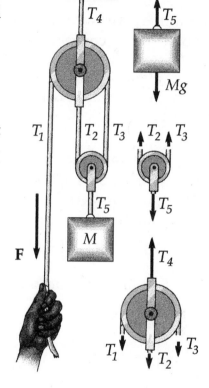

For mass M, $\Sigma F = 0$ becomes $T_5 - Mg = 0$:

$$T_5 = Mg \qquad \Diamond$$

Assume that the pulleys are frictionless, and that the tension is constant throughout each light, continuous rope. Therefore, $T_1 = T_2 = T_3$

For the small pulley,

$\Sigma F = 0$: $\qquad T_2 + T_3 - T_5 = 0$

$2T_2 = T_5$: $\qquad T_2 = T_5/2 = Mg/2$

$\qquad\qquad T_1 = T_2 = T_3 = Mg/2 \qquad \Diamond$

$\qquad\qquad F = T_1 = Mg/2 \qquad \Diamond$

For the large pulley,

$\Sigma F = 0$: $\qquad T_4 - T_1 - T_2 - T_3 = 0$

$\qquad\qquad T_4 = T_1 + T_2 + T_3 = 3Mg/2 \qquad \Diamond$

---

69. What horizontal force must be applied to the cart shown in Figure P5.69 so that the blocks remain stationary relative to the cart? Assume all surfaces, wheels, and pulley are frictionless. (**Hint:** Note that the force exerted by the string accelerates m_1.)

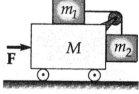

Figure P5.69

Solution

Draw separate free-body diagrams for blocks m_1 and m_2.

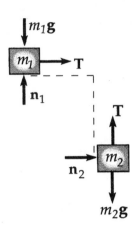

Remembering that normal forces are always perpendicular to the contacting surface, and always **push** on a body, draw n_1 and n_2 as shown. Note that m_2 should be in **contact** with the cart, and therefore does have a normal force from the cart. Remembering that ropes always **pull** on bodies in the direction of the rope, draw the tension force **T**. Finally, draw the gravitational force on each block, which always points downwards.

G: What can keep m_2 from falling? Only tension in the cord connecting it with m_1. This tension pulls forward on m_1 to accelerate that mass. We might guess that the acceleration is proportional to both m_2 and g and inversely proportional to m_1, so perhaps $a = m_2 g / m_1$. We should also expect the applied force to be proportional to the total mass of the system.

O: Use $\Sigma F = ma$ and the free-body diagrams above.

A: For m_2, $\qquad\qquad\qquad\qquad T - m_2 g = 0 \quad$ or $\quad T = m_2 g$
For m_1, $\qquad\qquad\qquad\qquad T = m_1 a \quad$ or $\quad a = T/m_1$
Substituting for T, we have $\qquad a = m_2 g / m_1$
For all 3 blocks, $\qquad\qquad\qquad F = (M + m_1 + m_2)a$
Therefore, $\qquad\qquad\qquad\qquad F = (M + m_1 + m_2)(m_2 g / m_1) \qquad\qquad \Diamond$

L: Even though this problem did not have a numerical solution, we were still able to rationalize the algebraic form of the solution. This technique does not always work, especially for complex situations, but often we can think through a problem to see if an equation for the solution makes sense based on the physical principles we know.

75. A van accelerates down a hill (Fig. P5.75), going from rest to 30.0 m/s in 6.00 s. During the acceleration, a toy ($m = 0.100$ kg) hangs by a string from the van's ceiling. The acceleration is such that the string remains perpendicular to the ceiling. Determine (a) the angle θ and (b) the tension in the string.

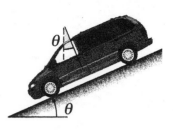

Figure P5.75

Solution The acceleration is obtained from

$$v_f = v_i + at: \qquad 30.0\ \text{m/s} = 0 + a(6.00\ \text{s})$$

$$a = 5.00\ \text{m/s}^2$$

The toy moves with the same acceleration as the van, 5.00 m/s² parallel to the hill. We take the x axis in this direction, so

$$a_x = 5.00\ \text{m/s}^2$$

and $\qquad a_y = 0$

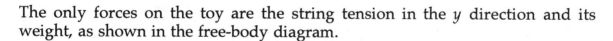

The only forces on the toy are the string tension in the y direction and its weight, as shown in the free-body diagram.

$$F_g = mg = (0.100\ \text{kg})(9.80\ \text{m/s}^2) = 0.980\ \text{N}$$

Using $\Sigma F_x = ma_x$: $\qquad (0.980\ \text{N})\sin\theta = (0.100\ \text{kg})(5.00\ \text{m/s}^2)$

(a) $\qquad\qquad\qquad\qquad \sin\theta = \dfrac{0.500}{0.980} \qquad \text{and} \qquad \theta = 30.7° \qquad\qquad ◊$

Using $\Sigma F_y = ma_y$: $\qquad +T - (0.980\ \text{N})\cos\theta = 0$

(b) $\qquad\qquad\qquad\qquad T = (0.980\ \text{N})\cos 30.7° = 0.843\ \text{N} \qquad\qquad ◊$

Chapter
6

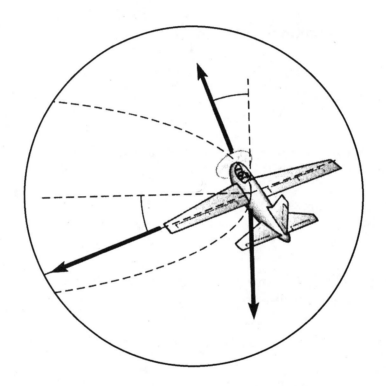

Circular Motion and
Other Applications of
Newton's Laws

CIRCULAR MOTION AND OTHER APPLICATIONS OF NEWTON'S LAWS

INTRODUCTION

In the previous chapter we introduced Newton's laws of motion and applied them to situations involving linear motion. In this chapter we shall apply Newton's laws of motion to circular motion. We shall also discuss the motion of an object when observed in an accelerated, or noninertial, frame of reference and the motion of an object through a viscous medium. Finally, we conclude this chapter with a brief discussion of the fundamental forces in nature.

EQUATIONS AND CONCEPTS

When an object of mass m moves uniformly in a circular path, the net force acting on the object is a centripetal force (directed toward the center of the circular path.)

$$F_r = ma_r = m\frac{v^2}{r} \tag{6.1}$$

When Newton's second law is applied to the motion of an object falling vertically through a viscous medium, the motion can be described by a differential equation. The constant, b, has a value which depends on the properties of the medium and the dimensions and shape of the object.

$$\frac{dv}{dt} = g - \frac{b}{m}v \tag{6.4}$$

As the resistive force approaches the weight, the acceleration approaches zero, and the object reaches a terminal speed, v_t. Equation 6.5 gives the speed as a function of time, where the object is released from rest at $t = 0$.

$$v = v_t\left(1 - e^{-t/\tau}\right) \tag{6.5}$$

where

$$v_t = \frac{mg}{b}$$

and

$$\tau = \frac{m}{b}$$

SUGGESTIONS, SKILLS, AND STRATEGIES

Section 6.4 deals with the motion of a body through a gas or liquid. If you covered this section in class, the following solution to Equation 6.3 (when the resistive force $\mathbf{R} = -b\mathbf{v}$) may be useful to know:

$$\frac{dv}{dt} = g - \frac{b}{m}v \tag{6.4}$$

In order to solve this equation, it is convenient to change variables. If we let $y = g - (b/m)v$, it follows that $dy = -(b/m)dv$. With these substitutions, Equation 6.4 becomes

$$-\left(\frac{m}{b}\right)\frac{dy}{dt} = y \qquad \text{or} \qquad \frac{dy}{y} = -\frac{b}{m}dt$$

Integrating this expression (now that the variables are separated) gives

$$\int \frac{dy}{y} = -\frac{b}{m} \int dt \qquad \text{or} \qquad \ln y = -\frac{b}{m}t + \text{constant}$$

This is equivalent to $y = (\text{const})\, e^{-bt/m} = g - (b/m)v$. Taking $v = 0$ at $t = 0$, we see $\text{const} = g$.

so $$v = \frac{mg}{b}(1 - e^{-bt/m}) = v_t(1 - e^{-t/\tau}) \qquad (6.5)$$

where $\tau = m/b$

REVIEW CHECKLIST

▷ Discuss Newton's universal law of gravity (the inverse-square law), and understand that it is an **attractive** force between two **particles** separated by a distance r.

▷ Apply Newton's second law to uniform and nonuniform circular motion.

▷ Remember that Newton's laws are only valid in constant velocity (inertial) frames of reference. When motion is described by an observer in an accelerated (noninertial) frame, the observer must invent fictitious forces which arise due to the acceleration of the reference frame.

▷ Recognize that motion of an object through a liquid or gas can involve resistive forces which have a complicated velocity dependence.

ANSWERS TO SELECTED CONCEPTUAL QUESTIONS

6. It has been suggested that rotating cylinders about 10 mi in length and 5 mi in diameter be placed in space and used as colonies. The purpose of the rotation is to simulate gravity for the inhabitants. Explain this concept for producing an effective gravity.

Answer

The centripetal force on the inhabitants is provided by the normal force exerted on them by the cylinder wall. If the rotation rate is adjusted to such a speed that this normal force is equal to their weight on Earth, the inhabitants would not be able to distinguish between this artificial gravity and normal gravity.

□   □   □   □

7. Why does a pilot tend to black out when pulling out of a steep dive?

Answer

When pulling out of a dive, blood leaves the pilot's head because the pilot's blood pressure is not great enough to compensate for both the gravitational force and the centripetal acceleration of the airplane. This loss of blood from the brain can cause the pilot to black out.

□   □   □   □

SOLUTIONS TO SELECTED END-OF-CHAPTER PROBLEMS

3.   A light string can support a stationary hanging load of 25.0 kg before breaking.   A 3.00-kg mass attached to the string rotates on a horizontal, frictionless table in a circle of radius 0.800 m.   What range of speeds can the mass have before the string breaks?

Solution

The string will break if the tension T exceeds the weight

$$m g = (25.0 \text{ kg})(9.80 \text{ m} / \text{s}^2) = 245 \text{ N}$$

As the 3.00-kg mass rotates in a horizontal circle, the tension provides the central force.

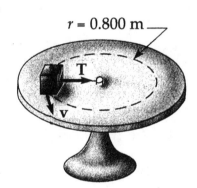

$r = 0.800$ m

From $\Sigma F = ma,$ $T = \dfrac{mv^2}{r}$

Then, $v \le \sqrt{\dfrac{rT}{m}}$ can be solved to yield

$$v \le \sqrt{\frac{(0.800 \text{ m})(245 \text{ N})}{(3.00 \text{ kg})}} = 8.08 \text{ m} / \text{s}$$

So the mass can have speeds between 0 and 8.08 m/s.   ◊

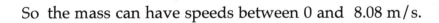

9. A coin placed 30.0 cm from the center of a rotating, horizontal turntable slips when its speed is 50.0 cm/s. (a) What provides the force in the radial direction when the coin is stationary relative to the turntable? (b) What is the coefficient of static friction between coin and turntable?

Solution

(a) The central force is provided by the force of static friction.

(b) The motion of the coin is shown in the upper diagram. The forces on the coin, shown in the free-body diagram below, are the normal force, the weight, and the force of static friction. The only force in the radial direction is **f**.

$v = 50.0 \text{ cm/s}$

ω

v

a

30.0 cm

n

f

mg

Therefore, $\qquad f = m \dfrac{v^2}{r}$

The normal force balances the weight, so $n - mg = 0$: $\qquad n = mg$

The frictional force, $\qquad\qquad\qquad\qquad\qquad f = \mu_s n = \mu_s mg$

The coin slips when the frictional force is exactly enough to cause the centripetal acceleration. $\qquad \mu_s mg = mv^2 / r$

$$v^2 / r g = \mu_s$$

Taking $r = 30.0 \text{ cm}$, $v = 50.0 \text{ cm / s}$, and $g = 980 \text{ cm / s}^2$,

$$\mu_s = \frac{(50.0 \text{ cm / s})^2}{(30.0 \text{ cm})(980 \text{ cm / s}^2)} = 0.0850 \qquad\qquad \Diamond$$

114

11. A crate of eggs is located in the middle of the flatbed of a pickup truck as the truck negotiates an unbanked curve in the road. The curve may be regarded as an arc of a circle of radius 35.0 m. If the coefficient of static friction between crate and truck is 0.600, how fast can the truck be moving without the crate sliding?

Solution

Call the mass of the egg crate m. The forces on it are its weight mg vertically down, the normal force $\mathbf{n}$ of the truck bed vertically up, and static friction $\mathbf{f}_s$ directed to oppose relative sliding motion of the crate over the truck bed. The friction force is directed radially inward. It is the only horizontal force on the crate, so it must provide the centripetal acceleration. When the truck has maximum speed, friction f_s will have its maximum value with $f_s = \mu_s n$.

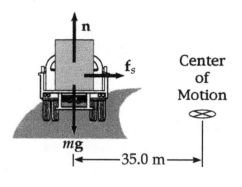

$\Sigma F_y = ma_y$ gives $n - mg = 0$ or $n = mg$

$\Sigma F_x = ma_x$ gives $f_s = ma_r$

From these two equations, $\mu_s n = mv^2/r$

$$\mu_s mg = mv^2/r$$

The mass divides out, leaving $v = \sqrt{\mu_s gr} = \sqrt{(0.600)(9.80 \text{ m} / \text{s}^2)(35.0 \text{ m})}$

$$v = 14.3 \text{ m} / \text{s} \hspace{3cm} \Diamond$$

15. Tarzan ($m = 85.0$ kg) tries to cross a river by swinging from a vine. The vine is 10.0 m long, and his speed at the bottom of the swing (as he just clears the water) is 8.00 m/s. Tarzan doesn't know that the vine has a breaking strength of 1 000 N. Does he make it safely across the river?

Solution

The forces acting on Tarzan are the force of gravity $m\mathbf{g}$ and the force of tension $\mathbf{T}$. At the lowest point in his motion, $\mathbf{T}$ is upward and $m\mathbf{g}$ is downward as in the free-body diagram.

$$F_g = mg = (85.0 \text{ kg})(9.80 \text{ m/s}^2) = 8.33 \times 10^2 \text{ N} \quad \text{(down)}$$

Thus, Newton's second law gives

$$T - mg = mv^2/r$$

1011-7-23-50
© 1950, Edgar Rice Burroughs, Inc. -
™ Reg. U.S. Pat. Off.,
Distributed by United Feature Syndicate, Inc.

Solving for T, with $v = 8.00$ m / s, $r = 10.0$ m, and $m = 85.0$ kg, gives

$$T = m\left(g + \frac{v^2}{r}\right) = (85.0 \text{ kg})\left(9.80 \text{ m / s}^2 + \frac{(8.00 \text{ m / s})^2}{10.0 \text{ m}}\right) = 1.38 \times 10^3 \text{ N} \qquad \lozenge$$

Since T **exceeds** the breaking strength of the vine (1 000 N), Tarzan **doesn't make it !** The vine breaks **before** he reaches the bottom of the swing. $\qquad \lozenge$

---

17. A 40.0-kg child sits in a swing supported by two chains, each 3.00 m long. If the tension in each chain at the lowest point is 350 N, find (a) the child's speed at the lowest point and (b) the force exerted by the seat on the child at the lowest point. (Neglect the mass of the seat.)

Solution

G: If the tension in each chain is 350 N at the lowest point, then the force of the seat on the child should be twice this force or 700 N. The child's speed is not as easy to determine, but somewhere between 0 and 10 m/s would be reasonable for the situation described.

O: We should first draw a free body diagram that shows the forces acting on the seat and apply Newton's laws to solve the problem.

A: We can see from the diagram that the only forces acting on the system of child+seat are the tension in the two chains and the weight of the boy:

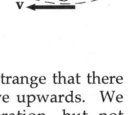

$$\Sigma F = 2T - mg = ma$$
$$\Sigma F = F_{net} = 2(350 \text{ N}) - (40.0 \text{ kg})(9.80 \text{ m} / \text{s}^2) = 308 \text{ N (up)}$$

Because $a = v^2/r$ is the centripetal acceleration,

$$v = \sqrt{\frac{F_{net}\,r}{m}} = \sqrt{\frac{(308 \text{ N})(3.00 \text{ m})}{40.0 \text{ kg}}} = 4.81 \text{ m} / \text{s} \qquad \Diamond$$

The child feels a normal force exerted by the seat equal to the total tension in the chains:

$$\mathbf{n} = 2(350 \text{ N}) = 700 \text{ N (upwards)} \qquad \Diamond$$

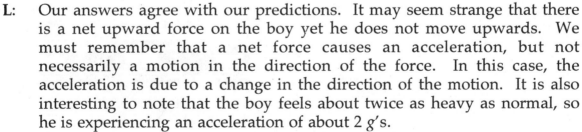

L: Our answers agree with our predictions. It may seem strange that there is a net upward force on the boy yet he does not move upwards. We must remember that a net force causes an acceleration, but not necessarily a motion in the direction of the force. In this case, the acceleration is due to a change in the direction of the motion. It is also interesting to note that the boy feels about twice as heavy as normal, so he is experiencing an acceleration of about 2 g's.

19. A pail of water is rotated in a vertical circle of radius 1.00 m. What must be the minimum speed of the pail at the top of the circle if no water is to spill out?

Solution

At the top of the vertical circle, the speed must be great enough so the central force is equal to or greater than the weight, mg, of the water.

That is,

$$\frac{mv^2}{r} \geq mg \qquad \text{or} \qquad v^2 \geq rg$$

At the minimum speed, $\qquad v_{min}^2 = rg$

so

$$v_{min} = \sqrt{rg} = \sqrt{(1.00 \text{ m})(9.80 \text{ m}/\text{s}^2)} = 3.13 \text{ m}/\text{s} \quad \lozenge$$

25. A 0.500-kg object is suspended from the ceiling of an accelerating boxcar as was seen in Figure 6.13. If $a = 3.00 \text{ m}/\text{s}^2$, find (a) the angle that the string makes with the vertical and (b) the tension in the string.

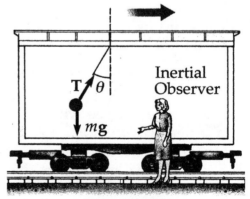

Figure 6.13

Solution

G: If the horizontal acceleration were zero, then the angle would be 0°, and if $a=g$, then the angle would be 45°, but since the acceleration is 3.00 m/s², a reasonable estimate of the angle is about 20°. Similarly, the tension in the string should be slightly more than the weight of the object, which is about 5 N.

O: We will apply Newton's second law to solve the problem.

A: The only forces acting on the suspended object are the force of gravity $m\mathbf{g}$ and the force of tension $\mathbf{T}$, as shown in the free-body diagram. Applying Newton's second law in the x and y directions,

$$\Sigma F_x = T\sin\theta = ma \qquad\qquad (1)$$

$$\Sigma F_y = T\cos\theta - mg = 0$$

or $\qquad T\cos\theta = mg \qquad\qquad\qquad\qquad (2)$

(a) Dividing equation (1) by (2), $\quad \tan\theta = \dfrac{a}{g} = \dfrac{3.00 \text{ m}/\text{s}^2}{9.80 \text{ m}/\text{s}^2} = 0.306$

Solving for θ, $\qquad\qquad\qquad\qquad \theta = 17.0° \qquad\qquad\qquad\qquad \lozenge$

(b) From equation (1), $\quad T = \dfrac{ma}{\sin\theta} = \dfrac{(0.500 \text{ kg})(3.00 \text{ m}/\text{s}^2)}{\sin(17.0°)} = 5.12 \text{ N} \quad \lozenge$

L: Our answers agree with our original estimates. This problem is very similar to Problem 5.30, so the same concept seems to apply to various situations.

27. A person stands on a scale in an elevator. As the elevator starts, the scale has a constant reading of 591 N. As the elevator later stops, the scale reading is 391 N. Assume the magnitude of the acceleration is the same during starting and stopping, and determine (a) the weight of the person, (b) the person's mass, and (c) the acceleration of the elevator.

Solution   The scale reads the upward normal force exerted by the floor on the passenger. The maximum force occurs during upward acceleration (when starting an upward trip or ending a downward trip). The minimum normal force occurs with downward acceleration. For each respective situation,

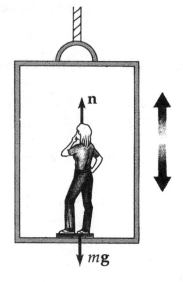

$\Sigma F_y = ma_y$ starting $+591 \text{ N} - mg = +ma$

stopping $+391 \text{ N} - mg = -ma$

(a)   These two simultaneous equations can be added to eliminate a and solve for mg.

$$+591 \text{ N} - mg + 391 \text{ N} - mg = 0 \quad \text{or} \quad 982 \text{ N} - 2mg = 0$$

$$w = mg = \frac{982 \text{ N}}{2} = 491 \text{ N} \qquad \Diamond$$

(b) $m = \dfrac{w}{g} = \dfrac{491 \text{ N}}{9.80 \text{ m} / \text{s}^2} = 50.1 \text{ kg}$ $\Diamond$

(c)   Substituting back gives $+591 \text{ N} - 491 \text{ N} = (50.1 \text{ kg})a$

$$a = \frac{100 \text{ N}}{50.1 \text{ kg}} = 2.00 \text{ m} / \text{s}^2 \qquad \Diamond$$

35. A small, spherical bead of mass 3.00 g is released from rest at $t = 0$ in a bottle of liquid shampoo. The terminal speed is observed to be $v_t = 2.00$ cm / s. Find (a) the value of the constant b in Equation 6.4, (b) the time τ it takes to reach $0.630v_t$, and (c) the value of the resistive force when the bead reaches terminal speed.

Solution

(a) The speed v varies with time according to Equation 6.4,

$$v = \frac{mg}{b}\left(1 - e^{-bt/m}\right) = v_t\left(1 - e^{-bt/m}\right)$$

where $v_t = mg/b$ is the terminal speed.

Hence, $\qquad b = \dfrac{mg}{v_t} = \dfrac{\left(3.00\times10^{-3}\text{ kg}\right)\left(9.80\text{ m / s}^2\right)}{2.00\times10^{-2}\text{ m / s}} = 1.47\text{ N}\cdot\text{s / m} \qquad \lozenge$

(b) To find the time for v to reach $0.630v_t$, we substitute $v = 0.630v_t$ into Equation 6.4, giving

$$0.630v_t = v_t\left(1 - e^{-bt/m}\right) \qquad \text{or} \qquad 0.370 = e^{-(1.47t/0.00300)}$$

Solve for t by taking the log of each side of the equation:

$$\ln(0.370) = -\frac{1.47t}{3.00\times10^{-3}} \qquad \text{and} \qquad t = 2.03\times10^{-3}\text{ s} \qquad \lozenge$$

(c) At terminal speed, $\qquad R = v_t b = mg$

Therefore, $\qquad R = \left(3.00\times10^{-3}\text{ kg}\right)\left(9.80\text{ m / s}^2\right) = 2.94\times10^{-2}\text{ N} \quad \lozenge$

•

37. A motor boat cuts its engine when its speed is 10.0 m/s and coasts to rest. The equation governing the motion of the motorboat during this period is $v = v_i e^{-ct}$, where v is the speed at time t, v_i is the initial speed, and c is a constant. At $t = 20.0$ s, the speed is 5.00 m/s. (a) Find the constant c. (b) What is the speed at $t = 40.0$ s? (c) Differentiate the expression for $v(t)$ and thus show that the acceleration of the boat is proportional to the speed at any time.

Solution

(a) We must fit the equation $v = v_i e^{-ct}$ to the two data points:

Substituting $t = 0$ and $v = 10.0$ m/s,

$$10.0 \text{ m/s} = v_i e^0 = (v_i)(1) \qquad \text{so} \qquad v_i = 10.0 \text{ m/s}$$

Substituting $t = 20.0$ s and $v = 5.00$ m/s,

$$5.00 \text{ m/s} = (10.0 \text{ m/s})e^{-c(20.0 \text{ s})} \qquad \text{or} \qquad 0.500 = e^{c(20.0 \text{ s})}$$

$$\ln(0.500) = (-c)(20.0 \text{ s})$$

$$c = \frac{-\ln(0.500)}{20.0 \text{ s}} = 0.0347 \text{ s}^{-1} \qquad\qquad\qquad ◊$$

(b) At all times $\qquad v = (10.0 \text{ m/s})e^{-(0.0347 \text{ /s})t}$

At $t = 40.0$ s, $\qquad v = (10.0 \text{ m/s})e^{-(0.0347 \text{ /s})(40.0 \text{ s})} = 2.50 \text{ m/s} \qquad\qquad ◊$

(c) The acceleration is the rate-of-change of velocity:

$$a = \frac{dv}{dt} = \frac{d}{dt}v_i e^{-ct} = v_i\left(e^{-ct}\right)(-c)$$

$$= -c\left(v_i e^{-ct}\right) = -cv = \left(-0.0347 \text{ s}^{-1}\right)v$$

Thus, the acceleration is a negative constant times the speed. ◊

---

41. A hailstone of mass 4.80×10^{-4} kg falls through the air and experiences a net force given by $F = -mg + Cv^2$, where $C = 2.50 \times 10^{-5}$ kg / m. (a) Calculate the terminal speed of the hailstone. (b) Use Euler's method of numerical analysis to find the speed and position of the hailstone at 0.2-s intervals, taking the initial speed to be zero. Continue the calculation until the hailstone reaches 99% of terminal speed.

Solution

(a) Since the force is proportional to the acceleration of the hailstone, we can rewrite the given equation as

$$F = ma = m\frac{dv}{dt} = -mg + Cv^2 \qquad \text{or} \qquad \frac{dv}{dt} = -g + \frac{C}{m}v^2 \qquad (1)$$

At terminal speed, the left-hand term dv/dt is equal to zero, so we can then solve the equation $0 = -g + Cv^2/m$ for the velocity.

$$v = \sqrt{\frac{gm}{C}} = \sqrt{\frac{\left(9.80 \text{ m / s}^2\right)\left(4.80 \times 10^{-4} \text{ kg}\right)}{2.50 \times 10^{-5} \text{ kg / m}}} = 13.7 \text{ m / s} \qquad ◊$$

(b) In using euler's method, we begin with $x=0$ and $v=0$. At each interval, we can calculate new values for x and v with the equations $x'=x+v\Delta t$ and $v'=v+a\Delta t$. On the other hand, at any moment the acceleration must be found by equation (1) above:

$$a' = -g + \frac{C}{m}(v')^2 = \left(-9.80 \text{ m}/\text{s}^2\right) + \left(0.0521 \text{ m}^{-1}\right)(v')^2$$

Calculating these values in tabular form, we find that the hailstone reaches 99% of terminal velocity after 3.4 seconds.

Time elapsed (s) $t' = t + 0.20$ s	Position (m) $x' = x + (\Delta t)v$	Velocity (m/s) $v' = v + (\Delta t)a$	Acceleration (m/s²) $a' = (0.0521)(v')^2 - g$
0.00	0	0	−9.80
0.20	0	−1.96	−9.60
0.40	−0.392	−3.88	−9.02
0.60	−1.17	−5.68	−8.11

and so on, until...

3.00	−27.4	−13.5	−0.321
3.20	−30.1	−13.55	−0.230
3.40	−32.8	−13.60	−0.165

◊

53. Because the Earth rotates about its axis, a point on the equator experiences a centripetal acceleration of 0.0337 m/s², while a point at the poles experiences no centripetal acceleration. (a) Show that at the equator the gravitational force on an object (the true weight) must exceed the object's apparent weight. (b) What is the apparent weight at the equator and at the poles of a person having a mass of 75.0 kg? (Assume the Earth is a uniform sphere and take $g = 9.800$ m/s².)

Solution

G: Since the centripetal acceleration is a small fraction (~0.3%) of g, we should expect that a person would have an apparent weight that is just slightly less at the equator than at the poles due to the rotation of the Earth.

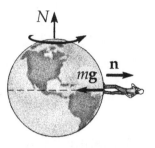

O: We will apply Newton's second law and the equation for centripetal acceleration.

A: (a) Let **n** represent the force exerted on the person by a scale, which is the "apparent weight." The true weight is $m\mathbf{g}$. Summing up forces on the object in the direction towards the Earth's center gives

$$mg - n = ma_c \qquad\qquad (1)$$

where
$$a_c = v^2/R_e = 0.0337 \text{ m / s}^2$$

is the centripetal acceleration directed toward the center of the Earth.
Thus, we see that $\quad n = m(g - a_c) < mg$
or $\qquad\qquad\qquad mg = n + ma_c > n \qquad\qquad (2) \qquad \Diamond$

(b) If $m = 75.0 \text{ kg}$ and $g = 9.800 \text{ m/s}^2$,

at the Equator: $\qquad n = m(g - a_c) = (75.0 \text{ kg})(9.800 \text{ m / s}^2 - 0.0337 \text{ m / s}^2)$

$\qquad\qquad\qquad\qquad n = 732.47 \text{ N} \qquad\qquad\qquad\qquad\qquad\qquad \Diamond$

at the Poles, $a_c = 0$: $\quad n = mg = (75.0 \text{ kg})(9.80 \text{ m / s}^2) = 735 \text{ N} \qquad\qquad \Diamond$

L: As we expected, the person does appear to weigh about 0.3% less at the equator than the poles. We might extend this problem to consider the effect of the earth's bulge on a person's weight. Since the earth is fatter at the equator than the poles, would you expect g to be less than 9.80 m/s^2 at the equator and slightly more at the poles?

63. An amusement park ride consists of a large vertical cylinder that spins about its axis fast enough that any person inside is held up against the wall when the floor drops away (Fig. P6.63). The coefficient of static friction between person and wall is μ_s, and the radius of the cylinder is R.   (a) Show that the maximum period of revolution necessary to keep the person from falling is $T = (4\pi^2 R \mu_s / g)^{1/2}$. (b) Obtain a numerical value for T if $R = 4.00$ m and $\mu_s = 0.400$. How many revolutions per minute does the cylinder make?

Solution

Figure P6.63

(a) The normal force of the wall pushes inward:

$$n = \frac{mv^2}{R} = \frac{m}{R}\left(\frac{2\pi R}{T}\right)^2 = \frac{4\pi^2 Rm}{T^2}$$

and the maximum force of friction balances the weight:

$$f_s = \mu_s n = mg$$

Therefore, $\mu_s n = mg$ becomes $\mu_s \dfrac{4\pi^2 Rm}{T^2} = mg$

Solving, $T^2 = \dfrac{4\pi^2 R \mu_s}{g}$ and $T = \sqrt{\dfrac{4\pi^2 R \mu_s}{g}}$ ◊

(b) $T = \left[\dfrac{4\pi^2(4.00 \text{ m})(0.400)}{9.80 \text{ m} / \text{s}^2}\right]^{1/2} = 2.54$ s   ◊

The angular speed is $\left(\dfrac{1 \text{ rev}}{2.54 \text{ s}}\right)\left(\dfrac{60 \text{ s}}{\text{min}}\right) = 23.6$ rev / min   ◊

69. A model airplane of mass 0.750 kg flies in a horizontal circle at the end of a 60.0-m control wire, with a speed of 35.0 m/s. Compute the tension in the wire if it makes a constant angle of 20.0° with the horizontal. The forces exerted on the airplane are the pull of the control wire, its own weight, and aerodynamic lift, which acts at 20.0° inward from the vertical as shown in Figure P6.69.

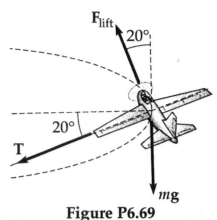

Figure P6.69

Solution   The plane's acceleration is toward the center of the circle of motion, so it is horizontal.   The radius of the circle of motion is $(60.0 \text{ m})\cos 20.0° = 56.4 \text{ m}$, and the acceleration is

$$a_c = \frac{v^2}{r} = \frac{(35 \text{ m / s})^2}{56.4 \text{ m}} = 21.7 \text{ m / s}^2$$

We can also calculate the weight of the airplane:

$$F_g = mg = (0.750 \text{ kg})(9.80 \text{ m / s}^2) = 7.35 \text{ N}$$

We define our axes for convenience.   In this case, two of the forces--one of them our force of interest-- are directed along the 20° lines.   We define the x-axis to be directed in the $(+\mathbf{T})$ direction, and the y-axis to be directed in the direction of lift.   With these definitions, the x component of the centripetal acceleration is

$$a_{cx} = a_c \cos(20°),$$

$\Sigma F_x = ma_x$ yields

$$T + F_g \sin 20.0° = ma_{cx}$$

Solving for T,

$$T = ma_{cx} - F_g \sin 20.0°$$

$$T = (0.750 \text{ kg})(21.7 \text{ m / s}^2)(\cos 20.0°) - (7.35 \text{ N})\sin 20.0°$$

and

$$T = (15.3 \text{ N}) - (2.50 \text{ N}) = 12.8 \text{ N} \qquad \lozenge$$

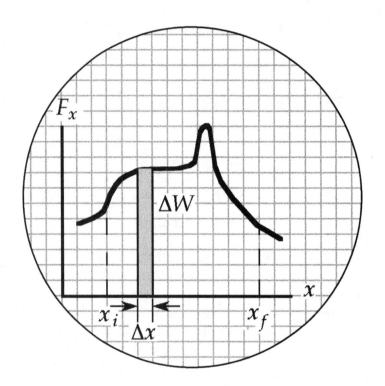

Work and Kinetic Energy

WORK AND KINETIC ENERGY

INTRODUCTION

In this chapter, we introduce the concepts of work and energy. Work is done by a force acting on an object when the point of application of that force moves through some distance and the force has a component along the line of motion. Kinetic energy is energy associated with the motion of an object. The concepts of work and energy can be applied to the dynamics of a mechanical system without resorting to Newton's laws. However, it is important to note that the work-energy concepts are based upon Newton's laws and therefore do not involve any new physical principles. In a complex situation, the "energy approach" can often provide a much simpler analysis than the direct application of Newton's second law.

EQUATIONS AND CONCEPTS

The work done on a body by a **constant** force **F** is defined to be the product of the displacement and the component of force in the direction of the displacement.

$$W = Fd\cos\theta \qquad (7.1)$$

It is convenient to express the work done by a constant force as the **dot product** (scalar product) **F · d**.

$$W = \mathbf{F} \cdot \mathbf{d} = Fd\cos\theta \qquad (7.2)$$

The dot product of any two vectors **A** and **B** is defined to be a scalar quantity whose magnitude is $AB\cos\theta$. (The product of the magnitudes of the two vectors and the cosine of the angle that is included between their directions.)

$$\mathbf{A} \cdot \mathbf{B} \equiv AB\cos\theta \qquad (7.3)$$

Note that work done by a force can be positive, negative, or zero depending on the value of θ. Work done by a force is positive if **F** has a component in the direction of **d** $(0 \le \theta < 90°)$; W is negative if the projection of **F** onto **d** is opposite to **d** $(90° < \theta < 180°)$. Finally, W is zero if **F** is perpendicular to **d** $(\theta = \pm 90°)$. Work is a **scalar** quantity with SI units of joules (J).

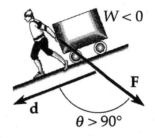

$$1\,\mathrm{J} = 1\,\mathrm{N}\cdot\mathrm{m}$$

The scalar product of two vectors **A** and **B** can be expressed in terms of the x, y and z components of the two vectors.

$$\mathbf{A} \cdot \mathbf{B} = A_x B_x + A_y B_y + A_z B_z \qquad (7.6)$$

If a force acting along x varies with position, and the body is displaced from x_i to x_f, the **work done by that force** is given by an integral expression. Graphically, the work done equals the area under the F_x versus x curve.

$$W = \int_{x_i}^{x_f} F_x \, dx \qquad (7.7)$$

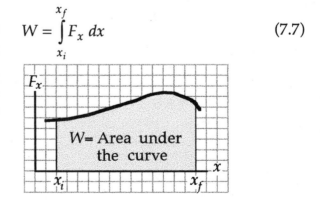

130

If a mass is connected to a spring of force constant k, **the work done by the spring force** $(-kx)$ as the mass undergoes an arbitrary displacement from x_i to x_f is given by Equation 7.11.

$$W_s = \tfrac{1}{2}kx_i^2 - \tfrac{1}{2}kx_f^2 \qquad (7.11)$$

Kinetic energy, K, is energy associated with the motion of an object.

$$K \equiv \tfrac{1}{2}mv^2 \qquad (7.14)$$

The **work-energy theorem** states that the work done by the resultant force on a body equals the **change** in kinetic energy of the body.

$$\Sigma W = K_f - K_i = \Delta K \qquad (7.15)$$

$$\Sigma W = \tfrac{1}{2}mv_f^2 - \tfrac{1}{2}mv_i^2 \qquad (7.16)$$

Note that if ΣW is positive, the kinetic energy increases; if ΣW is negative, the kinetic energy decreases. Therefore, the speed of a body will only change if there is net work done on it. (When $\Sigma W = 0$, the resultant work is zero, and $\Delta K = 0$.)

$$\tfrac{1}{2}mv_i^2 + \Sigma W = \tfrac{1}{2}mv_f^2$$

The loss in kinetic energy due to friction corresponds to the energy dissipated by the force of kinetic friction.

$$\Delta K_{\text{friction}} = -f_k d \qquad (7.17a)$$

131

The **average power** supplied by a force is the ratio of the work done by that force to the time interval over which it acts.

$$\overline{\mathscr{P}} \equiv \frac{W}{\Delta t}$$

The **instantaneous power** is equal to the limit of the average power as the time interval approaches zero.

$$\mathscr{P} = \frac{dW}{dt} = \mathbf{F} \cdot \mathbf{v} \qquad (7.18)$$

The SI unit of power is J/s, which is called a watt (W).

$$1\,\text{W} = 1\,\text{J}\,/\,\text{s}$$

$$1\,\text{kW} \cdot \text{h} = 3.60 \times 10^6\,\text{J}$$

The unit of power in the British engineering system is the horsepower.

$$1\,\text{hp} = 746\,\text{W}$$

When particle speeds are comparable to the speed of light, the equations of Newtonian mechanics must be replaced by more general equations. In particular, the relativistic form of the kinetic energy equation must be used.

$$K = mc^2 \left(\frac{1}{\sqrt{1 - (v/c)^2}} - 1 \right) \qquad (7.19)$$

SUGGESTIONS, SKILLS, AND STRATEGIES

There are two new mathematical skills you must learn. The first is the definition of the scalar (or dot) product, $\mathbf{A} \cdot \mathbf{B} \equiv AB\cos\theta$, where θ is the angle between $\mathbf{A}$ and $\mathbf{B}$. Since $\mathbf{A} \cdot \mathbf{B}$ is a scalar, then the order of product can be interchanged. That is, $\mathbf{A} \cdot \mathbf{B} = \mathbf{B} \cdot \mathbf{A}$. Furthermore, $\mathbf{A} \cdot \mathbf{B}$ can be positive, negative, or zero depending on the value of θ. (That is, $\cos\theta$ varies from -1 to $+1$.) If vectors are expressed in unit vector form, then the dot product is conveniently carried out using the multiplication table for unit vectors:

$$\mathbf{i} \cdot \mathbf{i} = \mathbf{j} \cdot \mathbf{j} = \mathbf{k} \cdot \mathbf{k} = 1; \qquad \mathbf{i} \cdot \mathbf{j} = \mathbf{i} \cdot \mathbf{k} = \mathbf{k} \cdot \mathbf{j} = 0$$

The second operation introduced in this chapter is the definite integral. In Section 7.4, it is shown that the work done by a **variable** force F_x in displacing a particle a small distance Δx is given by

$$\Delta W \approx F_x \Delta x$$

(ΔW equals the area of the shaded rectangle in the figure at the right.) The total work done by F_x as the particle is displaced from x_i to x_f is given approximately by the **sum** of such terms.

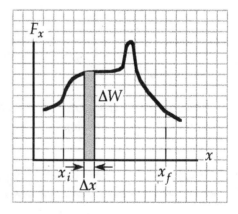

If we take such a sum, letting the widths of the displacements approach dx, the number of terms in the sum becomes very large and we get the actual work done:

$$W = \lim_{\Delta x \to 0} \sum_{x_i}^{x_f} F_x \, \Delta x = \int_{x_i}^{x_f} F_x \, dx$$

The quantity on the right is a definite integral, which graphically represents the **area under the** F_x **versus** x **curve**, as in the figure. Appendix B6 of the text represents a brief review of integration operations with some examples.

REVIEW CHECKLIST

▷ Define the work done by a constant force, and realize that work is a scalar. Describe the work done by a force which **varies** with position. In the one-dimensional case, note that the work done equals the area under the F_x versus x curve.

▷ Take the scalar or dot product of any two vectors **A** and **B** using the definition $\mathbf{A} \cdot \mathbf{B} \equiv AB\cos\theta$, or by writing **A** and **B** in unit vector form and using the multiplication table for unit vectors.

▷ Define the kinetic energy of an object of mass m moving with a speed v.

▷ Relate the work done by the net force on an object to the **change** in kinetic energy. The relation $\Sigma W = K_f - K_i = \Delta K$ is called the work-energy theorem, and is valid whether or not the (resultant) force is constant. That is, if we know the net work done on a particle as it undergoes a displacement, we also know the **change** in its kinetic energy. This is the most important concept in this chapter, so you must understand it thoroughly.

▷ Define the concepts of average power and instantaneous power (the time rate of doing work).

ANSWERS TO SELECTED CONCEPTUAL QUESTIONS

4. Can the kinetic energy of an object be negative? Explain.

Answer No. Kinetic energy = $mv^2/2$. Since v^2 is always positive, K is always positive.

5. (a) If the speed of a particle is doubled, what happens to its kinetic energy? (b) If the net work done on a particle is zero, what can be said about the speed?

Answer Kinetic energy, K, depends on the square of the velocity. Therefore if the speed is doubled, the kinetic energy will increase by a factor of four.

□　　□　　□　　□

8. One bullet has twice the mass of another bullet. If both bullets are fired so that they have the same speed, which has the greater kinetic energy? What is the ratio of the kinetic energies of the two bullets?

Answer The kinetic energy of the more massive bullet is twice that of the lower mass bullet.

□　　□　　□　　□

12. As a simple pendulum swings back and forth, the forces acting on the suspended mass are the force of gravity, the tension in the supporting cord, and air resistance. (a) Which of these forces, if any, does no work on the pendulum? (b) Which of these forces does negative work at all times during its motion? (c) Describe the work done by the force of gravity while the pendulum is swinging.

Answer (a) The tension in the supporting cord does no work, because the motion of the pendulum is always perpendicular to the cord, and therefore to the tension force. (b) The air resistance does negative work at all times, since the air resistance is always acting in a direction opposite to the motion. (c) The weight always acts downward; therefore, the work done by gravity is positive on the downswing, and negative on the upswing.

□　　□　　□　　□

SOLUTIONS TO SELECTED END-OF-CHAPTER PROBLEMS

5. A block of mass 2.50 kg is pushed 2.20 m along a frictionless horizontal table by a constant 16.0-N force directed 25.0° below the horizontal. Determine the work done by (a) the applied force, (b) the normal force exerted by the table, (c) the force of gravity. (d) Determine the total work done on the block.

Solution $W = Fd\cos\theta$

(a) By the applied force, $W_{app} = (16.0\text{ N})(2.20\text{ m})\cos(25.0°) = 31.9\text{ J}$ ◊

(b) By normal force, $W_n = nd\cos\theta = nd\cos(90°) = 0$ ◊

(c) By force of gravity, $W_g = F_g d\cos\theta = mgd\cos(90°) = 0$ ◊

(d) Net work done on the block: $W_{net} = W_{app} + W_n + W_g = 31.9\text{ J}$ ◊

---

7. A certain superhero, whose mass is 80.0 kg, is holding on to the free end of a 12.0-m rope, the other end of which is fixed to a tree limb above. He is able to get the rope in motion as only he knows how, eventually getting it to swing enough so that he can reach a ledge when the rope makes a 60.0° angle with the vertical. How much work was done by him against the force of gravity in this maneuver?

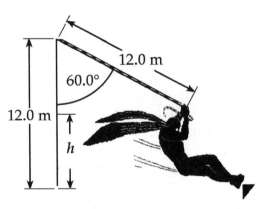

Solution The work done is force of gravity, times the vertical height.

The superhero's weight is $F_g = mg = (80.0\text{ kg})(9.80\text{ m / s}^2) = 784\text{ N}$

This height, h, is $h = 12.0\text{ m} - (12.0\text{ m})\cos 60.0° = 6.00\text{ m}$

Thus, $W = (784\text{ N})(6.00\text{ m}) = 4.70 \times 10^3\text{ J} = 4.70\text{ kJ}$ ◊

11. A force $\mathbf{F} = (6\mathbf{i} - 2\mathbf{j})$ N acts on a particle that undergoes a displacement $\mathbf{d} = (3\mathbf{i} + \mathbf{j})$ m. Find (a) the work done by the force on the particle and (b) the angle between $\mathbf{F}$ and $\mathbf{d}$.

Solution   Assume all numbers in this problem have 3 significant figures:

(a) $W = \mathbf{F} \cdot \mathbf{d} = \left((6\mathbf{i} - 2\mathbf{j})\ \mathrm{N}\right) \cdot \left((3\mathbf{i} - \mathbf{j})\ \mathrm{m}\right) = (6.00\ \mathrm{N})(3.00\ \mathrm{m}) + (-2.00\ \mathrm{N})(1.00\ \mathrm{m})$

$W = 18.0\ \mathrm{J} - 2.00\ \mathrm{J} = 16.0\ \mathrm{J}$ ◊

(b) $|\mathbf{F}| = \sqrt{F_x^2 + F_y^2} = \sqrt{6.00^2 + (-2.00)^2}\ \mathrm{N} = 6.32\ \mathrm{N}$

$|\mathbf{d}| = \sqrt{d_x^2 + d_y^2} = \sqrt{3.00^2 + 1.00^2}\ \mathrm{m} = 3.16\ \mathrm{m}$

$W = Fd\cos\theta$, so $\cos\theta = \dfrac{W}{Fd} = \dfrac{16.0\ \mathrm{J}}{(6.32\ \mathrm{N})\,(3.16\ \mathrm{m})} = 0.8012$

Thus, $\theta = \cos^{-1}(0.8012) = 36.8°$ ◊

17. A particle is subject to a force F_x that varies with position as in Figure P7.17. Find the work done by the force on the body as it moves (a) from $x = 0$ to $x = 5.00$ m, (b) from $x = 5.00$ m to $x = 10.0$ m, and (c) from $x = 10.0$ m to $x = 15.0$ m. (d) What is the total work done by the force over the distance $x = 0$ to $x = 15.0$ m?

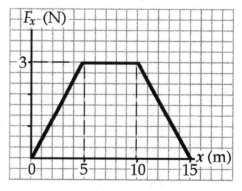

Figure P7.17

Solution The meaning of the equation $W = \int_{x_i}^{x_f} F_x dx$ is that W equals the area under the force-displacement curve.

(a) For the region $0 \le x \le 5.00$ m, $W = \dfrac{(3.00 \text{ N}) (5.00 \text{ m})}{2} = 7.50$ J ◊

(b) For the region $5.00 \text{ m} \le x \le 10.0$ m, $W = (3.00 \text{ N})(5.00 \text{ m}) = 15.0$ J ◊

(c) For the region $10.0 \text{ m} \le x \le 15.0$ m, $W = \dfrac{(3.00 \text{ N}) (5.00 \text{ m})}{2} = 7.50$ J ◊

(d) For the region $0 \le x \le 15.0$ m, $W = (7.50 \text{ J} + 7.50 \text{ J} + 15.0 \text{ J}) = 30.0$ J ◊

---

23. If it takes 4.00 J of work to stretch a Hooke's-law spring 10.0 cm from its unstressed length, determine the extra work required to stretch it an additional 10.0 cm.

Solution

G: We know that the force required to stretch a spring is proportional to the distance the spring is stretched, and since the work required is proportional to the force **and** to the distance, then $W \propto x^2$. This means if the extension of the spring is doubled, the work will increase by a factor of 4, so that for $x = 20$ cm, $W = 16$ J, requiring 12 J of additional work.

O: Let's confirm our answer using Hooke's law and the definition of work.

A: The linear spring force relation is given by Hooke's law: $F_s = -kx$
Integrating with respect to x, we find the work done by the spring is:

$$W_s = \int_{x_i}^{x_f} F \, dx = \int_{x_i}^{x_f} (-kx) \, dx = \tfrac{1}{2} k \left(x_f{}^2 - x_i{}^2 \right)$$

However, we want the work done **on** the spring, which is

$$W = -W_s = \frac{1}{2}k\left(x_f^2 - x_i^2\right)$$

We know the work for the first 10 cm, so we can find the force constant:

$$k = \frac{2W}{x_f^2 - x_i^2} = \frac{2(4.00 \text{ J})}{(0.100 \text{ m})^2 - 0} = 800 \text{ N / m}$$

Substituting for k, x_i and x_f, the extra work for the next step of extension is

$$W = \left(\tfrac{1}{2}\right)(800 \text{ N / m})\left[(0.200 \text{ m})^2 - (0.100 \text{ m})^2\right] = 12.0 \text{ J} \qquad \Diamond$$

L: Our calculated answer agrees with our prediction. It is helpful to remember that the force required to stretch a spring is proportional to the distance the spring is extended, but the work is proportional to the square of the extension.

---

33. A 40.0-kg box initially at rest is pushed 5.00 m along a rough, horizontal floor with a constant applied horizontal force of 130 N. If the coefficient of friction between the box and the floor is 0.300, find (a) the work done by the applied force, (b) the energy loss due to friction, (c) the work done by the normal force (d) the work done by gravity, (e) the change in kinetic energy of the box, and (f) the final speed of the box.

Solution

$\mu_k = 0.300$
$d = 5.00 \text{ m}$
$v_i = 0$

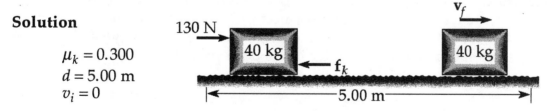

(a) The applied force and the motion are both **horizontal.**

Therefore, $W_F = \mathbf{F} \cdot \mathbf{d} = Fd\cos(0°) = (130 \text{ N})(5.00 \text{ m})(1) = 650 \text{ J} \qquad \Diamond$

(b) $f_k = \mu_k n = \mu_k mg = 0.300(40.0 \text{ kg})(9.80 \text{ m} / \text{s}^2) = 117.6 \text{ N}$

$W_{f_k} = \mathbf{f}_k \cdot \mathbf{d} = f_k d\cos(180°) = (117.6 \text{ N})(5.00 \text{ m})(-1) = -588 \text{ J}$ ◊

(c) Since the normal force is perpendicular to the motion,

$W_n = Fd\cos(90°) = (130 \text{ N})(5.00 \text{ m})(0) = 0$ ◊

(d) The force of gravity is also perpendicular to the motion, so $W_g = 0$ ◊

(e) $\Delta K = \Sigma W = W_F + W_{f_k} = 650 \text{ J} + (-588 \text{ J}) = 62.0 \text{ J}$ ◊

(f) $\frac{1}{2}mv_f^2 - \frac{1}{2}mv_i^2 = \Sigma W$

$v_f = \sqrt{\frac{2}{m}\left(\Sigma W + \frac{1}{2}mv_i^2\right)} = \sqrt{\left(\frac{2}{40.0 \text{ kg}}\right)\left[62.0 \text{ J} + \frac{1}{2}(40.0 \text{ kg})(0)^2\right]} = 1.76 \text{ m} / \text{s}$ ◊

---

35. A crate of mass 10.0 kg is pulled up a rough incline with an initial speed of 1.50 m/s. The pulling force is 100 N parallel to the incline, which makes an angle of 20.0° with the horizontal. The coefficient of kinetic friction is 0.400, and the crate is pulled 5.00 m. (a) How much work is done by gravity? (b) How much energy is lost because of friction? (c) How much work is done by the 100-N force? (d) What is the change in kinetic energy of the crate? (e) What is the speed of the crate after it has been pulled 5.00 m?

Solution

The force of gravity is (10.0 kg)(9.80 m/s²) = 98.0 N straight down, at an angle of (90.0° + 20.0°) = 110.0° with the motion. The work done by gravity is

(a) $$W_g = \mathbf{F} \cdot \mathbf{d} = (98.0 \text{ N})(5.00 \text{ m}) \cos(110.0°) = -168 \text{ J} \qquad \Diamond$$

(b) Setting the x-y axes parallel and perpendicular to the incline,

$$\Sigma F_y = ma_y: \qquad +n - (98.0 \text{ N})\cos(20.0°) = 0$$

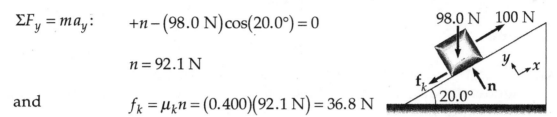

$$n = 92.1 \text{ N}$$

and $$f_k = \mu_k n = (0.400)(92.1 \text{ N}) = 36.8 \text{ N}$$

Therefore, $$W_k = \mathbf{f}_k \cdot \mathbf{d} = (36.8 \text{ N})(5.00 \text{ m}) \cos(180°) = -184 \text{ J}$$

and friction **losses** total 184 J $\qquad \Diamond$

(c) $$W = \mathbf{F} \cdot \mathbf{d} = 100 \text{ N } (5.00 \text{ m}) \cos(0°) = +500 \text{ J} \qquad \Diamond$$

(d) The normal force does zero work, because it is at 90° to the motion.

$$\Delta K = W_{net}$$

$$\Delta K = K_f - K_i = W_g + W_k + W_{applied \ force} + W_n$$

$$= -168 \text{ J} - 184 \text{ J} + 500 \text{ J} + 0 = 148 \text{ J} \qquad \Diamond$$

(e) Since $$K_f = K_i + W_{total}$$

$$\tfrac{1}{2}(10.0 \text{ kg}) \, v_f^2 = \tfrac{1}{2}(10.0 \text{ kg})(1.50 \text{ m / s})^2 + 148 \text{ J} = 159 \text{ J}$$

Thus, $$v_f = \sqrt{\frac{2(159 \text{ kg} \cdot \text{m}^2 / \text{s}^2)}{10.0 \text{ kg}}} = 5.65 \text{ m / s} \qquad \Diamond$$

37. A sled of mass m is given a kick on a frozen pond. The kick imparts to it an initial speed $v_i = 2.00 \text{ m / s}$. The coefficient of kinetic friction between the sled and the ice is $\mu_k = 0.100$. Utilizing energy considerations, find the distance the sled moves before it stops.

Solution

G: Since the sled's initial speed of 2 m/s (~ 4 mph) is reasonable for a moderate kick, we might expect the sled to travel several meters before coming to rest.

O: We could solve this problem using Newton's second law, but we are asked to use the work-kinetic energy theorem: $W = \Delta K = K_f - K_i$, where the only work done on the sled after the kick results from the friction between the sled and ice. (The weight and normal force both act at 90° to the motion, and therefore do no work on the sled.)

A: The work due to friction is $W = -f_k d$ where $f_k = \mu_k n = \mu_k mg$.

Since the final kinetic energy is zero, $W = \Delta K = 0 - K_i = -\frac{1}{2}mv_i^2$

So $\qquad -\mu_k mgd = -\frac{1}{2}mv_i^2$

Thus, $\quad d = \dfrac{mv_i^2}{2f_k} = \dfrac{mv_i^2}{2\mu_k \, mg} = \dfrac{v_i^2}{2\mu_k \, g} = \dfrac{(2.00 \text{ m / s})^2}{2(0.100)(9.80 \text{ m / s}^2)} = 2.04 \text{ m}$ ◊

L: The distance agrees with the prediction. It is interesting that the distance does not depend on the mass and is proportional to the square of the initial velocity. This means that a small car and a massive truck should be able to stop within the same distance if they both skid to a stop from the same initial speed. Also, doubling the speed requires 4 times as much stopping distance, which is consistent with advice given by transportation safety officers who suggest at least a 2 second gap between vehicles (as opposed to a fixed distance like 100 feet).

43. A 700-N Marine in basic training climbs a 10.0-m vertical rope at a constant speed in 8.00 s. What is his power output?

Solution

The marine must exert a 700 N upward force opposite the force of gravity to lift his body at constant speed. Then his muscles do work:

$$W = \mathbf{F} \cdot \mathbf{d} = (700\mathbf{j}\ \text{N}) \cdot (10.0\mathbf{j}\ \text{m}) = 7000\ \text{J}$$

The power he puts out is $\quad \overline{\mathcal{P}} = \dfrac{W}{\Delta t} = \dfrac{7000\ \text{J}}{8.00\ \text{s}} = 875\ \text{W}$ ◊

---

49. A compact car of mass 900 kg has an overall motor efficiency of 15.0%. (That is, 15.0% of the energy supplied by the fuel is delivered to the wheels of the car.) (a) If burning 1 gal of gasoline supplies 1.34×10^8 J of energy, find the amount of gasoline used by the car in accelerating from rest to 55.0 mi/h. Here you may ignore the effects of air resistance and rolling resistance. (b) How many such accelerations will 1 gal provide? (c) The mileage claimed for the car is 38.0 mi/gal at 55.0 mi/h. What power is delivered to the wheels (to overcome frictional effects) when the car is driven at this speed?

Solution

We first must convert units from miles per hour to SI units:

$$\left(\frac{55.0\ \text{mi}}{\text{h}}\right)\left(\frac{1609\ \text{m}}{\text{mi}}\right)\left(\frac{\text{h}}{3600\ \text{s}}\right) = 24.6\ \text{m}\,/\,\text{s}$$

143

(a) The engine must do work to provide kinetic energy to the chassis, according to

$$\Sigma W = K_f - K_i = \frac{1}{2}mv_f{}^2 - \frac{1}{2}mv_i{}^2$$

$$W_{\text{engine output}} = \frac{1}{2}(900 \text{ kg})(24.6 \text{ m / s})^2 - 0 = 2.72 \times 10^5 \text{ J}$$

This output is only 15.0% of the chemical energy the engine takes in; the other 85.0% is lost as heat.

Since
$$\text{efficiency} = \frac{\text{useful energy output}}{\text{total energy input}} = \frac{2.72 \times 10^5 \text{ J}}{\text{total energy input}}$$

$$\text{total energy input} = \frac{2.72 \times 10^5 \text{ J}}{0.150} = 1.82 \times 10^6 \text{ J}$$

and
$$\text{amt of gas} = (1.82 \times 10^6 \text{ J})\left(\frac{1 \text{ gal}}{1.34 \times 10^8 \text{ J}}\right) = 0.0135 \text{ gal} \qquad \lozenge$$

(b)
$$1 \text{ gal} = (n \text{ accel})(0.0135 \text{ gal / accel})$$

$$n = 73.8 \text{ accelerations} \qquad \lozenge$$

(c) Consider driving 38.0 miles with energy input 1.34×10^8 J and output work equal to $(0.150)(1.34 \times 10^8 \text{ J}) = 2.01 \times 10^7$ J. The time for this trip is $(38.0 \text{ mi})/(55.0 \text{ mi / h}) = 0.691 \text{ h} = 2490 \text{ s}$. So, the output power is

$$\overline{\mathcal{P}} = \frac{W}{t} = \frac{2.01 \times 10^7 \text{ J}}{2490 \text{ s}} = 8.08 \text{ kW} \qquad \lozenge$$

53. A proton in a high-energy accelerator moves with a speed of $c/2$. Using the work-kinetic energy theorem, find the work required to increase its speed to (a) $0.750c$, (b) $0.995c$.

Solution

G: Since particle accelerators have typical maximum energies on the order of a GeV ($1\,eV = 1.60\times10^{-19}\,J$), we could expect the work to be $\sim 10^{-10}\,J$.

O: The work-energy theorem is $W = \Delta K = K_f - K_i$ which for relativistic speeds ($v \sim c$) is:

$$W = \left(\frac{1}{\sqrt{1-v_f^2/c^2}} - 1\right)mc^2 - \left(\frac{1}{\sqrt{1-v_i^2/c^2}} - 1\right)mc^2$$

or (simplified),
$$W = \left(1/\sqrt{1-v_f^2/c^2} - 1/\sqrt{1-v_i^2/c^2}\right)mc^2$$

A: (a) $W = \left(\frac{1}{\sqrt{1-0.750^2}} - \frac{1}{\sqrt{1-0.500^2}}\right)(1.67\times10^{-27}\,kg)(3.00\times10^8\,m/s)^2$

$W = (1.512 - 1.155)(1.50\times10^{-10}\,J) = 5.37\times10^{-11}\,J$ ◊

(b) $W = \left(\frac{1}{\sqrt{1-0.995^2}} - \frac{1}{\sqrt{1-0.500^2}}\right)(1.67\times10^{-27}\,kg)(3.00\times10^8\,m/s)^2$

$W = (10.01 - 1.155)(1.50\times10^{-10}\,J) = 1.33\times10^{-9}\,J$ ◊

L: Even though these energies may seem like small numbers, we must remember that the proton has very small mass, so these input energies are comparable to the rest mass energy of the proton ($1.50 \times 10^{-10}\,J$). To produce a speed higher by 33%, the answer to part (b) is 25 times larger than the answer to part (a). Even with arbitrarily large accelerating energies, the particle will never reach or exceed the speed of light. This is a consequence of special relativity, which will be examined more closely in a later chapter.

59. A 4.00-kg particle moves along the x axis. Its position varies with time according to $x = t + 2.0t^3$, where x is in meters and t is in seconds. Find (a) the kinetic energy at any time t, (b) the acceleration of the particle and the force acting on it at time t, (c) the power being delivered to the particle at time t, and (d) the work done on the particle in the interval $t = 0$ to $t = 2.00$ s.

Solution Given that $m = 4.00$ kg and $x = t + 2.0t^3$, we find

(a) $\quad v = \dfrac{dx}{dt} = \dfrac{d}{dt}\left(t + 2.00t^3\right) = 1 + 6.00t^2$

$\quad K = \tfrac{1}{2}mv^2 = \tfrac{1}{2}(4.00 \text{ kg})\left(1 + 6.00t^2\right)^2 = \left(2.00 + 24.0t^2 + 72.0t^4\right) \text{J}$ ◊

(b) $\quad a = \dfrac{dv}{dt} = \dfrac{d}{dt}\left(1 + 6.00t^2\right) = 12.0t \text{ m} / \text{s}^2$ ◊

$\quad F = ma = (4.00 \text{ kg})\left(12.0t \text{ m} / \text{s}^2\right) = 48.0t \text{ N}$ ◊

(c) $\quad \mathcal{P} = \dfrac{dW}{dt} = \dfrac{dK}{dt} = \dfrac{d}{dt}\left(2.00 + 24.0t^2 + 72.0t^4\right) = \left(48.0t + 288t^3\right) \text{W}$ ◊

(d) At $t_i = 0$, $\qquad K_i = 2.00$ J

At $t_f = 2.00$ s, $\qquad K_f = \left[2.00 \text{ J} + \left(24.0 \text{ J} / \text{s}^2\right)(2.00 \text{ s})^2 + \left(72.0 \text{ J} / \text{s}^4\right)(2.00 \text{ s})^4\right]$

$\qquad K_f = 1250$ J

$W = K_f - K_i$, so $\quad W = K_f - K_i = 1.25 \times 10^3$ J ◊

Alternatively, for part (c) and (d) you can use:

$\mathcal{P} = Fv = 48.0t(1 + 6.00t^2)$ $\qquad W = \displaystyle\int_{t_i}^{t_f} \mathcal{P} \, dt = \int_0^2 (48.0t + 288t^3) \, dt$ etc.

63. A 2100-kg pile driver is used to drive a steel I-beam into the ground. The pile driver falls 5.00 m before cooming into contact with the beam, and it drives the beam 12.0 cm into the ground before coming to rest. Using energy considerations, calculate the average force the beam exerts on the pile driver while the pile driver is brought to rest.

Solution

G: Anyone who has hit their thumb with a hammer knows that the resulting force is greater than just the weight of the hammer, so we should also expect the force of the pile driver to be significantly greater than its weight: $F >> mg \sim 20$ kN. The force **on** the pile driver will be directed upwards.

O: The average force stopping the driver can be found from the work that results from the gravitational force starting its motion. The initial and final kinetic energies are zero.

A: Choose the initial point when the mass is elevated and the final point when it comes to rest again 5.12 m below. Two forces do work on the pile driver: gravity (weight) and the normal force exerted by the beam on the pile driver.

$$\Sigma W = K_f - K_i \quad \text{so that} \quad mgd_w \cos(0) + nd_n \cos(180°) = 0 - 0$$

where $\quad d_w = 5.12$ m, $\quad d_n = 0.120$ m, $\quad$ and $\quad m = 2100$ kg

In this situation, the weight vector is in the direction of motion and the beam exerts a force on the pile driver opposite the direction of motion.

$$(2100 \text{ kg})(9.80 \text{ m} / \text{s}^2)(5.12 \text{ m}) + n(0.120 \text{ m})(-1) = 0$$

Solve for n. $\quad n = \dfrac{1.05 \times 10^5 \text{ J}}{0.120 \text{ m}} = 878$ kN $\quad\quad\quad\quad$ ◊

L: The normal force is larger than 20 kN as we expected, and is actually about 43 times greater than the weight of the pile driver, which is why this machine is so effective.

Additional Calculation: Show that the work done by gravity on an object can be represented by mgh, where h is the vertical height that the object falls. Apply your results to the problem above.

By the figure to the right, where **d** is the path of the object, and h is the height that the object falls,

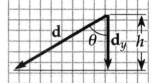

$$h = |\mathbf{d}_y| = d\cos\theta$$

Since $F = mg$, $\qquad mgh = Fd\cos\theta = \mathbf{F}\cdot\mathbf{d}$ $\qquad\qquad\qquad\qquad \Diamond$

In this problem, $mgh = n(d_n)$ $\quad$ so $\quad$ $n = mgh / d_n$

and $\qquad\qquad n = (2100 \text{ kg})(9.80 \text{ m} / \text{s}^2)(5.12 \text{ m})/(0.120 \text{ m}) = 878 \text{ kN} \ \Diamond$

---

69. A 200-g block is pressed against a spring of force constant 1.40 kN/m until the block compresses the spring 10.0 cm. The spring rests at the bottom of a ramp inclined at 60.0° to the horizontal. Use energy considerations, determine how far up the incline the block moves before it stops (a) if there is no friction between block and ramp and if (b) the coefficient of kinetic friction is 0.400.

Solution

We have two forms of energy involved: work done by gravity, and work done by the spring.

$$W_g = \mathbf{F}\cdot\mathbf{d} = mg\,d\cos 150° \qquad\qquad\qquad W_s = \tfrac{1}{2}kx_m^2$$

(a) Apply the work-energy theorem between the starting point and the point of maximum travel up the incline. The initial and final states of the block are stationary, so $\Delta K = 0$, and $\Sigma W = \Delta K$ becomes $W_g + W_s = 0$. Substituting known values into this equation,

$$mg\,d\cos 150° + \frac{1}{2}kx_m^2 = 0$$

and $\qquad d = \dfrac{-kx_m^2}{2mg\cos 150°} = \dfrac{(-1400\text{ N}/\text{m})\,(0.100\text{ m})^2}{2(0.200\text{ kg})(9.80\text{ m}/\text{s}^2)(-0.866)} = 4.12\text{ m}$ ◊

(b) We add a term for the friction's energy loss:

$$W_f = f_k d\cos 180° = -f_k d$$

Taking our vertical axis to be perpendicular to the incline, $\Sigma F_y = ma_y$:

$$+n - (1.96\text{ N})(\cos 60.0°) = 0 \qquad \text{and} \qquad n = 0.980\text{ N}$$

Then the force of friction is $f_k = \mu_k n$:

$$f_k = (0.400)(0.980\text{ N}) = 0.392\text{ N}$$

We again apply our energy equation $K_i + \Sigma W = K_f$, this time adding a friction term: $K_i + W_g + W_f + W_s = K_f$. Since K_i and K_f are both zero,

$$mgd\cos 150° + f_k d\cos 180° + \frac{1}{2}kx_m^2 = 0$$

$$0 + (1.96\text{ N})(d)(-0.866) + (0.392\text{ N})(d)(-1.00) + 7.00\text{ J} = 0$$

$$d = \frac{7.00\text{ J}}{1.70\text{ N} + 0.392\text{ N}} = 3.35\text{ m}$$ ◊

71. The ball launcher in a pinball machine has a spring that has a force constant of 1.20 N/cm (Fig. P7.71). The surface on which the ball moves is inclined 10.0° with respect to the horizontal. If the spring is initially compressed 5.00 cm, find the launching speed of a 100-g ball when the plunger is released. Friction and the mass of the plunger are negligible.

Figure P7.71

Solution   By the work-energy theorem., $\Sigma W = \Delta K$:

$$W_s + W_g = \Delta K$$

$$\tfrac{1}{2}kx^2 - mgd \sin(10.0°) = \tfrac{1}{2}m(v_f^2 - v_i^2)$$

Since $v_i = 0$, $v_f = \sqrt{\dfrac{kx^2}{m} - 2gd \sin(10.0°)}$

In this case, $x = d = 5.00 \text{ cm} = 0.00500 \text{ m}$

and $k = 1.20 \text{ N / cm} = 120 \text{ N / m}$

So, $v_f = \sqrt{\dfrac{(120 \text{ N/m})(0.0500 \text{ m})^2}{0.100 \text{ kg}} - 2\left(9.80 \text{ m/s}^2\right)(0.0500 \text{ m})(\sin 10.0°)}$

and $v_f = 1.68 \text{ m / s}$ ◊

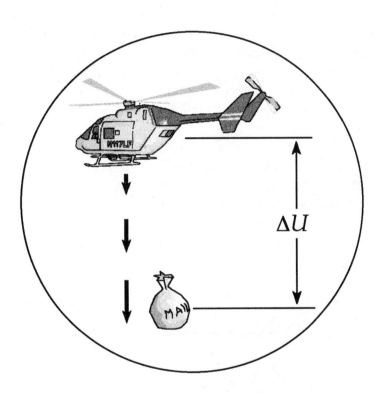

ΔU

Potential Energy and
Conservation of Energy

POTENTIAL ENERGY AND CONSERVATION OF ENERGY

INTRODUCTION

In Chapter 7 we introduced the concept of kinetic energy, which is the energy associated with the motion of an object. In this chapter we introduce another form of mechanical energy, **potential energy**, which is the energy associated with the position or configuration of an object. Potential energy can be thought of as stored energy that can either do work or be converted to kinetic energy.

The potential energy concept can be used only when dealing with a special class of forces called **conservative forces**. When only internal conservative forces, such as gravitational or spring forces, act within a system, the kinetic energy gained (or lost) by the system as its members change their relative positions is compensated by an equal energy loss (or gain) in potential energy. This is known as the **principle of conservation of mechanical energy**.

EQUATIONS AND CONCEPTS

The gravitational potential energy associated with an object at any point in space is the product of the object's weight and the vertical coordinate.

$$U_g \equiv mgy \tag{8.1}$$

The work done on any object by the gravitational force is equal to the change in the gravitational potential energy.

$$W_g = U_i - U_f = -\Delta U_g \tag{8.2}$$

The units of energy (kinetic and potential) are the same as the units of work — joules.

In calculating the work done by the gravitational force, remember that the difference in potential energy between two points is independent of the location of the origin. Choose an origin which is convenient to calculate U_i and U_f for a particular situation.

The quantity $\frac{1}{2}kx^2$ is referred to as the **elastic potential energy** stored in a spring which has been deformed a distance x from the equilibrium position.

$$U_s \equiv \tfrac{1}{2}kx^2 \tag{8.4}$$

A potential energy function U can be defined for a conservative force.

$$\Delta U = U_f - U_i = -\int_{x_i}^{x_f} F_x \, dx \tag{8.7}$$

The total mechanical energy of a system is the sum of the kinetic and potential energies.

$$E \equiv K + U \tag{8.9}$$

Note that both the gravitational force and the spring force satisfy the required properties of a conservative force. That is, the work done is path independent and is zero for any closed path.

The law of **conservation of mechanical energy** says that if only conservative forces act on a system, the sum of the kinetic and potential energies remains constant--or is conserved. According to this important conservation law, if the kinetic energy of the system increases by some amount, the potential energy must decrease by the same amount--and vice versa.

$$K_i + U_i = K_f + U_f \qquad (8.10)$$

If more than one conservative force acts on an object, then a potential energy term is associated with each force.

$$K_i + \Sigma U_i = K_f + \Sigma U_f \qquad (8.11)$$

If the only conservative force is the gravitational force, the equation for conservation of mechanical energy takes a special form.

$$\tfrac{1}{2}mv_i^2 + mgy_i = \tfrac{1}{2}mv_f^2 + mgy_f$$

Conservation of mechanical energy for a mass-spring system is similar.

$$\tfrac{1}{2}mv_i^2 + \tfrac{1}{2}kx_i^2 = \tfrac{1}{2}mv_f^2 + \tfrac{1}{2}kx_f^2$$

When nonconservative forces (e.g. friction) are present, the change in kinetic energy due to the **net force** can be calculated.

$$\int \mathbf{F}_{net} \cdot d\mathbf{x} = \Delta K$$

There is an important relationship between a conservative force and the potential energy of the system on which the force acts. The conservative force equals the negative derivative of the system's potential energy with respect to x.

$$F_x = -\frac{dU}{dx}$$

(8.16)

Mass and energy are conserved not separately, but as a single entity called **mass-energy**.

$$\Delta m = \frac{\Delta E}{c^2}$$

(8.18)

SUGGESTIONS, SKILLS, AND STRATEGIES

CHOOSING A ZERO LEVEL

In working problems involving gravitational potential energy, it is always necessary to choose a location at which the gravitational potential energy is zero. This choice is completely arbitrary because the important quantity is the **difference** in potential energy, and that difference is independent of the location of zero. It is often convenient, but not essential, to choose the surface of the Earth as the reference position for zero potential energy. In most cases, the statement of the problem suggests a convenient level to use.

CONSERVATION OF ENERGY

Take the following steps in applying the principle of conservation of energy:

- Define your system, which may consist of more than one object.

- Select a reference position for the zero point of gravitational potential energy.

- Determine whether or not nonconservative forces are present.

- If mechanical energy is conserved (that is, if only conservative forces are present), you can write the total initial energy at some point as the sum of the kinetic and potential energies at that point, $K_i + U_i$. Then, write an expression for the total final energy, $K_f + U_f$, at the final point of interest. Since mechanical energy is conserved, you can equate the two total energies and solve for the unknown.

- If nonconservative forces such as friction are present (and thus mechanical energy is not conserved), first write expressions for the total initial and total final energies. In this case, the difference between the two total energies is equal to the mechanical energy lost due to nonconservative force(s).

REVIEW CHECKLIST

▷ Recognize that the gravitational potential energy function, $U_g = mgy$, can be positive, negative, or zero, depending on the location of the reference level used to measure y. Be aware of the fact that although U depends on the origin of the coordinate system, the **change** in potential energy, $(U)_f - (U)_i$, is **independent** of the coordinate system used to define U.

▷ Understand that a force is said to be **conservative** if the work done by that force on a body moving between any two points is independent of the path taken. **Nonconservative** forces are those for which the work done on a particle moving between two points depends on the path. Account for nonconservative forces acting on a system using the work-energy theorem. In this case, the work done by all nonconservative forces equals the change in total mechanical energy of the system.

▷ Understand the distinction between kinetic energy (energy associated with motion), potential energy (energy associated with the position or configuration of a system), and the total mechanical energy of a system. State the law of conservation of mechanical energy, noting that mechanical energy is conserved when only conservative forces act on a system. This extremely powerful concept is most important in all areas of physics.

ANSWERS TO SELECTED CONCEPTUAL QUESTIONS

3. A bowling ball is suspended from the ceiling of a lecture hall by a strong cord. The bowling ball is drawn away from its equilibrium position and released from rest at the tip of the student's nose as in Figure Q8.3. If the student remains stationary, explain why she will not be struck by the ball on its return swing. Would the student be safe if she pushed the ball as she released it?

Figure Q8.3

Answer The total energy of the system (bowling ball) must be conserved. Since the ball initially has a potential energy mgh, and no kinetic energy, it cannot have any kinetic energy when returning to its initial position. Of course, air resistance will cause the ball to return to a point slightly below its initial position. On the other hand, if the ball is given a push, the demonstrator's nose will be in big trouble.

□   □   □   □

4. One person drops a ball from the top of a building, while another person at the bottom observes its motion. Will these two people agree on the value of the potential energy of the ball – Earth system? on its change in potential energy? on the kinetic energy of the ball?

Answer The two will not necessarily agree on the potential energy, since this depends on the origin--which may be chosen differently for the two observers. However, the two **must** agree on the value of the **change** in potential energy, which is independent of the choice of the reference frames. The two will also agree on the kinetic energy of the ball, assuming both observers are at rest with respect to each other, and hence measure the same velocity.

□   □   □   □

SOLUTIONS TO SELECTED END-OF-CHAPTER PROBLEMS

3. A 4.00-kg particle moves from the origin to position C, which has coordinates $x = 5.00$ m and $y = 5.00$ m. One force on it is the force of gravity acting in the negative y direction (Fig. P8.3). Using Equation 7.2, calculate the work done by gravity as the particle moves from O to C along (a) OAC, (b) OBC, (c) OC. Your results should all be identical. Why?

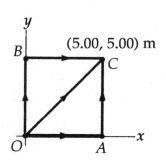

Figure P8.3

Solution

$$w = mg = (4.00 \text{ kg})(9.80 \text{ m} / \text{s}^2) = 39.2 \text{ N}$$

(a) $W_{OAC} = W_{OA} + W_{AC}$

$$= wd_{OA}\cos(270°) + wd_{AC}\cos(180°)$$

$$= (39.2 \text{ N})(5.00 \text{ m})(0) + (39.2 \text{ N})(5.00 \text{ m})(-1) = -196 \text{ J} \qquad \lozenge$$

(b) $W_{OBC} = W_{OB} + W_{BC}$

$$= (39.2 \text{ N})(5.00 \text{ m}) \cos 180° + (39.2 \text{ N})(5.00 \text{ m}) \cos 90° = -196 \text{ J} \qquad \lozenge$$

(c) $W_{OC} = wd_{OC}\cos 135° = (39.2 \text{ N})(5\sqrt{2} \text{ m})\left(-\dfrac{\sqrt{2}}{2}\right) = -196 \text{ J} \qquad \lozenge$

The results should all be the same since the gravitational force is conservative, and the work done by a conservative force is independant of the path. $\qquad \lozenge$

5. A force acting on a particle moving in the xy plane is given by $\mathbf{F} = (2y\mathbf{i} + x^2\mathbf{j})$ N, where x and y are in meters. The particle moves from the origin to a final position having coordinates $x = 5.00$ m and $y = 5.00$ m, as in Figure P8.3. Calculate the work done by $\mathbf{F}$ along (a) OAC, (b) OBC, (c) OC. (d) Is $\mathbf{F}$ conservative or nonconservative? Explain.

Solution   In the following integrals, remember $\mathbf{i}\cdot\mathbf{i} = \mathbf{j}\cdot\mathbf{j} = 1$ and $\mathbf{i}\cdot\mathbf{j} = 0$.

(a) $W_{OA} = \int_0^{5.00}(2y\mathbf{i} + x^2\mathbf{j})\cdot(\mathbf{i}\,dx) = \int_0^{5.00}2y\,dx = 2y\int_0^{5.00}dx = 2yx\Big]_{x=0,y=0}^{x=5.00,y=0} = 0$

$W_{AC} = \int_0^{5.00}(2y\mathbf{i} + x^2\mathbf{j})\cdot(\mathbf{j}\,dy) = \int_0^{5.00}x^2dy = x^2\int_0^{5.00}dy = x^2y\Big]_{x=5.00,y=0}^{x=5.00,y=5.00} = 125\text{ J}$

$W_{OAC} = 0 + 125\text{ J} = 125\text{ J}$ ◊

(b) $W_{OB} = \int_0^{5.00}(2y\mathbf{i} + x^2\mathbf{j})\cdot(\mathbf{j}\,dy) = \int_0^{5.00}x^2dy = x^2\int_0^{5.00}dy = x^2y\Big]_{x=0...}^{x=0...} = 0$

$W_{BC} = \int_0^{5.00}(2y\mathbf{i} + x^2\mathbf{j})\cdot(\mathbf{j}\,dx) = \int_0^{5.00}2y\,dx = 2y\int_0^{5.00}dx = 2(5.00)x\Big]_0^{5.00} = 50.0\text{ J}$

$W_{OBC} = 0 + 50.0\text{ J} = 50.0\text{ J}$ ◊

(c) $W_{OC} = \int(2y\mathbf{i} + x^2\mathbf{j})\cdot(\mathbf{i}\,dx + \mathbf{j}\,dy) = \int(2y\,dx + x^2\,dy)$

Since $x = y$ along OC, $dx = dy$ and

$W_{OC} = \int_0^{5.00}(2x + x^2)dx = 66.7\text{ J}$ ◊

(d) $\mathbf{F}$ is non-conservative since the work done is path dependent.   ◊

7. A single conservative force acts on a 5.00-kg particle. The equation $F_x = (2x + 4)$ N, where x is in meters, describes this force. As the particle moves along the x axis from $x = 1.00$ m to $x = 5.00$ m, calculate (a) the work done by this force, (b) the change in the potential energy of the system, and (c) the kinetic energy of the particle at $x = 5.00$ m if its speed at $x = 1.00$ m is 3.00 m / s.

Solution

We integrate the force with respect to distance, to find the work:

(a) Taking $F_x = (2.00x + 4.00)$ N, $x_i = 1.00$ m   and $x_f = 5.00$ m

We find $W_F = \int_{x_i}^{x_f} F_x \, dx = \int_{1.00 \text{ m}}^{5.00 \text{ m}} (2.00x + 4.00) dx$ N · m

and $W_F = x^2 + 4.00x \Big]_{x=1.00}^{x=5.00}$ N · m $= 40.0$ J   ◊

(b) The change in potential energy equals the negative of the work done by the conservative force.

$$\Delta U = -W_F = -40.0 \text{ J}$$ ◊

(c) When only conservative forces act, conservation of mechanical energy tells us that $\Delta K + \Delta U = 0$,

so the initial energy equals the final energy: $K_i + U_i = K_f + U_f$

Rearranging, $K_f = K_i - (U_f - U_i)$,   so $K_f = \tfrac{1}{2} m v_i^2 - \Delta U$

$$K_f = \left(\tfrac{1}{2}\right)(5.00 \text{ kg})(3.00 \text{ m / s})^2 - (-40.0 \text{ J}) = 62.5 \text{ J}$$ ◊

---

15. A bead slides without friction around a loop-the-loop (Fig. P8.15). If the bead is released from a height $h = 3.50R$, what is its speed at point A? How great is the normal force on it if its mass is 5.00 g?

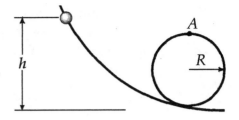

Figure P8.15

Solution

G: Since the bead is released above the top of the loop, it will have enough potential energy to reach point A and still have excess kinetic energy. The energy of the bead at point A will be proportional to h and g. If it is moving relatively slowly, the track will exert an upward force on the bead, but if it is whipping around fast, the normal force will push it toward the center of the loop.

O: The speed at the top can be found from the conservation of energy, and the normal force can be found from Newton's second law.

A: We define the bottom of the loop as the zero level for the gravitational potential energy.

Since $v_i = 0$, $\qquad E_i = K_i + U_i = 0 + mgh = mg(3.50\ R)$

The total energy of the bead at point A can be written as

$$E_A = K_A + U_A = \tfrac{1}{2}mv_A{}^2 + mg(2R)$$

Since mechanical energy is conserved, $E_i = E_A$, and we get

$$\tfrac{1}{2}mv_A{}^2 + mg(2R) = mg(3.50R)$$

$$v_A{}^2 = 3.00gR \qquad \text{or} \qquad v_A = \sqrt{3.00gR} \qquad \Diamond$$

161

To find the normal force at the top, we may construct a free-body diagram as shown, where we assume that **n** is downward, like *mg*. Newton's second law gives $\Sigma F = ma_c$, where a_c is the centripetal acceleration.

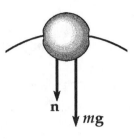

$$n + mg = \frac{mv_A{}^2}{R} = \frac{m(3.00gR)}{R} = 3.00mg$$

$$n = 3.00mg - mg = 2.00mg$$

$$n = 2.00(5.00 \times 10^{-3} \text{ kg})(9.80 \text{ m / s}^2) = 0.0980 \text{ N} \quad \text{downward} \qquad \Diamond$$

L: Our answer represents the speed at point *A* as proportional to the square root of the product of *g* and *R*, but we must not think that simply increasing the diameter of the loop will increase the speed of the bead at the top. In general, the speed will increase with increasing release height, which for this problem was defined in terms of the radius. The normal force may seem small, but it is twice the weight of the bead.

---

17. A block of mass 0.250 kg is placed on top of a light vertical spring of constant $k = 5000$ N/m and is pushed downward so that the spring is compressed 0.100 m. After the block is released, it travels upward and then leaves the spring. To what maximum height above the point of release does it rise?

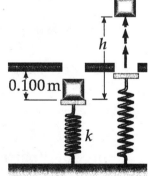

Solution   In both the initial and final states, the block is not moving. Therefore, the initial and final energies are:

$$E_i = K_i + U_i = 0 + (U_g + U_s)_i = 0 + \left(0 + \tfrac{1}{2}kx^2\right)$$

$$E_f = K_f + U_f = 0 + (U_g + U_s)_f = 0 + (mgh + 0)$$

Since $E_i = E_f$, $mgh = \frac{1}{2}kx^2$

and $h = \dfrac{kx^2}{2mg} = \dfrac{(5000 \text{ N} / \text{m})(0.100 \text{ m})^2}{2(0.250 \text{ kg})(9.80 \text{ m} / \text{s}^2)} = 10.2 \text{ m}$ ◊

---

21. Two masses are connected by a light string passing over a light frictionless pulley as shown in Figure P8.21. The 5.00-kg mass is released from rest. Using the law of conservation of energy, (a) determine the speed of the 3.00-kg mass just as the 5.00-kg mass hits the ground and (b) find the maximum height to which the 3.00-kg mass rises above the ground.

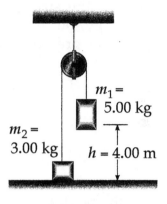

Figure P8.21

Solution

(a) Choose the initial point before release and the final point just before the larger mass hits the floor. The total energy of the two objects together remains constant, and the work-energy theorem for the two objects 1 and 2 is

$$(K_1 + K_2 + U_1 + U_2)_i + W_{app} - f_k d = (K_1 + K_2 + U_1 + U_2)_f$$

At the initial point, K_{1i}, K_{2i}, and U_{2i} are zero; W_{app} is zero, energy added to the system by nonconservative forces is zero, and U_{1f} is zero. Further, the initial height h of mass 1 is the same as the final height of mass 2, and the final velocity of both blocks is the same, so:

$$\left(0 + 0 + m_1 gh + 0\right) + 0 = \left(\frac{1}{2}m_1 v_f^2 + \frac{1}{2}m_2 v_f^2 + 0 + m_2 gh\right)$$

Rearranging, $\frac{1}{2}v_f^2(m_1 + m_2) = m_1 gh - m_2 gh$

Thus,
$$v_f = \sqrt{\frac{2gh(m_1 - m_2)}{m_1 + m_2}}$$

$$v_f = \sqrt{\frac{2(9.80 \text{ m} / \text{s}^2)(4.00 \text{ m})(5.00 \text{ kg} - 3.00 \text{ kg})}{5.00 \text{ kg} + 3.00 \text{ kg}}} = 4.43 \text{ m} / \text{s} \quad \Diamond$$

(b) Now the string goes slack. The 5.00-kg mass loses all its mechanical energy, but the 3.00-kg mass becomes a projectile. Take the initial point at the previous final point, and the new final point at its maximum height:

$$(K + U_g)_i = (K + U_g)_f \quad \text{or} \quad \tfrac{1}{2}mv_i^2 + mgh_i = 0 + mgh_f$$

Solving for h_f,
$$h_f = \frac{v_i^2}{2g} + h_i = \frac{(4.43 \text{ m} / \text{s})^2}{2(9.80 \text{ m} / \text{s}^2)} + 4.00 \text{ m} = 5.00 \text{ m} \quad \Diamond$$

This is 1.00 m higher than the height of the 5.00-kg mass when it was released.

---

31. The coefficient of friction between the 3.00-kg block and the surface in Figure P8.31 is 0.400. The system starts from rest. What is the speed of the 5.00-kg ball when it has fallen 1.50 m?

Solution

G: Assuming that the block does not reach the pulley within the 1.50 m distance, a reasonable speed for the ball might be somewhere between 1 and 10 m/s based on common experience.

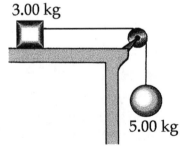

3.00 kg

5.00 kg

Figure P8.31

O: We could solve this problem by using $\Sigma F = ma$ to give a pair of simultaneous equations in the unknown acceleration and tension; then we would have to solve a motion problem to find the final speed. We may find it easier to solve using the work-energy theorem.

A: For objects A (block) and B (ball), the work-energy theorem is

$$(K_A + K_B + U_A + U_B)_i + W_{app} - f_k d = (K_A + K_B + U_A + U_B)_f$$

Choose the initial point before release and the final point after each block has moved 1.50 m. For the 3.00-kg block, choose $U_g = 0$ at the tabletop. For the 5.00-kg ball, take the zero level of gravitational energy at the final position. So $K_{Ai} = K_{Bi} = U_{Ai} = U_{Af} = U_{Bf} = 0$. Also, since the only external forces are gravity and friction, $W_{app} = 0$.

We now have $0 + 0 + 0 + m_B g y_{Bi} - fd = \frac{1}{2}m_A v_f^2 + \frac{1}{2}m_B v_f^2 + 0 + 0$

where the frictional force is $f_k = \mu_k n = \mu_k m_A g$ and does negative work since the force opposes the motion. Since all of the variables are known except for v_f, we can substitute and solve for the final speed.

$$(5.00 \text{ kg})(9.80 \text{ m}/\text{s}^2)(1.50 \text{ m}) - (0.400)(3.00 \text{ kg})(9.80 \text{ m}/\text{s}^2)(1.50 \text{ m})$$
$$= \frac{1}{2}(3.00 \text{ kg})v_f^2 + \frac{1}{2}(5.00 \text{ kg})v_f^2$$

$$73.5 \text{ J} - 17.6 \text{ J} = \frac{1}{2}(8.00 \text{ kg})v_f^2 \quad \text{or} \quad v_f = \sqrt{\frac{2(55.9 \text{ J})}{8.00 \text{ kg}}} = 3.74 \text{ m}/\text{s} \qquad \Diamond$$

L: The final speed seems reasonable based on our expectation. This speed must also be less than if the rope were cut and the ball simply fell, in which case its final speed would be

$$v_f' = \sqrt{2gy} = \sqrt{2(9.80 \text{ m}/\text{s}^2)(1.50 \text{ m})} = 5.42 \text{ m}/\text{s}$$

33. A 5.00-kg block is set into motion up an inclined plane with an initial speed of 8.00 m/s (Fig. P8.33). The block comes to rest after traveling 3.00 m along the plane, which is inclined at an angle of 30.0° to the horizontal. For this motion determine (a) the change in the block's kinetic energy, (b) the change in its potential energy, and (c) the frictional force exerted on it (assumed to be constant). (d) What is the coefficient of kinetic friction?

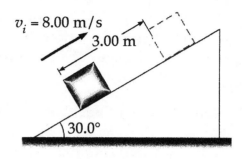

Figure P8.33

Solution

(a) $\quad \Delta K = K_f - K_i = \frac{1}{2}mv_f^2 - \frac{1}{2}mv_i^2 = 0 - \frac{1}{2}(5.00 \text{ kg})(8.00 \text{ m / s})^2 = -160 \text{ J}$ ◊

(b) $\quad \Delta U_g = U_{gf} - U_{gi} = mgy_f - 0$

$\quad \Delta U_g = (5.00 \text{ kg})(9.80 \text{ m / s}^2)(3.00 \text{ m}) \sin(30.0°) = 73.5 \text{ J}$ ◊

(c) Conservation of energy gives $\quad (K_i + U_{gi}) + W_{app} - f_k d = (K_f + U_{gf})$

Substituting for the energy terms, $\quad \left(\frac{1}{2}mv_i^2 + 0\right) - f_k d = \left(0 + mg(d \sin 30.0°)\right)$

Isolating the friction term, $\quad f_k = \left(mv_i^2 / 2d\right) - mg \sin(30.0°)$

Thus, $\quad f_k = \dfrac{(5.00 \text{ kg})(8.00 \text{ m / s})^2}{2(3.00 \text{ m})} - (5.00 \text{ kg})(9.80 \text{ m / s}^2)\left(\frac{1}{2}\right) = 28.8 \text{ N}$ ◊

(d) The forces perpendicular to the incline must add to zero:

$\quad \Sigma F_y = 0: \qquad\qquad +n - mg \cos 30.0° = 0$

$\qquad\qquad\qquad\qquad\qquad n = (5.00 \text{ kg})(9.80 \text{ m / s}^2) \cos 30.0° = 42.4 \text{ N}$

Now, $F_k = \mu_k n$ gives $\qquad \mu_k = f_k / n = 28.8 \text{ N} / 42.4 \text{ N} = 0.679$ ◊

41. The potential energy of a two - particle system separated by a distance r is given by $U(r) = A/r$, where A is a constant. Find the radial force $\mathbf{F}_r$ that each particle exerts on the other.

Solution

The force is the negative derivative of the potential energy with respect to distance:

$$F_r = -\frac{dU}{dr} = -\frac{d}{dr}\left(Ar^{-1}\right) = -A(-1)r^{-2} = \frac{A}{r^2} \qquad \lozenge$$

This describes an inverse-square-law force between the two particles.

---

49. The expression for the kinetic energy of a particle moving with speed v is given by Equation 7.19, which can be written as $K = \gamma mc^2 - mc^2$ where $\gamma = [1 - (v/c)^2]^{-1/2}$. The term γmc^2 is the total energy of the particle, and the term mc^2 is its rest energy. A proton moves with a speed of $0.990c$, where c is the speed of light. Find (a) its rest energy, (b) its total energy, and (c) its kinetic energy.

Solution

(a) $mc^2 = \left(1.67 \times 10^{-27}\ \text{kg}\right)(3.00 \times 10^8\ \text{m}/\text{s})^2 = 1.50 \times 10^{-10}\ \text{J} \qquad \lozenge$

(b) $\gamma = \dfrac{1}{\sqrt{1 - v^2/c^2}} = \dfrac{1}{\sqrt{1 - 0.990^2}} = 7.09$

$\gamma mc^2 = 7.09\ (1.50 \times 10^{-10}\ \text{J}) = 1.07 \times 10^{-9}\ \text{J} \qquad \lozenge$

(c) $K = \gamma mc^2 - mc^2 = (7.09 - 1)\ 1.50 \times 10^{-10}\ \text{J} = 9.15 \times 10^{-10}\ \text{J} \qquad \lozenge$

---

53. The particle described in Problem 52 (Fig. P8.52) is released from rest at A, and the surface of the bowl is rough. The speed of the particle at B is 1.50 m/s. (a) What is its kinetic energy at B? (b) How much mechanical energy is lost owing to friction as the particle moves from A to B? (c) Is it possible to determine μ from these results in any simple manner? Explain.

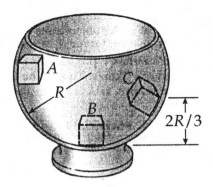

Solution

Figure P8.52

Take $U = 0$ at B. Since $v_i = 0$ at A,

$$K_A = 0 \quad \text{and} \quad U_A = mgR$$

(a) Since $v_B = 1.50$ m/s, and $m = 200$ g,

$$K_B = \tfrac{1}{2}mv_B^2 = \tfrac{1}{2}(0.200 \text{ kg})(1.50 \text{ m/s})^2 = 0.225 \text{ J} \qquad \Diamond$$

(b) At A, $\quad E_i = K_A + U_A = 0 + mgR = (0.200 \text{ kg})(9.80 \text{ m/s}^2)(0.300 \text{ m}) = 0.588 \text{ J}$

At B, $\quad E_f = K_B + U_B = 0.225 \text{ J} + 0$

By the energy equation, $\quad E_i - f_k d = E_f$

The energy lost is: $\quad f_k d = E_i - E_f = 0.588 \text{ J} - 0.225 \text{ J} = 0.363 \text{ J} \qquad \Diamond$

(c) Even though the energy lost is known, both the normal force and the friction force change with position as the block slides on the inside of the bowl. Therefore, there is no easy way to find μ.

57. A 10.0-kg block is released from point A in Figure P8.57. The track is frictionless except for the portion between B and C, which has a length of 6.00 m. The block travels down the track, hits a spring of force constant $k = 2250$ N/m, and compresses the spring 0.300 m from its equilibrium position before coming to rest momentarily. Determine the coefficient of kinetic friction between the block and the rough surface betwen B and C.

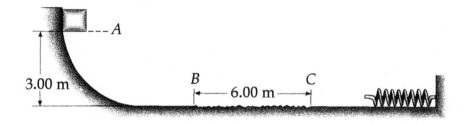

Figure P8.57

Solution

G: We should expect the coefficient of friction to be somewhere between 0 and 1 since this is the range of typical μ_k values. It is possible that μ_k could be greater than 1, but it can never be less than 0.

O: The easiest way to solve this problem is by considering the energy changes experienced by the block between the point of release (initial) and the point of full compression of the spring (final). Recall that the change in potential energy (gravitational and elastic) plus the change in kinetic energy must equal the work done on the block by non-conservative forces. Choose the gravitational energy to be zero along the flat portion of the track.

A: Putting the energy equation into symbols: $\Delta U_g + \Delta U_s + \Delta K = W_f$

Expanding into specific variables: $\left(-mgy_A\right) + \left(\frac{1}{2}kx_s^2\right) + 0 = -f_k d_{BC}$

The friction is $f_k = \mu_k mg$: $mgy_A - \frac{1}{2}kx^2 = \mu_k mgd$

Solving for the unknown variable μ_k:
$$\mu_k = \frac{y_A}{d} - \frac{kx^2}{2mgd}$$

Substituting: $\mu_k = \dfrac{3.00 \text{ m}}{6.00 \text{ m}} - \dfrac{(2250 \text{ N / m})(0.300 \text{ m})^2}{2(10.0 \text{ kg})(9.80 \text{ m / s}^2)(6.00 \text{ m})} = 0.328$ ◊

L: Our calculated value seems reasonable based on the friction data in Table 5.2. The most important aspect to solving these energy problems is considering how the energy is transferred from the initial to final energy states and remembering to subtract the energy resulting from any non-conservative forces (like friction).

---

61. A 20.0-kg block is connected to a 30.0-kg block by a string that passes over a frictionless pulley. The 30.0-kg block is connected to a spring that has negligible mass and a force constant of 250 N/m, as shown in Figure P8.61. The spring is unstretched when the system is as shown in the figure, and the incline is frictionless. The 20.0-kg block is pulled 20.0 cm down the incline (so that the 30.0-kg block is 40.0 cm above the floor) and is released from rest. Find the speed of each block when the 30.0-kg block is 20.0 cm above the floor (that is, when the spring is unstretched).

Solution   Let x be the distance the spring is stretched from equilibrium ($x = 0.200$ m), which corresponds to the upward displacement of the 30.0-kg mass. Also let $U_g = 0$ be measured from the lowest position of the 20.0-kg mass when the system is released from rest. Finally, define v as the speed of both blocks at the moment the spring passes through its unstretched position. Since all forces are conservative, conservation of energy yields $\Delta K + \Delta U_s + \Delta U_g = 0$.

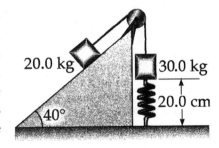

Figure P8.61

Solving for each variable,

$$\Delta K = \frac{1}{2}(m_1 + m_2)v^2 - 0 = \frac{1}{2}(50.0 \text{ kg})v^2 = (25.0 \text{ kg})v^2$$

$$\Delta U_s = 0 - \frac{1}{2}kx^2 = -\frac{1}{2}(250 \text{ N/m})(0.200 \text{ m})^2 = -5.00 \text{ N} \cdot \text{m}$$

$$\Delta U_g = (m_2 \sin \theta - m_1)gx = \left[(20.0 \text{ kg})\sin 40.0° - 30.0 \text{kg}\right](9.80 \text{ m} / \text{s}^2)(0.200 \text{ m})$$
$$= -33.6 \text{ N} \cdot \text{m}$$

Substituting into the energy equation,

$$(25.0 \text{ kg})v^2 - (5.00 \text{ N} \cdot \text{m}) - (33.6 \text{ N} \cdot \text{m}) = 0$$

Solving for v gives $\qquad v = 1.24 \text{ m} / \text{s}$ ◊

---

63. A block of mass 0.500 kg is pushed against a horizontal spring of negligible mass until the spring is compressed a distance of Δx (Fig. P8.63). The spring constant is 450 N/m. When it is released, the block travels along a frictionless, horizontal surface to point B, at the bottom of a vertical circular track of radius $R = 1.00$ m, and continues to move up the track. The speed of the block at the bottom of the track is

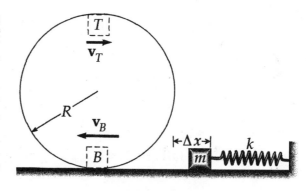

Figure P8.63

$v_B = 12.0$ m/s, and the block experiences an average frictional force of 7.00 N while sliding up the track. (a) What is Δx? (b) What speed do you predict for the block at the top of the track? (c) Does the block actually reach the top of the track, or does it fall off before reaching the top?

Solution

Energy is conserved in the firing of the block. Therefore,

(a) $\frac{1}{2}kx^2 = \frac{1}{2}mv^2$, or $\qquad\qquad \frac{1}{2}(450 \text{ N / m})(\Delta x)^2 = \frac{1}{2}(0.500 \text{ kg})(12.0 \text{ m / s})^2$

Thus, $\qquad\qquad\qquad\qquad\qquad \Delta x = 0.400 \text{ m}$ ◊

(b) Between bottom and top, $\quad (K + U_g)_B - f_k d = (K + U_g)_T$

$$\left(\tfrac{1}{2}mv_B^2 + mgh_B\right) - f(\pi R) = \left(\tfrac{1}{2}mv_T^2 + mgh_T\right)$$

Rearranging terms, $\qquad\qquad v_T^2 = v_B^2 + 2g(h_B - h_T) - 2f\pi R/m$

Substituting known variables, this becomes

$$v_T^2 = (12.0 \text{ m/s})^2 + 2\left(9.80 \text{ m/s}^2\right)(0 - 2.00 \text{ m}) - \frac{2(7.00 \text{ N})(\pi)(1.00 \text{ m})}{0.500 \text{ kg}}$$

$$v_T = \sqrt{16.8 \text{ m}^2/\text{s}^2} = 4.10 \text{ m / s} \qquad ◊$$

(c) The block falls if $a_c < g$. $\qquad a_c = \dfrac{v_T^2}{R} = \dfrac{(4.10 \text{ m / s})^2}{1.00 \text{ m}} = 16.8 \text{ m / s}^2$

Therefore, $a_c > g$. Some downward normal force is required along with the block's weight to constitute the central force, and the block stays on the track. ◊

69. A ball at the end of a string whirls around in a vertical circle. If the ball's total energy remains constant, show that the tension in the string at the bottom is greater than the tension at the top by a value six times the weight of the ball.

Solution

Applying Newton's second law at the bottom (b) and top (t) of the circular path gives

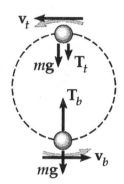

$$(1) \qquad T_b - mg = \frac{mv_b^2}{R}$$

$$(2) \qquad -T_t - mg = -\frac{mv_t^2}{R}$$

Subtracting (1) and (2) gives

$$(3) \qquad T_b = T_t + 2mg + \frac{m(v_b^2 - v_t^2)}{R}$$

Also, energy must be conserved; that is, $\Delta K + \Delta U = 0$

So,

$$\tfrac{1}{2}m\left(v_b^2 - v_t^2\right) + \left(0 - 2mgR\right) = 0$$

or $\qquad (4) \qquad \dfrac{v_b^2 - v_t^2}{R} = 4g$

Substituting (4) into (3) gives $\qquad T_b = T_t + 6mg \qquad\qquad \Diamond$

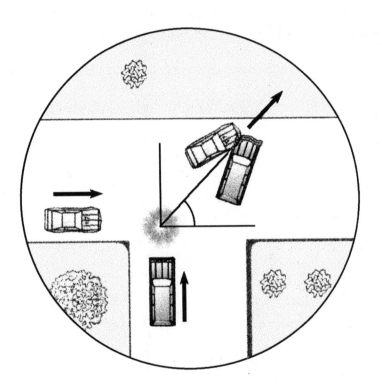

Linear Momentum and Collisions

LINEAR MOMENTUM AND COLLISIONS

INTRODUCTION

In this chapter we introduce the concept of **momentum**, a term that is used in describing objects in motion. Momentum is defined as the product of mass and velocity. The concept of momentum leads us to a second conservation law, that of conservation of momentum. This law is especially useful for treating problems that involve collisions between objects and for analyzing rocket propulsion. The concept of the center of mass of a system of particles is also introduced, and we shall see that the motion of a system of particles can be represented by the motion of one representative particle located at the center of mass.

EQUATIONS AND CONCEPTS

The **linear momentum p** of a particle is defined as the product of its mass m with its velocity **v**. This equation is equivalent to three component scalar equations, one along each of the coordinate axes.

$$\mathbf{p} \equiv m\mathbf{v} \tag{9.1}$$

The time rate of change of the linear momentum of a particle is equal to the resultant force acting on the particle. This is Newton's second law for a particle.

$$\Sigma \mathbf{F} = \frac{d\mathbf{p}}{dt} = \frac{d(m\mathbf{v})}{dt} \tag{9.3}$$

When two particles interact with each other (but are otherwise isolated from their surroundings), Newton's third law **tells us that the force of particle 1 on particle 2 is equal and opposite to the force of particle 2 on particle 1.** Since force is the time rate of change of momentum (Newton's second law), one finds that the total momentum of the isolated pair of particles is conserved.

$$\mathbf{p}_{tot} = \mathbf{p}_1 + \mathbf{p}_2 = \text{constant} \tag{9.4}$$

In general, when the **external force** acting on a system of particles is zero, the **total linear momentum of the system is conserved.** This important statement is known as the **law of conservation of momentum.** It is especially useful in treating problems involving collisions between two bodies.

$$\mathbf{p}_{1i} + \mathbf{p}_{2i} = \mathbf{p}_{1f} + \mathbf{p}_{2f} \tag{9.5}$$

The **impulse** of a force $\mathbf{F}$ acting on a particle **equals the change in momentum of the particle.** This is known as the impulse-momentum theorem.

$$\mathbf{I} \equiv \int_{t_i}^{t_f} \mathbf{F}\,dt = \Delta\mathbf{p} \tag{9.9}$$

To calculate impulse, we usually define a **time-averaged force $\mathbf{F}$** which would give the same impulse to the particle as the actual time-varying force over the time interval Δt.

$$\overline{\mathbf{F}} \equiv \frac{1}{\Delta t}\int_{t_i}^{t_f} \mathbf{F}\,dt \tag{9.10}$$

The average force can be thought of as the constant force that would give the same change in momentum over the time interval Δt as the applied impulse.

$$I = \bar{F}\Delta t \qquad (9.11)$$

In the impulse approximation of Equation 9.9, the force $\mathbf{F}$ acts for a short time and is much larger than any other force present. This approximation is usually made in collision problems, where the force is the contact force between the particles during the collision.

It is useful to consider two particular types of collisions that can occur between two bodies. An **elastic collision** is one in which **both linear momentum and kinetic energy are conserved.** An **inelastic collision** is one in which **only linear momentum is conserved.** A perfectly inelastic collision is an inelastic collision in which the two bodies stick together after the collision. In any type of collision, momentum is conserved. Note that when we say that the momentum is conserved, we are speaking of the momentum of the **entire system:** The momentum of any particle in a collision may change, but the momentum of the total system remains unchanged.

$$\left.\begin{array}{l} \mathbf{p}_1 + \mathbf{p}_2 = \text{const} \\[2mm] K_1 + K_2 = \text{const} \end{array}\right\} \quad \text{(Elastic collision)}$$

$$\mathbf{p}_1 + \mathbf{p}_2 = \text{const} \quad \text{(Inelastic collision)}$$

When two particles moving along a straight line collide and stick together (perfectly inelastic collision), the common velocity after the collision can be calculated in terms of the two mass values and the two initial velocities.

$$\mathbf{v}_f = \frac{m_1 \mathbf{v}_{1i} + m_2 \mathbf{v}_{2i}}{m_1 + m_2} \qquad (9.14)$$

When two particles undergo a perfectly elastic collision, both momentum and kinetic energy are conserved. When such a collision occurs, **the relative velocity before the collision equals the negative of the relative velocity of the two particles following the collision.**

$$v_{1i} - v_{2i} = -\left(v_{1f} - v_{2f}\right) \qquad (9.19)$$

When the masses and initial speeds of both particles are known, the final speeds can be calculated.

$$v_{1f} = \left(\frac{m_1 - m_2}{m_1 + m_2}\right)v_{1i} + \left(\frac{2m_2}{m_1 + m_2}\right)v_{2i} \qquad (9.20)$$

$$v_{2f} = \left(\frac{2m_1}{m_1 + m_2}\right)v_{1i} + \left(\frac{m_2 - m_1}{m_1 + m_2}\right)v_{2i} \qquad (9.21)$$

An important special use occurs **when the second particle (m_2, the "target") is initially at rest.** Remember the appropriate algebraic signs (designating direction) must be included for v_{1i} and v_{2i}.

$$v_{1f} = \left(\frac{m_1 - m_2}{m_1 + m_2}\right)v_{1i} \qquad (9.22)$$

$$v_{2f} = \left(\frac{2m_1}{m_1 + m_2}\right)v_{1i} \qquad (9.23)$$

The **x coordinate of the center of mass of n particles** with individual coordinates of $x_1, x_2, x_3 \ldots$ and masses of $m_1, m_2, m_3 \ldots$ is given by Equation 9.28. The y and z coordinates of the center of mass are defined by similar expressions. The center of mass of a homogeneous, symmetric body must lie on an axis of symmetry.

$$x_{CM} \equiv \frac{\sum_i m_i x_i}{\sum_i m_i} \tag{9.28}$$

The center of mass for a collection of particles can be located by its position vector.

$$\mathbf{r}_{CM} \equiv \frac{\sum_i m_i \mathbf{r}_i}{M} \tag{9.30}$$

For an extended object, the center of mass can be calculated by integrating over the total length, area, or volume which includes the total mass M.

$$\mathbf{r}_{CM} = \frac{1}{M} \int \mathbf{r} \, dm \tag{9.33}$$

In this expression for the **velocity of the center of mass of a system of particles**, $\mathbf{v}_i$ is the velocity of the ith particle and M is the total mass of the system.

$$\mathbf{v}_{CM} = \frac{\sum_i m_i \mathbf{v}_i}{M} \tag{9.34}$$

The **total momentum of a system of particles** is equal to the total mass M multiplied by the velocity of the center of mass.

$$\mathbf{p}_{tot} = M\mathbf{v}_{CM} = \text{constant} \tag{9.39}$$

$$(\text{when } \Sigma \mathbf{F}_{ext} = 0)$$

179

The **acceleration of the center of mass of a system of particles** depends on the value of the acceleration for each of the individual particles.

$$a_{CM} = \frac{1}{M} \sum_i m_i a_i \qquad (9.36)$$

Newton's second law applied to a system of particles says that the **resultant external force** acting on the system **equals the time rate of change of the total momentum.** This form of Newton's second law must be used when the mass of the system changes. If the net external force on the system is zero, then the total momentum of the system remains constant.

$$\Sigma F_{ext} = M a_{CM} = \frac{d p_{tot}}{dt} \qquad (9.38)$$

The principle behind the operation of a rocket is the law of conservation of momentum as applied to the rocket and its ejected fuel. If a rocket moves in the absence of gravity and ejects fuel with an exhaust velocity v_e, **its change in velocity** is proportional to the exhaust velocity, where M_i and M_f refer to its initial and final mass values for the rocket.

$$v_f - v_i = v_e \ln\left(\frac{M_i}{M_f}\right) \qquad (9.41)$$

The thrust on a rocket increases as the exhaust speed increases and as the burn rate increases.

$$\text{Thrust} = \left| v_e \frac{dM}{dt} \right| \qquad (9.42)$$

SUGGESTIONS, SKILLS, AND STRATEGIES

The following procedure is recommended when dealing with problems involving collisions between two objects:

- Set up a coordinate system and define your velocities with respect to that system. That is, objects moving in the direction selected as the positive direction of the x axis are considered as having a positive velocity and negative if moving in the negative x direction. It is convenient to have the x axis coincide with one of the initial velocities.

- In your sketch of the coordinate system, draw all velocity vectors with labels and include all the given information.

- Write expressions for the momentum of each object before and after the collision. (In two-dimensional collision problems, write expressions for the x and y components of momentum before and after the collision.) Remember to include the appropriate signs for the velocity vectors.

- Now write expressions for the **total** momentum **before** and **after** the collision and equate the two. (For two-dimensional collisions, this expression should be written for the momentum in both the x and y directions.) It is important to emphasize that it is the momentum of the **system** (the two colliding objects) that is conserved, not the momentum of the individual objects.

- If the collision is **inelastic**, you should then proceed to solve the momentum equations for the unknown quantities.

- If the collision is **elastic**, kinetic energy is also conserved, so you can equate the total kinetic energy before the collision to the total kinetic energy after the collision. This gives an additional relationship between the various velocities. The conservation of kinetic energy for elastic collisions leads to the expression $v_{1i} - v_{2i} = -(v_{1f} - v_{2f})$, which is often easier to use in solving elastic collision problems than is an expression for conservation of kinetic energy.

REVIEW CHECKLIST

▷   The impulse of a force acting on a particle during some time interval equals the **change** in momentum of the particle, and the impulse equals the area under the force-time graph.

▷   The momentum of any isolated system (one for which the net external force is zero) is conserved, regardless of the nature of the forces between the masses which comprise the system.

▷   There are two types of collisions that can occur between two particles, namely elastic and inelastic collisions. Recognize that a **perfectly** inelastic collision is an inelastic collision in which the colliding particles stick together after the collision, and hence move as a composite particle.

▷   The conservation of linear momentum applies not only to head-on collisions (one-dimensional), but also to glancing collisions (two- or three-dimensional). For example, in a two-dimensional collision, the total momentum in the x direction is conserved and the total momentum in the y direction is conserved.

▷   The equations for momentum and kinetic energy can be used to calculate the final velocities in a two-body head-on elastic collision; and to calculate the final velocity and the change of kinetic energy in a two-body system for a completely inelastic collision.

ANSWERS TO SELECTED CONCEPTUAL QUESTIONS

4. If two particles have equal momenta, are their kinetic energies necessarily equal? Explain.

Answer   No, the kinetic energies might not be equal. Equal momenta means that the product of mass and velocity are equal. Thus, if the mass of one particle is increased by a factor of α, then its velocity will be reduced by a factor of $1/\alpha$. But the kinetic energy, $K = mv^2/2$, takes that reduced velocity into account twice, while taking the increased mass into account only once. Therefore, a more massive particle will have a lower kinetic energy.

□   □   □   □

7. Explain how linear momentum is conserved when a ball bounces from a floor.

Answer   The ball's downward momentum increases as it accelerates downward. A larger upward momentum change occurs when it touches the floor and rebounds. The outside forces of gravity and the normal force inject impulses to change its momentum.

If we think of ball-and-Earth-together as our system, these forces are internal and do not change the total momentum. It is conserved, if we neglect the curvature of the Earth's orbit. As the ball falls down, the Earth lurches up to meet it, on the order of 10^{25} times more slowly. Then, ball and Earth bounce off each other and separate. In other words, while you dribble a ball, you also dribble the Earth.

□   □   □   □

15. A sharpshooter fires a rifle while standing with the butt of the gun against his shoulder. If the forward momentum of a bullet is the same as the backward momentum of the gun, why is it not as dangerous to be hit by the gun as by the bullet?

Answer It is the product mv which is the same for both the bullet and the gun. The bullet has a large velocity and a small mass, while the gun has a small velocity and a large mass. However, the bullet carries much more kinetic energy than the gun, and inflicts that energy on a much smaller surface area. Thus, it is more likely to bruise (in the case of rubber bullets), break, and pierce whatever it hits.

□   □   □   □

20. Does the center of mass of a rocket in free space accelerate? Explain. Can the speed of a rocket exceed the exhaust speed of the fuel? Explain.

Answer The center of mass of a **rocket, plus its exhaust** does not accelerate: momentum must be conserved. However, if you consider the "rocket" to be the mechanical system plus the unexpended fuel, then it becomes obvious that the center of mass of that "rocket" does accelerate.

Ultimately, a rocket is no more complicated than two rubber balls that are pressed together, and then released. One springs in one direction with one velocity; the other springs in the other direction with a velocity equal to

$$v_2 = v_1(m_1 \, / \, m_2)$$

In a similar manner, the mass of the rocket times its speed must equal the mass of all parts of its exhaust, times their speed. After a sufficient quantity of fuel has been exhausted the ratio m_1/m_2 will be greater than 1; then the speed of the rocket **can** exceed the speed of the exhaust.

□   □   □   □

184

SOLUTIONS TO SELECTED END-OF-CHAPTER PROBLEMS

9. An estimated force-time curve for a baseball struck by a bat is shown in Figure P9.9. From this curve, determine (a) the impulse delivered to the ball, (b) the average force exerted on the ball, and (c) the peak force exerted on the ball.

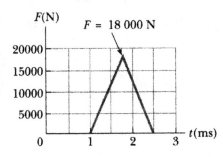

Figure P9.9

Solution

(a) $I = \int F\,dt =$ area under the F-t graph

$$= \frac{1}{2}(18000\ \text{N})(2.50\times 10^{-3}\,\text{s} - 1.00\times 10^{-3}\,\text{s}) = 13.5\ \text{N}\cdot\text{s} \qquad \Diamond$$

(b) $\overline{F} = \dfrac{\int F\,dt}{\Delta t} = \dfrac{13.5\ \text{N}\cdot\text{s}}{\left(2.50\times 10^{-3}\,\text{s} - 1.00\times 10^{-3}\,\text{s}\right)} = 9000\ \text{N}$ \qquad $\Diamond$

(c) From the graph, $F_{max} = 18000\ \text{N}$ \qquad $\Diamond$

11. A 3.00-kg steel ball strikes a wall with a speed of 10.0 m/s at an angle of 60.0° with the surface. It bounces off with the same speed and angle (Fig. P9.11). If the ball is in contact with the wall for 0.200 s, what is the average force exerted on the ball by the wall?

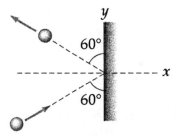

Figure P9.11

Solution

G: If we think about the angle as a variable and consider the limiting cases, then the force should be zero when the angle is 0° (no contact between the ball and the wall). When the angle is 90° the force will be its maximum and can be found from the momentum-impulse equation, so that $F = \Delta p / \Delta t < 300$ N, and the force on the ball must be directed to the left.

O: Use the momentum-impulse equation to find the force, and carefully consider the direction of the velocity vectors by defining up and to the right as positive.

A: $\Delta \mathbf{p} = \mathbf{F} \Delta t$

$$\Delta p_y = m\left(v_{fy} - v_{iy}\right) = m(v\cos 60.0° - v\cos 60.0°) = 0$$

So the wall does not exert a force on the ball in the y direction.

$$\Delta p_x = m\left(v_{fx} - v_{ix}\right) = m(-v\sin 60.0° - v\sin 60.0°) = -2mv\sin 60.0°$$

$$\Delta p_x = -2(3.00 \text{ kg})(10.0 \text{ m / s})(0.866) = -52.0 \text{ kg} \cdot \text{m / s}$$

$$\overline{\mathbf{F}} = \frac{\Delta \mathbf{p}}{\Delta t} = \frac{\Delta p_x \mathbf{i}}{\Delta t} = \frac{-52.0\mathbf{i} \text{ kg} \cdot \text{m / s}}{0.200 \text{ s}} = -260\mathbf{i} \text{ N} \qquad \Diamond$$

L: The force is to the left and has a magnitude less than 300 N as expected.

17. A 10.0-g bullet is fired into a stationary block of wood (m = 5.00 kg). The relative motion of the bullet stops inside the block. The speed of the bullet-plus-wood combination immediately after the collision is measured as 0.600 m/s. What was the original speed of the bullet?

Solution

G: A reasonable speed of a bullet should be somewhere between 100 and 1000 m/s.

O: We can find the initial speed of the bullet from conservation of momentum. We are told that the block of wood was originally stationary.

A: Since there is no external force on the block and bullet system, the total momentum of the system is constant so that $\Delta \mathbf{p} = 0$

$$\mathbf{p}_{1i} + \mathbf{p}_{2i} = \mathbf{p}_{1f} + \mathbf{p}_{2f}$$

$$(0.0100 \text{ kg})\mathbf{v}_{1i} + 0 = (0.0100 \text{ kg})(0.600 \text{ m}/\text{s})\mathbf{i} + (5.00 \text{ kg})(0.600 \text{ m}/\text{s})\mathbf{i}$$

$$\mathbf{v}_{1i} = \frac{(5.01 \text{ kg})(0.600 \text{ m}/\text{s})\mathbf{i}}{0.0100 \text{ kg}} = 301\mathbf{i} \text{ m}/\text{s} \qquad \Diamond$$

L: The speed seems reasonable, and is in fact just under the speed of sound in air (343 m/s at 20° C).

19. A 45.0-kg girl is standing on a plank that has a mass of 150 kg. The plank, originally at rest, is free to slide on a frozen lake, which is a flat, frictionless supporting surface. The girl begins to walk along the plank at a constant speed of 1.50 m/s relative to the plank. (a) What is her speed relative to the ice surface? (b) What is the speed of the plank relative to the ice surface?

Solution   Let $\mathbf{v}_g$ = velocity of the girl relative to the ice

$\mathbf{v}_{gp}$ = velocity of the girl relative to the plank

$\mathbf{v}_p$ = velocity of the plank relative to the ice

The girl and the plank exert forces on each other, but the ice isolates them from outside horizontal forces. Therefore, the net momentum is zero for the combined girl plus plank system.

$$0 = m_g\mathbf{v}_g + m_p\mathbf{v}_p$$

Further, the relation among relative speeds can be written:

$$\mathbf{v}_g = \mathbf{v}_{gp} + \mathbf{v}_p$$

$$\mathbf{v}_g = \left(1.50\mathbf{i} + \mathbf{v}_p\right) \text{m/s}$$

We substitute: $0 = \left(45.0 \text{ kg}\right)\left(1.50\mathbf{i} \text{ m/s} + \mathbf{v}_p\right) + \left(150 \text{ kg}\right)\mathbf{v}_p$

$$\left(195 \text{ kg}\right)\mathbf{v}_p = -\left(-45.0 \text{ kg}\right)\left(1.50\mathbf{i} \text{ m/s}\right)$$

(b) $\mathbf{v}_p = -0.346\mathbf{i} \text{ m/s}$ ◊

(a) $\mathbf{v}_g = 1.50\mathbf{i} - 0.346\mathbf{i} \text{ m/s} = 1.15\mathbf{i} \text{ m/s}$ ◊

25. A neutron in a reactor makes an elastic head-on collision with the nucleus of a carbon atom initially at rest. (a) What fraction of the neutron's kinetic energy is transferred to the carbon nucleus? (b) If the initial kinetic energy of the neutron is 1.60×10^{-13} J, find its final kinetic energy and the kinetic energy of the carbon nucleus after the collision. (The mass of the carbon nucleus is about 12.0 times greater than the mass of the neutron.)

Solution

(a) This is a perfectly elastic head-on collision, so we use:

$$v_{1i} - v_{2i} = -\left(v_{1f} - v_{2f}\right)$$

Let object 1 be the neutron, object 2 be the carbon nucleus: $m_2 = 12.0 m_1$

Since $v_{2i} = 0$, $\qquad\qquad\qquad\qquad\qquad v_{2f} = v_{1i} + v_{1f}$

Now, by conservation of momentum, $\qquad m_1 v_{1i} + m_2 v_{2i} = m_1 v_{1f} + m_2 v_{2f}$

or $\qquad\qquad\qquad\qquad\qquad\qquad m_1 v_{1i} = m_1 v_{1f} + 12.0 m_1 v_{2f}$

Substituting our velocity equation, $\qquad v_{1i} = v_{1f} + 12.0(v_{1i} + v_{1f})$

We solve $-11 v_{1i} = 13 v_{1f}$: $\qquad\qquad v_{1f} = -(11.0/13.0) v_{1i}$

and $\qquad\qquad\qquad\qquad v_{2f} = v_{1i} - \left(\dfrac{11.0}{13.0}\right) v_{1i} = \left(\dfrac{2.00}{13.0}\right) v_{1i}$

The neutron's original kinetic energy is $\frac{1}{2} m_1 v_{1i}^2$. The carbon's final kinetic energy is

$$\tfrac{1}{2} m_2 v_{2f}^2 = \tfrac{1}{2}(12.0 m_1)\left(\frac{2.00}{13.0}\right)^2 v_{1i}^2 = \left(\frac{48.0}{169}\right)\left(\tfrac{1}{2}\right) m_1 v_{1i}^2$$

So, $\qquad (48.0/169) = 0.284 = 28.4\%$ of the total energy is transferred. ◊

(b)   For the carbon nucleus, $K_{2f} = (0.284)(1.60 \times 10^{-13} \text{ J}) = 45.4 \text{ fJ}$ ◊

The collision is perfectly elastic, so the neutron retains the rest of the energy,

$$K_{1f} = 1.60 \times 10^{-13} \text{ J} - 0.454 \times 10^{-13} \text{ J} = 1.15 \times 10^{-13} \text{ J} = 115 \text{ fJ}$$ ◊

---

27.   A 12.0-g bullet is fired into a 100-g wooden block initially at rest on a horizontal surface. After impact, the block slides 7.50 m before coming to rest. If the coefficient of friction between block and surface is 0.650, what was the speed of the bullet immediately before impact?

Solution   The collision between picture (1) and picture (2) is **totally inelastic**, and momentum is conserved:

$$m_1 v_1 = (m_1 + m_2) v_2$$

In the sliding process between pictures (2) and (3), the change in kinetic energy is equal to the energy lost due to friction:

$$\tfrac{1}{2}(m_1 + m_2){v_2}^2 = f_f L$$

$$\tfrac{1}{2}(m_1 + m_2){v_2}^2 = \mu(m_1 + m_2)gL$$

Solving for v_2: $v_2 = \sqrt{2\mu L g} = \sqrt{2(0.650)(7.50 \text{ m})(9.80 \text{ m}/\text{s}^2)} = 9.77 \text{ m/s}$

By conservation of momentum,

$$v_1 = \left(\frac{m_1 + m_2}{m_1}\right)v_2 = \left(\frac{112 \text{ g}}{12.0 \text{ g}}\right)9.77 \text{ m}/\text{s} = 91.2 \text{ m/s}$$ ◊

---

33. A billiard ball moving at 5.00 m/s strikes a stationary ball of the same mass. After the collision, the first ball moves at 4.33 m/s, at an angle of 30.0° with respect to the original line of motion. Assuming an elastic collision (and ignoring friction and rotational motion), find the struck ball's velocity after the collision.

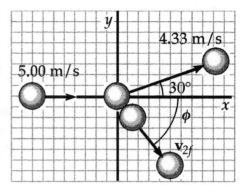

Solution

Call each mass m, and call $\mathbf{v}_{2f}$ the velocity of the second ball after the collision, as in the figure. We apply conservation of momentum:

In the x direction, $\qquad m(5.00 \text{ m / s}) = m(4.33 \text{ m / s})\cos 30.0° + mv_{2fx}$

Solving, $\qquad v_{2fx} = 1.25 \text{ m / s}$

In the y direction, $\qquad 0 = m(4.33 \text{ m / s})\sin 30.0° + mv_{2fy}$

Solving, $\qquad v_{2fy} = -2.17 \text{ m / s}$

$$\mathbf{v}_{2f} = (1.25\mathbf{i} - 2.17\mathbf{j}) \text{ m / s} \quad (\text{or } 2.50 \text{ m / s at } -60.0°) \quad \Diamond$$

35. A 3.00-kg mass with an initial velocity of 5.00**i** m/s collides with and sticks to a 2.00-kg mass with an initial velocity of –3.00**j** m/s. Find the final velocity of the composite mass.

Solution

Momentum is conserved, with both masses having the same final velocity:

$$m_1\mathbf{v}_{1i} + m_2\mathbf{v}_{2i} = m_1\mathbf{v}_{1f} + m_2\mathbf{v}_{2f}$$

$$(3.00 \text{ kg})(5.00\mathbf{i} \text{ m / s}) + (2.00 \text{ kg})(-3.00\mathbf{j} \text{ m / s}) = (3.00 \text{ kg} + 2.00 \text{ kg})\mathbf{v}_f$$

$$\mathbf{v}_f = \frac{15.0\mathbf{i} - 6.00\mathbf{j}}{5.00} \text{ m / s} = (3.00\mathbf{i} - 1.20\mathbf{j}) \text{ m / s} \qquad \Diamond$$

Related Calculation: Compute the kinetic energy both before and after the collision; show that kinetic energy is not conserved.

$$K_{1i} + K_{2i} = \frac{1}{2}(3.00 \text{ kg})(5.00 \text{ m / s})^2 + \frac{1}{2}(2.00 \text{ kg})(3.00 \text{ m / s})^2$$

$$K_{1i} + K_{2i} = 46.5 \text{ J}$$

$$K_{1f} + K_{2f} = \frac{1}{2}(5.00 \text{ kg})\left[(3.00 \text{ m / s})^2 + (1.20 \text{ m / s})^2\right] = 26.1 \text{ J} \qquad \Diamond$$

---

39. An unstable nucleus of mass 17.0×10^{-27} kg initially at rest disintegrates into three particles. One of the particles, of mass 5.00×10^{-27} kg, moves along the y axis with a velocity of 6.00×10^6 m / s. Another particle, of mass 8.40×10^{-27} kg, moves along the x axis with a speed of 4.00×10^6 m / s. Find (a) the velocity of the third particle and (b) the total kinetic energy increase in the process.

Solution

(a) With three particles, the total final momentum is $m_1\mathbf{v}_{1f} + m_2\mathbf{v}_{2f} + m_3\mathbf{v}_{3f}$, and it must be zero to equal the original momentum.

Assuming conservation of mass, the mass of the third particle is

$$m_3 = (17.0 - 5.00 - 8.40) \times 10^{-27} \text{ kg} = 3.60 \times 10^{-27} \text{ kg}$$

Because the total momentum is zero, $m_1\mathbf{v}_{1f} + m_2\mathbf{v}_{2f} + m_3\mathbf{v}_{3f} = 0$

Solving for $\mathbf{v}_{3f}$, $\mathbf{v}_{3f} = -\dfrac{m_1\mathbf{v}_{1f} + m_2\mathbf{v}_{2f}}{m_3}$

$$\mathbf{v}_{3f} = -\frac{(3.00\mathbf{j} + 3.36\mathbf{i}) \times 10^{-20} \text{ kg} \cdot \text{m / s}}{3.60 \times 10^{-27} \text{ kg}} = (-9.33\mathbf{i} - 8.33\mathbf{j}) \text{ Mm / s} \qquad \lozenge$$

(b) The original kinetic energy is zero. The final kinetic energy is $K = K_{1f} + K_{2f} + K_{3f}$.

$$K_{1f} = \tfrac{1}{2}\left(5.00 \times 10^{-27} \text{ kg}\right)\left(6.00 \times 10^6 \text{ m / s}\right)^2 = 9.00 \times 10^{-14} \text{ J}$$

$$K_{2f} = \tfrac{1}{2}\left(8.40 \times 10^{-27} \text{ kg}\right)\left(4.00 \times 10^6 \text{ m / s}\right)^2 = 6.72 \times 10^{-14} \text{ J}$$

$$K_{3f} = \tfrac{1}{2}\left(3.60 \times 10^{-27} \text{ kg}\right)\left(\left(9.33 \times 10^6 \text{ m / s}\right)^2 + \left(8.33 \times 10^6 \text{ m / s}\right)^2\right)$$
$$= 28.2 \times 10^{-14} \text{ J}$$

Thus $K = (9.00 \times 10^{-14} \text{ J}) + (6.72 \times 10^{-14} \text{ J}) + (28.2 \times 10^{-14} \text{ J}) = 4.39 \times 10^{-13} \text{ J} \qquad \lozenge$

41. A uniform piece of sheet steel is shaped as shown in Figure P9.41. Compute the x and y coordinates of the center of mass of the piece.

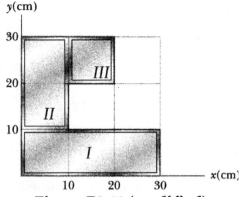

Figure P9.41 (modified)

Solution

G: By inspection, it appears that the center of mass is located at about

$$(12\,\mathbf{i} + 13\,\mathbf{j})\ \text{cm}$$

O: Think of the sheet as composed of three sections, and consider the mass of each section to be at the geometric center of that section. Define the mass per unit area to be σ, and number the rectangles as shown. We can then calculate the mass and identify the center of mass of each section.

A: $m_I = (30.0\ \text{cm})(10.0\ \text{cm})\sigma$ $\qquad CM_I = (15.0\ \text{cm},\ 5.00\ \text{cm})$

$m_{II} = (10.0\ \text{cm})(20.0\ \text{cm})\sigma$ $\qquad CM_{II} = (5.00\ \text{cm},\ 20.0\ \text{cm})$

$m_{III} = (10.0\ \text{cm})(10.0\ \text{cm})\sigma$ $\qquad CM_{III} = (15.0\ \text{cm},\ 25.0\ \text{cm})$

The overal center of mass is at a point defined by the vector equation

$$\mathbf{r}_{CM} \equiv \left(\Sigma m_i \mathbf{r}_i\right)/\Sigma m_i$$

Substituting the appropriate values, $\mathbf{r}_{CM}$ is calculated to be:

$$\mathbf{r}_{CM} = \frac{\sigma\left[(300)(15.0\mathbf{i} + 5.00\mathbf{j}) + (200)(5.00\mathbf{i} + 20.0\mathbf{j}) + (100)(15.0\mathbf{i} + 25.0\mathbf{j})\ \text{cm}^3\right]}{\sigma\left(300\ \text{cm}^2 + 200\ \text{cm}^2 + 100\ \text{cm}^2\right)}$$

$$\mathbf{r}_{CM} = \frac{(45.0\mathbf{i} + 15.0\mathbf{j} + 10.0\mathbf{i} + 40.0\mathbf{j} + 15.0\mathbf{i} + 25.0\mathbf{j})}{6.00}\ \text{cm}$$

$$\mathbf{r}_{CM} = (11.7\mathbf{i} + 13.3\mathbf{j})\ \text{cm} \qquad\qquad \Diamond$$

L: The coordinates are close to our eyeball estimate. In solving this problem, we could have chosen to divide the original shape some other way, but the answer would be the same. This problem also shows that the center of mass can lie outside the boundary of the object.

---

47. Romeo (77.0 kg) entertains Juliet (55.0 kg) by playing his guitar from the rear of their boat at rest in still water, 2.70 m away from Juliet who is in the front of the boat. After the serenade, Juliet carefully moves to the rear of the boat (away from shore) to plant a kiss on Romeo's cheek. How far does the 80.0-kg boat move toward the shore it is facing?

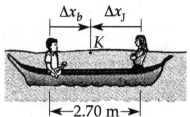

Solution Neglecting any viscous drag forces of the water on the boat, no outside forces act on the boat-plus-lovers system, so its momentum is conserved at zero and its center of mass stays fixed: $x_{CM,\,i} = x_{CM,\,f}$

Define K to be the point where they kiss, and Δx_J and Δx_b as shown in the figure. Since Romeo moves with the boat (and thus $\Delta x_{Romeo} = \Delta x_b$), let m_b be the combined mass of Romeo and the boat.

Then the center of mass equations is
$$\frac{m_J \Delta x_{Ji} + m_b \Delta x_{bi}}{m_J + m_b} = \frac{m_J \Delta x_{Jf} + m_b \Delta x_{bf}}{m_J + m_b}$$

which reduces to
$$m_J \Delta x_J + m_b \Delta x_b = 0$$

Let $+x$ point away from the shore:
$$(55.0 \text{ kg})\Delta x_J + (77.0 \text{ kg} + 80.0 \text{ kg})\Delta x_b = 0$$

and
$$\Delta x_J = -2.85\Delta x_b$$

As Juliet moves away from shore, the boat and Romeo glide toward the shore until the original 2.70 m gap between them is closed: $\Delta x_J - \Delta x_b = 2.70 \text{ m}$

Substituting, we find $\Delta x_b = -0.700 \text{ m}$, or 0.700 m towards the shore ◊

195

49. A 2.00-kg particle has a velocity of $(2.00\mathbf{i} - 3.00\mathbf{j})$ m/s, and a 3.00-kg particle has a velocity of $(1.00\mathbf{i} + 6.00\mathbf{j})$ m/s. Find (a) the velocity of the center of mass and (b) the total momentum of the system.

Solution

Use $\quad \mathbf{v}_{CM} = \dfrac{m_1 \mathbf{v}_1 + m_2 \mathbf{v}_2}{m_1 + m_2} \quad$ and $\quad \mathbf{p}_{CM} = (m_1 + m_2)\mathbf{v}_{CM}:$

(a) $\quad \mathbf{v}_{CM} = \dfrac{(2.00 \text{ kg})\left[(2.00\mathbf{i} - 3.00\mathbf{j}) \text{ m/s}\right] + (3.00 \text{ kg})\left[(1.00\mathbf{i} + 6.00\mathbf{j}) \text{ m/s}\right]}{(2.00 \text{ kg} + 3.00 \text{ kg})}$

$\quad \mathbf{v}_{CM} = (1.40\mathbf{i} + 2.40\mathbf{j}) \text{ m/s} \hfill \Diamond$

(b) $\quad \mathbf{p}_{CM} = (2.00 \text{ kg} + 3.00 \text{ kg})\left[(1.40\mathbf{i} + 2.40\mathbf{j}) \text{ m/s}\right] = (7.00\mathbf{i} + 12.0\mathbf{j}) \text{ kg} \cdot \text{m/s} \hfill \Diamond$

---

51. The first stage of a Saturn V space vehicle consumes fuel and oxidizer at the rate of 1.50×10^4 kg/s, with an exhaust speed of 2.60×10^3 m/s. (a) Calculate the thrust produced by these engines. (b) Find the initial acceleration of the vehicle on the launch pad if its initial mass is 3.00×10^6 kg. [**Hint:** You must include the force of gravity to solve part (b).]

Solution

G: The thrust must be at least equal to the weight of the rocket ($\cong 30$ MN); otherwise the launch will not be successful! However, since a Saturn V rocket accelerates rather slowly compared to the acceleration of falling objects, the thrust should be less than about twice the rocket's weight so that $0 < a < g$.

O: Use Newton's second law to find the force and acceleration from the changing momentum.

A: (a) The thrust, F, is equal to the time rate of change of momentum as fuel is exhausted from the rocket.

$$F = \frac{dp}{dt} = \frac{d}{dt}(mv_e)$$

Since v_e is a constant exhaust velocity, so

$$F = v_e(dm/dt)$$

where $dm/dt = 1.50 \times 10^4$ kg / s and $v_e = 2.60 \times 10^3$ m / s

$$F = (2.60 \times 10^3 \text{ m / s})(1.50 \times 10^4 \text{ kg / s}) = 39.0 \text{ MN} \qquad \Diamond$$

(b) Applying $\Sigma F = ma$,

$$(3.90 \times 10^7 \text{ N}) - (3.00 \times 10^6 \text{ kg})(9.80 \text{ m / s}^2) = (3.00 \times 10^6 \text{ kg})a$$

$$a = \frac{(3.90 \times 10^7 \text{ N}) - (29.4 \times 10^6 \text{ N})}{3.00 \times 10^6 \text{ kg}} = 3.20 \text{ m / s}^2 \text{ up} \qquad \Diamond$$

L: As expected, the thrust is slightly greater than the weight of the rocket, and the acceleration is about $0.3g$, so the answers appear to be reasonable. This kind of rocket science is not so complicated after all!

---

59. An 80.0-kg astronaut is working on the engines of his ship, which is drifting through space with a constant velocity. The astronaut, wishing to get a better view of the Universe, pushes against the ship and much later finds himself 30.0 m behind the ship and at rest with respect to it. Without a thruster, the only way to return to the ship is to throw his 0.500-kg wrench directly away from the ship. If he throws the wrench with a speed of 20.0 m/s relative to the ship, how long does it take the astronaut to reach the ship?

Solution

G: We hope the momentum of the wrench provides enough recoil so that the astronaut can reach the ship before he loses life support! We might expect the elapsed time to be on the order of several minutes based on the description of the situation.

O: No external force acts on the system (astronaut plus wrench), so the total momentum is constant. Since the final momentum (wrench plus astronaut) must be zero, we have

$$\text{Final momentum} = \text{Initial momentum} = 0.$$

A: In equation form, $\quad m_w v_w + m_a v_a = 0$

Thus, $\quad v_a = -m_w v_w / m_a$

$$v_a = -\frac{(0.500 \text{ kg})(20.0 \text{ m} / \text{s})}{80.0 \text{ kg}} = -0.125 \text{ m} / \text{s}$$

At this speed, the time he takes to travel to the ship is

$$t = \frac{30.0 \text{ m}}{0.125 \text{ m} / \text{s}} = 240 \text{ s} = 4.00 \text{ minutes} \qquad \lozenge$$

L: The astronaut is fortunate that the wrench gave him sufficient momentum to return to the ship in a reasonable amount of time! In this problem, we were told that the astronaut was not drifting away from the ship when he threw the wrench. However, this is not possible since he did not encounter an external force that would reduce his velocity away from the ship (there is no air friction beyond earth's atmosphere). If this were a real-life situation, the astronaut would have to throw the wrench hard enough to overcome his momentum caused by his original push away from the ship, and then some.

71. A 5.00-g bullet moving with an initial speed of 400 m/s is fired into and passes through a 1.00-kg block, as in Figure P9.71. The block, initially at rest on a frictionless, horizontal surface, is connected to a spring of force constant 900 N/m. If the block moves 5.00 cm to the right after impact, find (a) the speed at which the bullet emerges from the block and (b) the energy lost in the collision.

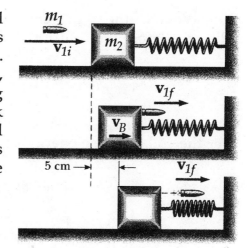

**Figure P9.71
(modified)**

Solution

First find the initial velocity of the block, using conservation of energy during compression of the spring. Assume that the bullet has passed completely through the block before the spring has started to compress. Note that conservation of momentum does not apply here.

$$\tfrac{1}{2}m_2 v_B^{\,2} = \tfrac{1}{2}kx^2 \quad \text{becomes} \quad \tfrac{1}{2}(1.00 \text{ kg})v_B^{\,2} = \tfrac{1}{2}(900 \text{ N}/\text{m})(0.0500 \text{ m})^2$$

$$\text{yielding} \quad v_B = \sqrt{2.25} \text{ m}/\text{s} = 1.50 \text{ m}/\text{s}$$

(a) When the bullet collides with the block, it is the momentum that is conserved:

$$m_1 v_{1i} + m_2 v_{2i} = m_1 v_{1f} + m_2 v_B$$

so $\quad v_{1f} = (m_1 v_{1i} - m_2 v_B)/m_1$

$$v_{1f} = \frac{\left(5.00 \times 10^{-3} \text{ kg}\right)(400 \text{ m}/\text{s}) - (1.00 \text{ kg})(1.50 \text{ m}/\text{s})}{5.00 \times 10^{-3} \text{ kg}} = 100 \text{ m}/\text{s} \qquad \lozenge$$

(b)   We use the work-energy theorem to find the energy lost in the collision. Before the collision, the block is motionless, and the bullet's energy is:

$$K_1 = \frac{1}{2}mv_{1i}^{\,2} = \frac{1}{2}(0.00500 \text{ kg})(400 \text{ m / s})^2 = 400 \text{ J}$$

After the collision, the energy is:

$$K_2 = \frac{1}{2}m_1v_{1f}^{\,2} + \frac{1}{2}m_2v_B^{\,2}$$

$$K_2 = \frac{1}{2}(0.00500 \text{ kg})(100 \text{ m / s})^2 + \frac{1}{2}(1.00 \text{ kg})(1.50 \text{ m / s})^2 = 26.1 \text{ J}$$

Therefore $|\Delta K| = |K_2 - K_1| = |26.1 \text{ J} - 400 \text{ J}| = 374 \text{ J}$ ◊

Chapter 10

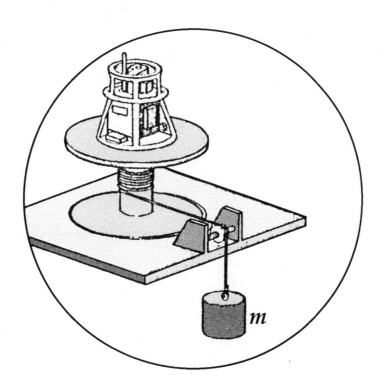

m

Rotation of a Rigid Object About a Fixed Axis

ROTATION OF A RIGID OBJECT
ABOUT A FIXED AXIS

INTRODUCTION

In dealing with the rotation of an object, analysis is greatly simplified by assuming the object to be rigid. A rigid object is defined as one that is nondeformable or, to say the same thing another way, one in which the distances between all pairs of particles remain constant. In this chapter, we treat the rotation of a rigid object about a fixed axis, commonly referred to as **pure rotational motion**.

EQUATIONS AND CONCEPTS

When a particle moves along a circular path of radius r, the distance traveled by the particle is called the arc length, s. The radial line from the center of the path to the particle sweeps out an angle, θ.

$$\theta = \frac{s}{r} \qquad\qquad (10.1b)$$

The angle θ is the ratio of two lengths (arc length to radius) and hence is a dimensionless quantity. However, it is common practice to refer to the angle as being in units of radians. In calculations, the relationship between radians and degrees is the following:

$$\theta \,(\text{rad}) = \frac{\pi}{180°}\,\theta\,(\text{deg})$$

The **average angular speed** $\overline{\omega}$ of a particle or body rotating about a fixed axis equals the ratio of the angular displacement $\Delta\theta$ to the time interval Δt, where θ is measured in radians.

$$\overline{\omega} = \frac{\Delta\theta}{\Delta t} \qquad (10.3)$$

The **instantaneous angular speed** ω is defined as the limit of the average angular velocity as Δt approaches zero.

$$\omega = \frac{d\theta}{dt} \qquad (10.4)$$

The **average angular acceleration** $\overline{\alpha}$ of a rotating body is defined as the ratio of the change in angular velocity to the time interval Δt.

$$\overline{\alpha} = \frac{\Delta\omega}{\Delta t} \qquad (10.5)$$

The **instantaneous angular acceleration** equals the limit of the average angular acceleration as Δt approaches zero.

$$\alpha = \frac{d\omega}{dt} \qquad (10.6)$$

If a particle or body rotates about a fixed axis with **constant** angular acceleration, we can apply the **equations of rotational kinematics.**

$$\omega_f = \omega_i + \alpha t \qquad (10.7)$$

$$\theta_f = \theta_i + \omega_i t + \tfrac{1}{2}\alpha t^2 \qquad (10.8)$$

$$\omega_f^2 = \omega_i^2 + 2\alpha\left(\theta_f - \theta_i\right) \qquad (10.9)$$

If a rigid body rotates about a fixed axis, the linear speed of any point on the body a distance r from the axis of rotation is related to the angular speed through the relation $v = r\omega$. Similarly, the tangential acceleration of any point on the body is related to the angular acceleration through the relation $a_t = r\alpha$. Note that **every point on the body has the same ω and α, but not every point has the same v and a_t.**

$$v = r\omega \quad (10.10)$$

$$a_t = r\alpha \quad (10.11)$$

The **moment of inertia of a system of particles** is defined by Equation 10.15, where m_i is the mass of the ith particle and r_i is its distance from a specified axis. Note that I has SI units of kg·m^2.

$$I = \sum_i m_i r_i^2 \quad (10.15)$$

The **kinetic energy** of a rigid body rotating with an angular speed ω about some axis is proportional to the square of the angular speed. Note that it does not represent a new form of energy. It is simply a convenient form for representing rotational kinetic energy.

$$K_R = \tfrac{1}{2}I\omega^2 \quad (10.16)$$

The **torque** τ due to an applied force has a magnitude given by the product of the force and its moment arm d, where d equals the **perpendicular distance** from the rotation axis to the line of action of **F**. Torque is a measure of the ability of a force to rotate a body about a specified axis. Note that the torque depends on the axis of rotation, which must be specified when τ is evaluated.

$$\tau \equiv rF \sin \phi = Fd \qquad (10.19)$$

The **net torque** acting on a rigid body about some axis is equal to the product of the moment of inertia and angular acceleration, where I is the moment of inertia about the axis of rotation. This is only true for a plane laminar body or for the case when the axis of rotation is a principal axis.

$$\Sigma \tau = I\alpha \qquad (10.21)$$

If a net torque τ acts on a rigid body, the **power supplied to the body** at any instant is proportional to the angular speed.

$$\mathcal{P} = \tau \omega \qquad (10.23)$$

The **work-energy theorem** says that the net work done by external forces in rotating a rigid body about a fixed axis equals the change in the body's rotational kinetic energy.

$$\Sigma W = \tfrac{1}{2} I \omega_f^2 - \tfrac{1}{2} I \omega_i^2 \qquad (10.24)$$

SUGGESTIONS, SKILLS, AND STRATEGIES

CALCULATING MOMENTS OF INERTIA

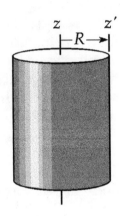

You should know how to calculate the moment of inertia of a system of particles about a specified axis. The technique is straightforward, and consists of applying $I = \Sigma m_i r_i^2$, where m_i is the mass of the ith particle and r_i is the distance from the axis of rotation to the particle.

Once the moment of inertia about an axis through the center of mass I_{CM} is known, you can easily evaluate the moment of inertia about any axis parallel to the axis through the center of mass using the **parallel axis theorem** :

$$I = I_{CM} + MD^2$$

where D is the distance between the two axes.

For example, the moment of inertia of a solid cylinder about an axis through its center (the z axis in the figure) is given by

$$I_z = \tfrac{1}{2}MR^2$$

Hence, the moment of inertia about the z' axis located a distance $D = R$ from the z axis is

$$I_{z'} = I_z + MR^2 = \tfrac{1}{2}MR^2 + MR^2 = \tfrac{3}{2}MR^2$$

REVIEW CHECKLIST

▷ Quantitatively, the angular displacement, speed, and acceleration for a rigid body system in rotational motion are related to the distance traveled, tangential speed, and tangential acceleration. The linear quantity is calculated by multiplying the angular quantity by the radius arm for an object or point in that system.

▷ If a body rotates about a fixed axis, every particle on the body has the same angular speed and angular acceleration. For this reason, rotational motion can be simply described using these quantities. The formulas which describe angular motion are analogous to the corresponding set of formulas pertaining to linear motion.

▷ Calculate the moment of inertia I of a system of particles or a rigid body about a specific axis. Note that the value of I depends on (a) the mass distribution and (b) the axis about which the rotation occurs. The parallel-axis theorem is useful for calculating I about an axis parallel to one that goes through the center of mass.

▷ Understand the concept of torque associated with a force, noting that the torque associated with a force has a magnitude equal to the force times the moment arm. Furthermore, note that the value of the torque depends on the origin about which it is evaluated.

▷ Recognize that the work-energy theorem can be applied to a rotating rigid body. That is, the net work done on a rigid body rotating about a fixed axis equals the change in its rotational kinetic energy.

ANSWERS TO SELECTED CONCEPTUAL QUESTIONS

1. What is the angular speed of the second hand of a clock? What is the direction of ω as you view a clock hanging vertically? What is the magnitude of the angular acceleration vector α of the second hand?

Answer   The second hand of a clock turns at one revolution per minute, so

$$\omega = \frac{2\pi \text{ rad}}{60 \text{ s}} = 0.105 \text{ rad / s}$$

The motion is clockwise, so the direction of the vector angular velocity is away from you. It turns steadily, so ω is constant, and α is zero.

☐   ☐   ☐   ☐

15.   If you see an object rotating, is there necessarily a net torque acting on it?

Answer   An object rotates with constant angular momentum when zero total torque acts on it. For example, consider the Earth; it rotates at a constant rate of once per day, but there is no net torque acting on it.

☐   ☐   ☐   ☐

SOLUTIONS TO SELECTED END-OF-CHAPTER PROBLEMS

1.   A wheel starts from rest and rotates with constant angular acceleration and reaches an angular speed of 12.0 rad/s in 3.00 s. Find (a) the magnitude of the angular acceleration of the wheel and (b) the angle (in radians) through which it rotates in this time.

Solution

(a) $\alpha = \dfrac{\omega_f - \omega_i}{t} = \dfrac{(12.0 - 0) \text{ rad / s}}{3.00 \text{ s}} = 4.00 \text{ rad / s}^2$ ◊

(b) $\theta_f = \omega_i t + \frac{1}{2}\alpha t^2 = \frac{1}{2}\left(4.00 \text{ rad / s}^2\right)(3.00 \text{ s})^2 = 18.0 \text{ rad}$ ◊

5. An electric motor rotating a grinding wheel at 100 rev/min is switched off. Assuming constant negative acceleration of magnitude $2.00 \text{ rad}/\text{s}^2$, (a) how long does it take the wheel to stop? (b) Through how many radians does it turn during the time found in part (a)?

Solution We are given $\omega_f = 0,$ and $\alpha = -2.00 \text{ rad}/\text{s}^2$

and convert

$$\omega_i = \left(100 \frac{\text{rev}}{\text{min}}\right)\left(2\pi \frac{\text{rad}}{\text{rev}}\right)\left(\frac{1 \text{ min}}{60.0 \text{ s}}\right) = 10.47 \text{ rad}/\text{s}$$

(a) $\omega_f = \omega_i + \alpha t$, so

$$t = \frac{\omega_f - \omega_i}{\alpha} = \frac{0-(10.5 \text{ rad}/\text{s})}{-2.00 \text{ rad}/\text{s}^2} = 5.24 \text{ s} \qquad \lozenge$$

(b) $\omega_f^2 - \omega_i^2 = 2\alpha(\theta_f - \theta_i)$, so $\theta_f - \theta_i = (\omega_f^2 - \omega_i^2)/2\alpha$

Thus,

$$\theta_f - \theta_i = \frac{0-(10.5 \text{ rad}/\text{s})^2}{2(-2.00 \text{ rad}/\text{s}^2)} = 27.4 \text{ rad} \qquad \lozenge$$

Note also in part (b) that since a constant acceleration is acting for time t,

$$\theta_f = \overline{\omega} t = \left(\frac{10.5 \text{ rad}/\text{s} + 0 \text{ rad}/\text{s}}{2}\right)(5.24 \text{ s}) = 27.4 \text{ rad} \qquad \lozenge$$

---

13. A racing car travels on a circular track of radius 250 m. If the car moves with a constant linear speed of 45.0 m/s, find (a) its angular speed and (b) the magnitude and direction of its acceleration.

Solution (a) $\omega = v/r = (45.0 \text{ m}/\text{s})/250 \text{ m} = 0.180 \text{ rad}/\text{s} \qquad \lozenge$

(b) With no change in speed, the car has no tangential acceleration. The acceleration is centripetal and acts toward the center.

$$a_c = v^2/r = (45.0 \text{ m}/\text{s})^2/250 \text{ m} = 8.10 \text{ m}/\text{s}^2 \qquad \lozenge$$

---

15. A wheel 2.00 m in diameter lies in a vertical plane and rotates with a constant angular acceleration of 4.00 rad / s^2. The wheel starts at rest at $t = 0$, and the radius vector at point P on the rim makes an angle of 57.3° with the horizontal at this time. At $t = 2.00$ s, find (a) the angular speed of the wheel, (b) the linear speed and acceleration of the point P, and (c) the angular position of the point P.

Solution $r = 1.00$ m, $\alpha = 4.00$ rad / s^2, $\omega_i = 0$, and $\theta_i = 57.3° = 1$ rad

(a) $\omega_f = \omega_i + \alpha t = 0 + \alpha t$

At $t = 2.00$ s, $\omega_f = \left(4.00 \text{ rad / s}^2\right)(2.00 \text{ s}) = 8.00 \text{ rad / s}$ ◊

(b) $v = r\omega = (1.00 \text{ m})(8.00 \text{ rad / s}) = 8.00 \text{ m / s}$ ◊

$a_c = r\omega^2 = (1.00 \text{ m})(8.00 \text{ rad / s})^2 = 64.0 \text{ m / s}^2$ ◊

$a_t = r\alpha = (1.00 \text{ m})\left(4.00 \text{ rad / s}^2\right) = 4.00 \text{ m / s}^2$ ◊

The magnitude of the total acceleration is:

$$a = \sqrt{a_c^2 + a_t^2} = \sqrt{\left(64.0 \text{ m / s}^2\right)^2 + \left(4.00 \text{ m / s}^2\right)^2} = 64.1 \text{ m / s}^2 \quad ◊$$

The direction of the total acceleration vector makes an angle ϕ with respect to the radius to point P:

$$\phi = \tan^{-1}(a_t / a_c) = \tan^{-1}\left(4.00 \text{ m / s}^2 / 64.0 \text{ m / s}^2\right) = 3.58° \quad ◊$$

(c) $\theta_f = \theta_i + \omega_i t + \frac{1}{2}\alpha t^2 = (1.00 \text{ rad}) + \frac{1}{2}\left(4.00 \text{ rad / s}^2\right)(2.00 \text{ s})^2 = 9.00 \text{ rad}$ ◊

θ_f is the total angle through which point P has passed, and is greater than one revolution. The position of point P is found by subtracting one revolution. Therefore, P is at $9.00 \text{ rad} - 2\pi \text{ rad} = 2.72 \text{ rad}$ ◊

19. A disk 8.00 cm in radius rotates at a constant rate of 1200 rev/min about its central axis. Determine (a) its angular speed, (b) the linear speed at a point 3.00 cm from its center, (c) the radial acceleration of a point on the rim, and (d) the total distance a point on the rim moves in 2.00 s.

Solution   (a) $\omega = (2\pi \, \text{rad} / \text{rev}) \left(\dfrac{1200 \, \text{rev} / \text{min}}{60 \, \text{s} / \text{min}} \right) = 126 \, \text{rad} / \text{s}$ ◊

(b) $v = \omega R = (126 \, \text{rad} / \text{s})(0.0300 \, \text{m}) = 3.77 \, \text{m} / \text{s}$ ◊

(c) $a_c = \omega^2 R = (126 \, \text{rad} / \text{s})^2 (0.0800 \, \text{m}) = 1.26 \times 10^3 \, \text{m} / \text{s}^2$ ◊

(d) $s = R\theta = R\omega t = (8.00 \times 10^{-2} \, \text{m})(126 \, \text{rad} / \text{s})(2.00 \, \text{s}) = 20.1 \, \text{m}$ ◊

---

25. The four particles in Figure P10.25 are connected by rigid rods of negligible mass. The origin is at the center of the rectangle. If the system rotates in the xy plane about the z axis with an angular speed of 6.00 rad/s, calculate (a) the moment of inertia of the system about the z axis and (b) the rotational energy of the system.

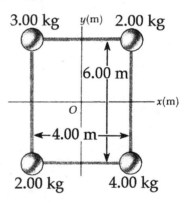

Figure P10.25

Solution   (a) All four particles are at a distance r from the z axis, so

$$r^2 = (3.00 \, \text{m})^2 + (2.00 \, \text{m})^2 = 13.00 \, \text{m}^2$$

Thus, $I_z = \sum m_i r_i^2 = (3.00 \, \text{kg})(13.00 \, \text{m}^2) + (2.00 \, \text{kg})(13.00 \, \text{m}^2)$

$$+ (4.00 \, \text{kg})(13.00 \, \text{m}^2) + (2.00 \, \text{kg})(13.00 \, \text{m}^2)$$

$I_z = 143 \, \text{kg} \cdot \text{m}^2$ ◊

(b) $K_R = \frac{1}{2} I_z \omega^2 = \frac{1}{2}(143 \, \text{kg} \cdot \text{m}^2)(6.00 \, \text{rad} / \text{s})^2 = 2.57 \, \text{kJ}$ ◊

---

33. Find the net torque on the wheel in Figure P10.33 about the axle through O if $a = 10.0$ cm and $b = 25.0$ cm.

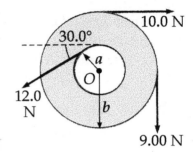

Figure P10.33

Solution

G: By examining the magnitudes of the forces and their respective lever arms, it appears that the wheel will rotate clockwise, and the net torque appears to be about 5 N·m.

O: To find the net torque, we add the individual torques, remembering to apply the convention that a torque producing clockwise rotation is negative and a counterclockwise rotation is positive.

A: $\sum \tau = \sum Fd = +(12.0 \text{ N})(0.100 \text{ m}) - (10.0 \text{ N})(0.250 \text{ m}) - (9.00 \text{ N})(0.250 \text{ m})$

$\sum \tau = -3.55 \text{ N·m}$ ◊

(This is 3.55 N·m into the plane of the page, a clockwise torque)

L: The resulting torque has a reasonable magnitude and produces clockwise rotation as expected. Note that the 30° angle was not required for the solution since each force acted perpendicular to its lever arm. The 10-N force is to the right, but its torque is negative – that is, clockwise, just like the torque of the downward 9-N force.

37. A model airplane having a mass of 0.750 kg is tethered by a wire so that it flies in a circle 30.0 m in radius. The airplane engine provides a net thrust of 0.800 N perpendicular to the tethering wire. (a) Find the torque the net thrust produces about the center of the circle. (b) Find the angular acceleration of the airplane when it is in level flight. (c) Find the linear acceleration of the airplane tangent to its flight path.

Solution

(a) $\tau = Fd = (0.800 \text{ N})(30.0 \text{ m}) = 24.0 \text{ N} \cdot \text{m}$ ◊

(b) $I = mr^2 = (0.750 \text{ kg})(30.0 \text{ m})^2 = 675 \text{ kg} \cdot \text{m}^2$

$\Sigma \tau = I\alpha$, so the angular acceleration is:

$$\alpha = \frac{\Sigma \tau}{I} = \frac{24.0 \text{ N} \cdot \text{m}}{675 \text{ kg} \cdot \text{m}^2} = 0.0356 \text{ rad} / \text{s}^2$$ ◊

(c) $a = r\alpha = (30.0 \text{ m})(0.0356 / \text{s}^2) = 1.07 \text{ m} / \text{s}^2$ ◊

We could also find this linear acceleration from $\Sigma F = ma$:

$$a = \frac{\Sigma F}{m} = \frac{0.800 \text{ N}}{0.750 \text{ kg}} = 1.07 \text{ m} / \text{s}^2$$

45. A weight of 50.0 N is attached to the free end of a light string wrapped around a reel with a radius of 0.250 m and a mass of 3.00 kg. The reel is a solid disk, free to rotate in a vertical plane about the horizontal axis passing through its center. The weight is released 6.00 m above the floor. (a) Determine the tension in the string, the acceleration of the mass, and the speed with which the weight hits the floor. (b) Find the speed calculated in part (a), using the principle of conservation of energy.

Solution

G: Since the rotational inertia of the reel will slow the fall of the weight, we should expect the downward acceleration to be less than g. If the reel did not rotate, the tension in the string would be equal to the weight of the object; and if the reel disappeared, the tension would be zero. Therefore, $T < mg$ for the given problem. With similar reasoning, the final speed must be less than if the weight were to fall freely:

$$v_f < \sqrt{2g\Delta y} \cong 11 \text{ m / s}$$

O: We can find the acceleration and tension using the rotational form of Newton's second law. The final speed can be found from the kinematics equation stated above and from conservation of energy. Free-body diagrams will greatly assist in analyzing the forces.

A: (a) Use $\Sigma\tau = I\alpha$ to find T and a.

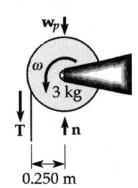

First find I for the reel, which we assume to be a uniform disk.

$$I = \tfrac{1}{2}MR^2 = \tfrac{1}{2}3.00 \text{ kg}(0.250 \text{ m})^2 = 0.0938 \text{ kg} \cdot \text{m}^2$$

The forces on it are shown, including a normal force exerted by its axle. From the diagram, we can see that the tension is the only unbalanced force causing the wheel to rotate.

$\Sigma\tau = I\alpha$ becomes

$$n(0) + F_g(0) + T(0.250 \text{ m}) = \left(0.0938 \text{ kg} \cdot \text{m}^2\right)(a/0.250 \text{ m}) \qquad (1)$$

where we have applied $a_t = r\alpha$ to the point of contact between string and pulley.

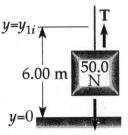

The falling weight has mass

$$m = \frac{F_g}{g} = \frac{50.0 \text{ N}}{9.80 \text{ m}/\text{s}^2} = 5.10 \text{ kg}$$

For this mass, $\Sigma F_y = ma_y$ becomes

$$+T - 50.0 \text{ N} = (5.10 \text{ kg})(-a) \qquad (2)$$

Note that since we have defined upwards to be positive, the minus sign shows that its acceleration is downward. We now have our two equations in the unknowns T and a for the two connected objects. Substituting T from equation (2) into equation (1), we have

$$\left[50.0 \text{ N} - (5.10 \text{ kg})a\right](0.250 \text{ m}) = \left(0.0938 \text{ kg} \cdot \text{m}^2\right)(a/0.250 \text{ m})$$

$$12.5 \text{ N} \cdot \text{m} - (1.28 \text{ kg} \cdot \text{m})a = (0.375 \text{ kg} \cdot \text{m})a$$

$$12.5 \text{ N} \cdot \text{m} = a(1.65 \text{ kg} \cdot \text{m}) \qquad \text{or} \qquad a = 7.57 \text{ m}/\text{s}^2$$

and $\qquad T = 50.0 \text{ N} - 5.10 \text{ kg}\left(7.57 \text{ m}/\text{s}^2\right) = 11.4 \text{ N} \qquad\qquad ◊$

For the motion of the weight,

$$v_f^2 = v_i^2 + 2a\left(x_f - x_i\right) = 0^2 + 2\left(7.57 \text{ m}/\text{s}^2\right)(6.00 \text{ m})$$

$$v_f = 9.53 \text{ m}/\text{s} \text{ (down)} \qquad\qquad ◊$$

(b) The work-energy theorem can take account of multiple objects more easily than Newton's second law. Like your bratty cousins, the work-energy theorem grows between visits; now it reads:

$$\left(K_1 + K_2 + U_{g1} + U_{g2}\right)_i = \left(K_1 + K_2 + U_{g1} + U_{g2}\right)_f$$

$$0 + 0 + m_1 g y_{1i} + 0 = \tfrac{1}{2} m_1 v_{1f}^2 + \tfrac{1}{2} I_2 \omega_{2f}^2 + 0 + 0$$

Now note that $\omega = v / r$ as the string unwinds from the reel. Making substitutions,

$$50.0 \text{ N}(6.00 \text{ m}) = \tfrac{1}{2}(5.10 \text{ kg}) v_f^2 + \tfrac{1}{2}\left(0.0938 \text{ kg} \cdot \text{m}^2\right)\left(\frac{v_f}{0.250 \text{ m}}\right)^2$$

$$300 \text{ N} \cdot \text{m} = \tfrac{1}{2}(5.10 \text{ kg}) v_f^2 + \tfrac{1}{2}(1.50 \text{ kg}) v_f^2$$

$$v_f = \sqrt{\frac{2(300 \text{ N} \cdot \text{m})}{6.60 \text{ kg}}} = 9.53 \text{ m / s} \qquad \Diamond$$

L: As we should expect, both methods give the same final speed for the falling object, but the energy method is simpler. The acceleration is less than g, and the tension is less than the object's weight as we predicted. Now that we understand the effect of the reel's moment of inertia, this problem solution could be applied to solve other real-world pulley systems with masses that should not be ignored.

47. This problem describes one experimental method of determining the moment of inertia of an irregularly shaped object such as the payload for a satellite. Figure P10.47 shows a mass m suspended by a cord wound around a spool of radius r, forming part of a turntable supporting the object. When the mass is released from rest, it descends through a distance h, acquiring a speed v. Show that the moment of inertia I of the equipment (including the turntable) is

$$mr^2\left(2gh\,/\,v^2 - 1\right).$$

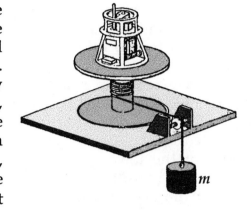

Figure P10.47

Solution

If the friction is negligible, then energy is conserved as the counterweight unwinds. Each point on the cord moves at a linear speed of $v = \omega r$, where r is the radius of the drum. The energy conservation equation gives us:

$$\left(K_1 + K_2 + U_{g1} + U_{g2}\right)_i + W_{nc} = \left(K_1 + K_2 + U_{g1} + U_{g2}\right)_f$$

Solving, we have

$$0 + 0 + mgh + 0 + 0 = \tfrac{1}{2}mv^2 + \tfrac{1}{2}I\omega^2 + 0 + 0$$

$$mgh = \tfrac{1}{2}mv^2 + \tfrac{1}{2}\frac{Iv^2}{r^2}$$

$$2mgh - mv^2 = I\frac{v^2}{r^2}$$

and finally,

$$I = mr^2\left(\frac{2gh}{v^2} - 1\right)$$

◊

49. (a) A uniform, solid disk of radius R and mass M is free to rotate on a frictionless pivot through a point on its rim (Fig. P10.49). If the disk is released from rest in the position shown by the blue circle, what is the speed of its center of mass when the disk reaches the position indicated by the dashed circle? (b) What is the speed of the lowest point on the disk in the dashed position? (c) Repeat part (a) using a uniform hoop.

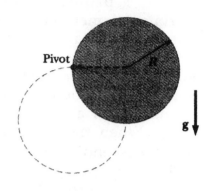

Figure P10.49

Solution

We cannot use the equation $\omega_f{}^2 - \omega_i{}^2 = 2\alpha(\pi/2)$ to find ω_f, because α is not constant. Instead, we use conservation of energy. To identify the change in gravitational energy, think of the height through which the center of mass falls. Using the parallel axis theorem, the moment of inertia of the disk about the pivot point on the disk is:

$$I = I_c + MD^2 = \tfrac{1}{2}MR^2 + MR^2 = \tfrac{3}{2}MR^2$$

The pivot point is fixed, so the kinetic energy is entirely rotational around the pivot:

$$(K+U)_i = (K+U)_f$$

$$MgR = \tfrac{1}{2}\left(\tfrac{3}{2}MR^2\right)\omega^2 + 0$$

Solving for ω,
$$\omega = \sqrt{\frac{4g}{3R}}$$

(a) At the center of mass, $\quad\quad\quad\quad\quad\quad v = R\omega = 2\sqrt{\dfrac{Rg}{3}}$ ◊

(b) At the lowest point on the rim, $\quad v = 2R\omega = 4\sqrt{\dfrac{Rg}{3}}$ ◊

(c) For a hoop, $\quad\quad\quad\quad\quad\quad\quad I_c = MR^2 \quad$ and $\quad I_{rim} = 2MR^2$

By conservation of energy then, $\quad MgR = \dfrac{1}{2}\left(2MR^2\right)\omega^2 + 0$

So $\quad\quad\quad\quad\quad\quad\quad\quad\quad\quad\quad \omega = \sqrt{\dfrac{g}{R}}$

and the center of mass moves at $\quad v_{CM} = R\omega = \sqrt{gR}$ ◊

---

55. A 4.00-m length of light nylon cord is wound around a uniform cylindrical spool of radius 0.500 m and mass 1.00 kg. The spool is mounted on a frictionless axle and is initially at rest. The cord is pulled from the spool with a constant acceleration of magnitude 2.50 m/s². (a) How much work has been done on the spool when it reaches an angular speed of 8.00 rad/s? (b) Assuming that there is enough cord on the spool, how long does it take the spool to reach this angular speed? (c) Is there enough cord on the spool?

Solution

(a) $W = \Delta K_R = \dfrac{1}{2}I\omega_f^2 - \dfrac{1}{2}I\omega_i^2 = \dfrac{1}{2}I\left(\omega_f^2 - \omega_i^2\right) \quad$ where $\quad I = \dfrac{1}{2}mR^2$

$W = \left(\dfrac{1}{2}\right)\left(\dfrac{1}{2}\right)(1.00 \text{ kg})(0.500 \text{ m})^2\left[(8.00 \text{ rad/s})^2 - 0\right] = 4.00 \text{ J}$ ◊

(b) $\omega_f = \omega_i + \alpha t$ where $\alpha = a/r = (2.50 \text{ m}/\text{s}^2)/(0.500 \text{ m}) = 5.00 \text{ rad}/\text{s}^2$

$t = (\omega_f - \omega_i)/\alpha = (8.00 \text{ rad}/\text{s} - 0)/(5.00 \text{ rad}/\text{s}^2) = 1.60 \text{ s}$ ◊

(c) $\theta_f = \theta_i + \omega_i t + \frac{1}{2}\alpha t^2 = 0 + 0 + \frac{1}{2}(5.00 \text{ rad}/\text{s}^2)(1.60 \text{ s})^2 = 6.40 \text{ rad}$

The length pulled from the spool is $s = r\theta = (0.500 \text{ m})(6.40 \text{ rad}) = 3.20 \text{ m}$

When the spool reaches an angular velocity of 8.00 rad/s, 1.60 s will have elapsed and 3.20 m of cord will have been removed from the spool. Our answer is **yes**.   ◊

---

59. A long, uniform rod of length L and mass M is pivoted about a horizontal, frictionless pin passing through one end. The rod is released from rest in a vertical position, as in Figure P10.59. At the instant the rod is horizontal, find (a) its angular speed, (b) the magnitude of its angular acceleration, (c) the x and y components of the acceleration of its center of mass, and (d) the components of the reaction force at the pivot.

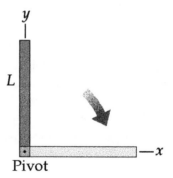

Figure P10.59

Solution   (a) Since only conservative forces are acting on the bar, use conservation of energy:

$\Delta K + \Delta U = 0$

$K_f - K_i + U_f - U_i = 0$

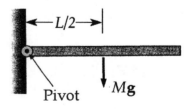

Take the zero level of potential energy at the level of the pivot, and assume the mass of the bar to be located at the center of mass. Under these conditions $U_f = 0$ and $U_i = mgL / 2$. Using the equation above,

$$\left(\tfrac{1}{2}I\omega_f^2 - 0\right) + \left(0 - \tfrac{1}{2}mgL\right) = 0$$

and

$$\omega_f = \sqrt{mgL/I}$$

For a bar rotating about an axis through one end, $I = mL^2 / 3$.

Therefore, $\omega_f = \sqrt{(mgL)/\left(\tfrac{1}{3}mL^2\right)} = \sqrt{3g/L}$ ◊

(b) $\sum \tau = I\alpha$: $mg(L/2) = \left(\tfrac{1}{3}mL^2\right)\alpha$,   and $\alpha = 3g/2L$ ◊

(c) $a_x = a_c = r\omega_f^2 = \left(\dfrac{L}{2}\right)\left(\dfrac{3g}{L}\right) = \dfrac{3g}{2}$ ◊

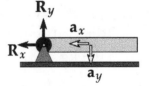

Since this is **centripetal** acceleration, it is directed along the **negative** horizontal:

$$a_y = a_t = r\alpha = \frac{L}{2}\alpha = \frac{3g}{4}$$ ◊

(d) Using $\sum \mathbf{F} = m\mathbf{a}$, we have

$R_x = ma_x = 3mg/2$ in the **negative** direction   ◊

$R_y - mg = -ma_y$ so $R_y = m\left(g - a_y\right) = m(g - 3g/4) = mg/4$ ◊

71. Two blocks, as shown in Figure P10.71, are connected by a string of negligible mass passing over a pulley of radius 0.250 m and moment of inertia I. The block on the frictionless incline is moving upward with a constant acceleration of 2.00 m / s². (a) Determine T_1 and T_2, the tensions in the two parts of the string. (b) Find the moment of inertia of the pulley.

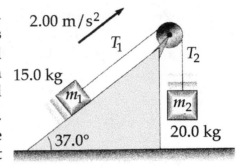

Figure P10.71

Solution

G: In earlier problems, we assumed that the tension in a string was the same on either side of a pulley. Here we see that the moment of inertia changes that assumption, but we should still expect the tensions to be similar in magnitude (about the weight of each mass ~ 150 N), and $T_2 > T_1$ for the pulley to rotate clockwise as shown.

If we knew the mass of the pulley, we could calculate its moment of inertia, but since we only know the acceleration, it is difficult to estimate I. We at least know that I must have units of kg·m², and a 50 – cm disk probably has a mass less than 10 kg, so I is probably less than 0.3 kg·m².

O: For each block, we know its mass and acceleration, so we can use Newton's second law to find the net force, and from it the tension. The difference in the two tensions causes the pulley to rotate, so this net torque and the resulting angular acceleration can be used to find the pulley's moment of inertia.

A: (a) Apply $\Sigma F = ma$ to each block to find each string tension.

The forces acting on the 15-kg block are its weight, the normal support from the incline, and T_1. Taking the positive x axis as directed up the incline , $\Sigma F_x = ma_x$ yields:

$$-(m_1 g)_x + T_1 = m_1(+a)$$

Substituting known values, and solving for T_1,

$$-(15.0 \text{ kg})(9.80 \text{ m / s}^2)\sin 37° + T_1 = (15.0 \text{ kg})(2.00 \text{ m / s}^2)$$

$$T_1 = 118 \text{ N} \qquad \qquad \lozenge$$

Similarly, for the counterweight, we have $\Sigma F_y = ma_y$, or

$$T_2 - m_2 g = m_2(-a)$$

$$T_2 - (20.0 \text{ kg})(9.80 \text{ m / s}^2) = (20.0 \text{ kg})(-2.00 \text{ m / s}^2)$$

So, $\quad T_2 = 156 \text{ N} \qquad \qquad \lozenge$

(b)   Now for the pulley, $\Sigma \tau = r(T_2 - T_1) = I\alpha$. We may choose to call clockwise positive. The angular acceleration is:

$$\alpha = \frac{a}{r} = \frac{2.00 \text{ m / s}^2}{0.250 \text{ m}} = 8.00 \text{ rad / s}^2$$

$\Sigma \tau = I\alpha$: $(+118 \text{ N})(0.250 \text{ m}) - (156 \text{ N})(0.250 \text{ m}) = I(-8.00 \text{ rad / s}^2)$

$$I = \frac{9.38 \text{ N} \cdot \text{m}}{8.00 \text{ rad / s}^2} = 1.17 \text{ kg} \cdot \text{m}^2 \qquad \qquad \lozenge$$

L: The tensions are close to the weight of each mass and $T_2 > T_1$ as expected. However, the moment of inertia for the pulley is about 4 times greater than expected. Unless we made a mistake in solving this problem, our result means that the pulley has a mass of 37.4 kg (about 80 lb), which means that the pulley is probably made of a dense material, like steel. This is certainly not a problem where the mass of the pulley can be ignored since the pulley has more mass than the combination of the two blocks!

73. As a result of friction, the angular speed of a wheel changes with time according to the relationship

$$\omega = \frac{d\theta}{dt} = \omega_0 e^{-\sigma t}$$

where ω_0 and σ are constants. The angular speed changes from 3.50 rad/s at $t = 0$ to 2.00 rad/s at $t = 9.30$ s. Use this information to determine σ and ω_0. Then determine (a) the magnitude of the angular acceleration at $t = 3.00$ s, (b) the number of revolutions the wheel makes in the first 2.50 s, and (c) the number of revolutions it makes before coming to rest.

Solution   When $t = 0$, $\omega = 3.50$ rad/s   and $e^0 = 1$

so $\omega = \omega_0 e^{-\sigma t}$ becomes $\omega_{t=0} = \omega_0 e^{-\sigma(t=0)} = 3.50$ rad/s

gives $\omega_0 = 3.50$ rad/s   ◊

We now calculate σ: $2.00 \text{ rad/s} = (3.50 \text{ rad/s}) e^{-\sigma(9.30 \text{ s})}$

$0.571 = e^{-\sigma(9.30 \text{ s})}$

$\left[\ln(0.571) = -5.60\right] = \left[\ln\left(e^{-9.30\sigma}\right) = -9.30\sigma\right]$

and $\sigma = 0.0602 \text{ s}^{-1}$ ◊

(a)   At all times, $\alpha = \frac{d\omega}{dt} = \frac{d}{dt}\left[\omega_0 e^{-\sigma t}\right] = -\sigma\omega_0 e^{-\sigma t}$

At $t = 3.00$ s, $\alpha = -\left(0.0602 \text{ s}^{-1}\right)(3.50 \text{ rad/s}) e^{-1.81} = -0.176 \text{ rad/s}^2$ ◊

(b)   From the given equation, $d\theta = \omega_0 e^{-\sigma t} dt,$

and

$$\theta = \int_{0\,s}^{2.50\,s} \omega_0 e^{-\sigma t} dt = \frac{\omega_0}{-\sigma} e^{-\sigma t} \Big|_{0\,s}^{2.50\,s} = \frac{\omega_0}{-\sigma}\left(e^{-2.50\sigma} - 1\right)$$

Substituting and solving, $\theta = -58.2(0.860 - 1)\ \text{rad} = 8.12\ \text{rad}$

or $\theta = (8.12\ \text{rad})\left(\dfrac{1\ \text{rev}}{2\pi\ \text{rad}}\right) = 1.29\ \text{rev}$ ◊

(c)   The motion continues to a finite limit, as ω approaches zero and t goes to infinity. From part (b), the total angular displacement is

$$\theta = \frac{\omega_0}{-\sigma} e^{-\sigma t} \Big|_{0}^{\infty} = \frac{\omega_0}{-\sigma}\left(e^{-\infty} - e^{0}\right) = \frac{\omega_0}{-\sigma}(0 - 1) = \frac{\omega_0}{-\sigma}$$

Substituting, $\theta = 58.2\ \text{rad}$

or $\theta = \left(\dfrac{1\ \text{rev}}{2\pi\ \text{rad}}\right)(58.2\ \text{rad}) = 9.26\ \text{rev}$ ◊

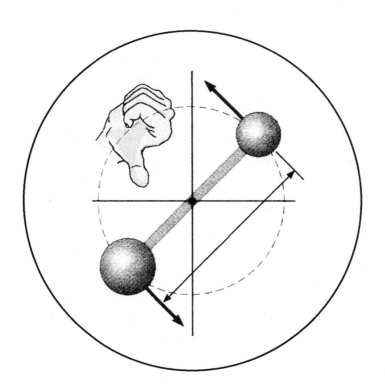

Rolling Motion and Angular Momentum

ROLLING MOTION
AND ANGULAR MOMENTUM

INTRODUCTION

In the previous chapter we learned how to treat the rotation of a rigid body about a fixed axis. This chapter deals in part with the more general case, where the axis of rotation is not fixed in space. We begin by describing the rolling motion of an object. Next, we define a vector product, a convenient mathematical tool for expressing such quantities as torque and angular momentum. The central point of this chapter is to develop the concept of the angular momentum of a system of particles, a quantity that plays a key role in rotational dynamics.

EQUATIONS AND CONCEPTS

If a uniform body of circular cross section rolls on a rough surface without slipping, the **speed and acceleration of the center of mass** are simply related to the angular speed and angular acceleration.

$$v_{CM} = R\omega \qquad (11.1)$$

$$a_{CM} = R\alpha \qquad (11.2)$$

The total kinetic energy of a rigid body rolling on a rough surface can be expressed as the sum of the rotational kinetic energy about the center of mass and the translational kinetic energy of the center of mass.

$$K = \tfrac{1}{2}I_{CM}\omega^2 + \tfrac{1}{2}Mv_{CM}^{\;2} \qquad (11.4)$$

If a body rolls down an incline **without slipping**, one can use conservation of energy to find the **velocity of the center of mass** as the body falls through a vertical distance h, starting from rest. From this expression, one can also find the acceleration of the center of mass.

$$v_{CM} = \sqrt{\frac{2gh}{1 + \dfrac{I_{CM}}{MR^2}}}$$

(11.6)

The **torque** acting on a particle whose vector position is $\mathbf{r}$ can be expressed as $\mathbf{r} \times \mathbf{F}$, where $\mathbf{F}$ is the external force acting on the particle. Torque depends on the choice of the origin and has SI units of N·m.

$$\tau \equiv \mathbf{r} \times \mathbf{F}$$

(11.7)

The **cross product** of any two vectors $\mathbf{A}$ and $\mathbf{B}$ is a vector $\mathbf{C}$ whose magnitude is given by $AB \sin \theta$ and whose direction is perpendicular to the plane formed by $\mathbf{A}$ and $\mathbf{B}$. The sense of $\mathbf{C}$ can be determined from the right-hand rule.

$$\mathbf{C} = \mathbf{A} \times \mathbf{B}$$

(11.8)

$$C \equiv AB \sin \theta$$

(11.9)

The **angular momentum** of a particle whose linear momentum is $\mathbf{p}$ and whose vector position is $\mathbf{r}$ is defined as $\mathbf{L} = \mathbf{r} \times \mathbf{p}$. The SI unit of angular momentum is $kg \cdot m^2 / s$. Note that both the magnitude and direction of $\mathbf{L}$ depend on the choice of origin.

$$\mathbf{L} \equiv \mathbf{r} \times \mathbf{p}$$

(11.15)

If the same origin is used to define **L** and τ, then the **torque** on the particle **equals the time rate of change of its angular momentum.** This expression is the rotational analog of Newton's second law, $\Sigma \mathbf{F} = d\mathbf{p}/dt$, and is the basic equation for treating rotating rigid bodies and rotating particles.

$$\Sigma \tau = \frac{d\mathbf{L}}{dt} \qquad (11.19)$$

The angular momentum of a system of particles is obtained by taking the vector sum of the individual angular momenta about some point in an inertial frame. The individual momenta may change with time, which can change the total angular momentum. However, the total angular momentum of the system will only change if a net **external** torque acts on the system. In fact, **the net torque acting on a system of particles equals the time rate of change of the total angular momentum.**

$$\Sigma \tau_{ext} = \frac{d\mathbf{L}}{dt} \qquad (11.20)$$

The **magnitude of the angular momentum of a rigid body** in the form of a plane lamina rotating in the x-y plane about a **fixed axis** (the z axis) is given by the product $I\omega$, where I is the moment of inertia about the axis of rotation and ω is the angular speed.

$$L_z = I\omega \qquad (11.21)$$

The **law of conservation of angular momentum** states that if the resultant external torque acting on a system is zero, the total angular momentum is constant. This follows from Equation 11.20.

If $\qquad \Sigma\tau_{ext} = \dfrac{d\mathbf{L}}{dt} = 0 \qquad$ (11.24)

$$\mathbf{L} = \text{constant} \qquad (11.25)$$

If a zero net torque acts on a body rotating about a fixed axis, and the moment of inertia changes from I_i to I_f, then the conservation of angular momentum can be used to find the final angular speed in terms of the initial angular speed.

$$I_i\omega_i = I_f\omega_f = \text{constant} \qquad (11.27)$$

SUGGESTIONS, SKILLS, AND STRATEGIES

The operation of the vector or cross product is used for the first time in this chapter. (Recall that the angular momentum $\mathbf{L}$ of a particle is defined as $\mathbf{L} = \mathbf{r} \times \mathbf{p}$, while torque is defined by the expression $\tau = \mathbf{r} \times \mathbf{F}$.) Let us briefly review the cross-product operation and some of its properties.

If $\mathbf{A}$ and $\mathbf{B}$ are any two vectors, their cross product, written as $\mathbf{A} \times \mathbf{B}$, is also a vector $\mathbf{C}$. That is,

$$\mathbf{C} = \mathbf{A} \times \mathbf{B}$$

where the magnitude of $\mathbf{C}$ is given by $C = |\mathbf{C}| = AB\sin\theta$, and θ is the angle between $\mathbf{A}$ and $\mathbf{B}$ as in the figure below.

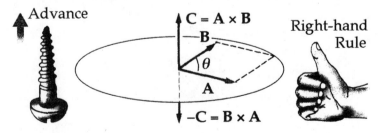

The direction of **C** is perpendicular to the plane formed by **A** and **B**, and its sense is determined by the right-hand rule. You should practice this rule for various choices of vector pairs. Note that **B** × **A** is directed **opposite** to **A** × **B**. That is, **A** × **B** = −**B** × **A**.

This follows from the right-hand rule. You should not confuse the cross product of two vectors, which is a vector quantity, with the dot product of two vectors, which is a scalar quantity. (Recall that the dot product is defined as **A·B** = $AB \cos \theta$.)

Note that the cross product of any vector with itself is zero. That is, **A** × **A** = 0 since $\theta = 0°$, and $\sin(0) = 0$.

Very often, vectors will be expressed in unit vector form, and it is convenient to make use of the multiplication table for unit vectors. Note that **i**, **j**, and **k** represent a set of mutually orthogonal vectors as shown below.

$$\mathbf{i} \times \mathbf{i} = \mathbf{j} \times \mathbf{j} = \mathbf{k} \times \mathbf{k} = 0$$

$$\mathbf{i} \times \mathbf{j} = -\mathbf{j} \times \mathbf{i} = \mathbf{k}$$

$$\mathbf{j} \times \mathbf{k} = -\mathbf{k} \times \mathbf{j} = \mathbf{i}$$

$$\mathbf{k} \times \mathbf{i} = -\mathbf{i} \times \mathbf{k} = \mathbf{j}$$

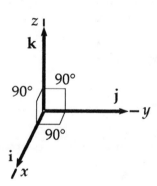

For example, if **A** = 3.00**i** + 5.00**j** and **B** = 4.00**j**, then we can take the cross-product:

$$\mathbf{A} \times \mathbf{B} = (3.00\mathbf{i} + 5.00\mathbf{j}) \times (4.00\mathbf{j}) = 3.00\mathbf{i} \times 4.00\mathbf{j} + 5.00\mathbf{j} \times 4.00\mathbf{j} = 12.0\mathbf{k}$$

REVIEW CHECKLIST

▷ Define the cross product (magnitude and direction) of any two vectors, **A** and **B**, and state the various properties of the cross product.

▷ Define the angular momentum **L** of a particle moving with a velocity **v** relative to a specified point, and the torque τ acting on the particle relative to that point. Note that both **L** and τ are quantities which depend on the choice of the origin since they involve the vector position **r** of the particle. (That is, $\mathbf{L} = \mathbf{r} \times \mathbf{p}$ and $\tau = \mathbf{r} \times \mathbf{F}$.)

▷ Apply the conservation of angular momentum principle to a body rotating about a fixed axis, in which the moment of inertia changes due to a change in the mass distribution.

▷ Describe the center of mass motion of a rigid body which undergoes both rotation about some axis and translation in space. Note that for pure rolling motion of an object such as a sphere or cylinder, the total kinetic energy can be expressed as the sum of a rotational kinetic energy about the center of mass plus the translational energy of the center of mass.

ANSWERS TO SELECTED CONCEPTUAL QUESTIONS

9. In a tape recorder, the tape is pulled past the read-and-write heads at a constant speed by the drive mechanism. Consider the reel from which the tape is pulled — as the tape is pulled off it, the radius of the roll of remaining tape decreases. How does the torque on the reel change with time? How does the angular speed of the reel change with time? If the tape mechanism is suddenly turned on so that the tape is quickly pulled with a great force, is the tape more likely to break when pulled from a nearly full reel or a nearly empty reel?

Answer   We can assume fairly accurately that the driving motor will run at a constant angular speed, and at a constant torque. Therefore, as the radius of the takeup reel increases, the tension in the tape will decrease.

$$T = \tau_{const} / R_{takeup} \tag{1}$$

$$\tau_{source} = FR_{source} = \tau_{const} R_{source} / R_{takeup} \tag{2}$$

As the radius of the source reel decreases, given a decreasing tension, the torque in the source reel will **decrease** even faster.

This torque will be partly absorbed by friction in the feed heads (which we assume to be small); some will be absorbed by friction in the source reel. Another small amount of the torque will be absorbed by the increasing angular velocity of the source reel.   However, in the case of a sudden jerk on the tape, the changing angular velocity of the source reel becomes important. If the source reel is full, then the moment of inertia will be large, and the tension in the tape will be large. If it is nearly empty, then the angular acceleration will be large, instead. Thus, the tape will be more likely to break when the source reel is nearly **full**. One sees the same effect in the case of paper towels; it is easier to snap a towel free when the roll is new than when it is nearly empty.

□   □   □   □

17. Two solid spheres — a large, massive sphere and a small sphere with low mass — are rolled down a hill. Which sphere reaches the bottom of the hill first? Next, a large, low-density sphere and a small, high-density sphere having the same mass are rolled down the hill. Which one reaches the bottom first in this case?

Answer   The rate of rolling depends on what fraction of the original gravitational potential energy appears as translational kinetic energy and what fraction appears as rotational kinetic energy. The moment of inertia, I, of a sphere increases as the square of the radius. The square of the angular velocity, ω^2, decreases as the square of the radius. Thus, when we multiply these together, we find that the rotational kinetic energy is independent of radius. Since the radial dependence has vanished, the only remaining dependence of the rotational kinetic energy is on mass and velocity. The translational kinetic energy also depends only on these two quantities. Thus, regardless of the mass and radius of the spheres, the same fraction of translational kinetic energy will always result, and all solid spheres will roll down the hill at the same rate.

□   □   □   □

19. Why do tightrope walkers carry a long pole to help themselves keep their balance?

Answer   The long pole increases the tightrope walker's moment of inertia, and therefore decreases his angular acceleration, under any given torque. That gives him more time to respond, and in essence, aids his reflexes.

□   □   □   □

22. If global warming occurs over the next century, it is likely that some polar ice will melt and the water will be distributed closer to the equator. How would this change the moment of inertia of the Earth? Would the length of the day (one revolution) increase or decrease?

Answer   The Earth already bulges slightly at the Equator, and is slightly flat at the poles. If more mass moved towards the Equator, it would essentially move the mass to a greater distance from the axis of rotation, and increase the moment of inertia. Since conservation of angular momentum requires that $\omega_z I_z$ = const, an increase in the moment of inertia would decrease the angular velocity, and slow down the spinning of the Earth. Thus, the length of each day would increase.

□   □   □   □

SOLUTIONS TO SELECTED END-OF-CHAPTER PROBLEMS

1. A cylinder of mass 10.0 kg rolls without slipping on a horizontal surface. At the instant its center of mass has a speed of 10.0 m/s, determine (a) the translational kinetic energy of its center of mass, (b) the rotational energy about its center of mass, and (c) its total energy.

Solution

(a) $$K_{trans} = \tfrac{1}{2}mv_{CM}{}^2 = \tfrac{1}{2}(10.0 \text{ kg})(10.0 \text{ m}/\text{s})^2 = 500 \text{ J} \qquad \lozenge$$

(b) Call the radius of the cylinder R. The moment of inertia is

$$I_{CM} = \tfrac{1}{2}mR^2$$

An observer at the center sees the rough surface and the circumference of the cylinder moving at 10.0 m/s, so the angular speed of the cylinder is:

$$\omega = v_{CM}/R = (10.0 \text{ m}/\text{s})/R$$

so $$K_{rot} = \tfrac{1}{2}I_{CM}\omega^2 = \left(\tfrac{1}{2}\right)\left[\tfrac{1}{2}(10.0 \text{ kg})R^2\right]\left(\frac{10.0 \text{ m}/\text{s}}{R}\right)^2 = 250 \text{ J} \qquad \lozenge$$

(c) $$K_{total} = K_{trans} + K_{rot} = 500 \text{ J} + 250 \text{ J} = 750 \text{ J} \qquad \lozenge$$

5. (a) Determine the acceleration of the center of mass of a uniform solid disk rolling down an incline making an angle θ with the horizontal. Compare this acceleration with that of a uniform hoop. (b) What is the minimum coefficient of friction required to maintain pure rolling motion for the disk?

Solution

G: The acceleration of the disk will depend on the angle of the incline. In fact, it should be proportional to $g \sin \theta$ since the disk should not accelerate when the incline angle is zero, and since $a = g$ when the angle is 90° and the disk can fall freely. The acceleration of the disk should also be greater than a hoop since the mass of the disk is closer to its center, giving it less rotational inertia so that it can roll faster than the hoop. The required coefficient of friction is difficult to predict, but is probably between 0 and 1 since this is a typical range for μ.

O: We can find the acceleration by applying Newton's second law and considering both the linear and rotational motion. A free-body diagram (left-hand disk) and a motion diagram (right-hand disk) will greatly assist us in defining our variables and seeing how the forces are related.

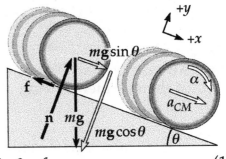

A: $\Sigma F_x = m a_x$ becomes $mg \sin \theta - f = m a_{CM}$ (1)

$\Sigma F_y = m a_y$ yields $n - mg \cos \theta = 0$ (2)

$\Sigma \tau = I_{CM} \alpha$ gives $f r = I_{CM} a_{CM} / r$ (3)

(a) For a disk, $\left(I_{CM} \right)_{disk} = \frac{1}{2} m r^2$

From (3) we find $f = \left(\frac{1}{2} m r^2 \right) \left(a_{CM} \right) / r^2 = \frac{1}{2} m a_{CM}$

Substituting this into (1), $\qquad$ $mg\sin\theta - \dfrac{1}{2}ma_{CM} = ma_{CM}$

so that $\qquad$ $(a_{CM})_{\text{disk}} = \dfrac{2}{3}g\sin\theta$ $\qquad$ ◊

For a hoop, $\qquad$ $(I_{CM})_{\text{hoop}} = mr^2$

From (3), $\qquad$ $f = mr^2 a_{CM}/r^2 = ma_{CM}$

Substituting this into (1), $\qquad$ $mg\sin\theta - ma_{CM} = ma_{CM}$

so $\qquad$ $(a_{CM})_{\text{hoop}} = \dfrac{1}{2}g\sin\theta$

Therefore, $\qquad$ $\dfrac{(a_{CM})_{\text{disk}}}{(a_{CM})_{\text{hoop}}} = \dfrac{\frac{2}{3}g\sin\theta}{\frac{1}{2}g\sin\theta} = \dfrac{4}{3}$ $\qquad$ ◊

(b) From (2) we find $n = mg\cos\theta$: $\qquad$ $f = \mu n = \mu mg\cos\theta$

Likewise, from equation (1), $\qquad$ $f = (mg\sin\theta) - ma_{CM}$

Setting these equations equal, $\qquad$ $\mu mg\cos\theta = mg\sin\theta - \dfrac{2}{3}mg\sin\theta$

so $\qquad$ $\mu = \dfrac{1}{3}\left(\dfrac{\sin\theta}{\cos\theta}\right) = \dfrac{1}{3}\tan\theta$ $\qquad$ ◊

L: As expected, the acceleration of the disk is proportional to $g\sin\theta$ and is slightly greater than the acceleration of the hoop. The coefficient of friction result is similar to the result found for a block on an incline plane, where $\mu = \tan\theta$ (see Example 5.12). However, μ is not always between 0 and 1 as predicted. For angles greater than 72° the coefficient of friction must be larger than 1. For angles greater than 80° μ must be extremely large to make the disk roll without slipping.

11. Two vectors are given by $\mathbf{A} = -3\mathbf{i} + 4\mathbf{j}$ and $\mathbf{B} = 2\mathbf{i} + 3\mathbf{j}$. Find (a) $\mathbf{A} \times \mathbf{B}$ and (b) the angle between $\mathbf{A}$ and $\mathbf{B}$.

Solution

(a) $\mathbf{A} \times \mathbf{B} = (-3\mathbf{i} + 4\mathbf{j}) \times (2\mathbf{i} + 3\mathbf{j})$

$\mathbf{A} \times \mathbf{B} = (-6\mathbf{i} \times \mathbf{i}) - (9\mathbf{i} \times \mathbf{j}) + (8\mathbf{j} \times \mathbf{i}) + (12\mathbf{j} \times \mathbf{j}) = 0 - 9\mathbf{k} + 8(-\mathbf{k}) + 0 = -17\mathbf{k}$ ◊

(b) Since $|\mathbf{A} \times \mathbf{B}| = |AB \sin \theta|$,

$$\theta = \sin^{-1} \frac{|\mathbf{A} \times \mathbf{B}|}{AB} = \sin^{-1} \left(\frac{17}{\sqrt{3^2 + 4^2} \sqrt{2^2 + 3^2}} \right) = 70.6°$$ ◊

Related Calculation: To solidify your understanding, review taking the dot product of these same two vectors, and again find the angle between them. For these next calculations, assume all given values to be known to 3 significant figures.

$$\mathbf{A} \cdot \mathbf{B} = (-3.00\mathbf{i} + 4.00\mathbf{j}) \cdot (2.00\mathbf{i} + 3.00\mathbf{j}) = -6.00 + 12.0 = 6.00$$ ◊

$$\mathbf{A} \cdot \mathbf{B} = AB \cos \theta$$

$$\theta = \cos^{-1} \frac{\mathbf{A} \cdot \mathbf{B}}{AB} = \cos^{-1} \left(\frac{6.00}{\sqrt{3^2 + 4^2} \sqrt{2^2 + 3^2}} \right) = 70.6°$$ ◊

---

17. If $|\mathbf{A} \times \mathbf{B}| = \mathbf{A} \cdot \mathbf{B}$, what is the angle between $\mathbf{A}$ and $\mathbf{B}$?

Solution We are given the condition $|\mathbf{A} \times \mathbf{B}| = \mathbf{A} \cdot \mathbf{B}$. This says that

$$AB \sin \theta = AB \cos \theta \qquad \text{and} \qquad \tan \theta = 1$$

$$\theta = 45° \qquad \text{and} \qquad \theta = 225° \quad \text{both satisfy this condition.}$$ ◊

---

19. A light, rigid rod 1.00 m in length joins two particles — with masses 4.00 kg and 3.00 kg — at its ends. The combination rotates in the xy plane about a pivot through the center of the rod (Figure P11.19). Determine the angular momentum of the system about the origin when the speed of each particle is 5.00 m / s.

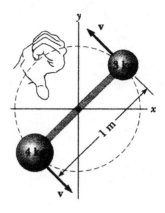

Figure P11.19

Solution

Taking the origin of the compound object to be the pivot, the angular speed and the moment of inertia are

$$\omega = v/r = (5.00 \text{ m / s})/0.500 \text{ m} = 10.0 \text{ rad / s}$$

$$I = \Sigma mr^2 = (4.00 \text{ kg})(0.500 \text{ m})^2 + (3.00 \text{ kg})(0.500 \text{ m})^2 = 1.75 \text{ kg} \cdot \text{m}^2$$

By the right-hand rule (shown in the modification to the figure), we find that this is directed out of the plane. So its angular momentum is

$$L = I\omega = (1.75 \text{ kg} \cdot \text{m}^2)(10.0 \text{ rad / s})$$

$$\mathbf{L} = 17.5 \text{ kg} \cdot \text{m}^2/ \text{s} \text{ out of the plane} \qquad \Diamond$$

Note: Alternatively, we could solve this using $\mathbf{L} = \sum m\mathbf{r} \times \mathbf{v}$:

$$L = (4.00 \text{ kg})(0.500 \text{ m})(5.00 \text{ m / s}) + (3.00 \text{ kg})(0.500 \text{ m})(5.00 \text{ m / s})$$

$$L = 10.0 \text{ kg} \cdot \text{m}^2/ \text{s} + 7.50 \text{ kg} \cdot \text{m}^2/ \text{s}$$

$$\mathbf{L} = 17.5 \text{ kg} \cdot \text{m}^2/ \text{s} \text{ out of the plane} \qquad \Diamond$$

21. The position vector of a particle of mass 2.00 kg is given as a function of time by $\mathbf{r} = (6.00\mathbf{i} + 5.00t\mathbf{j})$ m. Determine the angular momentum of the particle about the origin as a function of time.

Solution The velocity of the particle is

$$\mathbf{v} = \frac{d\mathbf{r}}{dt} = \frac{d}{dt}(6.00\mathbf{i} \text{ m} + 5.00t\mathbf{j} \text{ m}) = 5.00\mathbf{j} \text{ m / s}$$

The angular momentum is

$$\mathbf{L} = \mathbf{r} \times \mathbf{p} = m\mathbf{r} \times \mathbf{v} = (2.00 \text{ kg})(6.00\mathbf{i} \text{ m} + 5.00t\mathbf{j} \text{ m}) \times 5.00\mathbf{j} \text{ m / s}$$

$$= (60.0 \text{ kg} \cdot \text{m}^2/\text{ s})\mathbf{i} \times \mathbf{j} + (50.0t \text{ kg} \cdot \text{m}^2/\text{ s})\mathbf{j} \times \mathbf{j}$$

$$= 60.0\mathbf{k} \text{ kg} \cdot \text{m}^2/\text{ s} \quad \text{(constant in time)} \qquad \lozenge$$

25. A particle of mass m is shot with an initial velocity $\mathbf{v}_i$ and makes an angle θ with the horizontal, as shown in Figure P11.25. The particle moves in the gravitational field of the Earth. Find the angular momentum of the particle about the origin when the particle is (a) at the origin, (b) at the highest point of its trajectory, and (c) just before it hits the ground. (d) What torque causes its angular momentum to change?

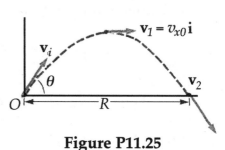

Figure P11.25

Solution

The angular momentum is $\mathbf{L} = \mathbf{r} \times m\mathbf{v}$.

(a) Since $\mathbf{r} = 0$, $\mathbf{L} = 0$ $\qquad \lozenge$

240

(b) At the highest point of the trajectory,

$$\mathbf{L} = \left(\frac{v_i^2 \sin 2\theta}{2g} \mathbf{i} + \frac{(v_i \sin \theta)^2}{2g} \mathbf{j} \right) \times mv_{xi} \mathbf{i} = \frac{-m(v_i \sin \theta)^2 v_i \cos \theta}{2g} \mathbf{k} \qquad \lozenge$$

(c) $\mathbf{L} = R\mathbf{i} \times m\mathbf{v}_2$, where $R = \dfrac{v_i^2 \sin 2\theta}{g}$

$$\mathbf{L} = m\left[R\mathbf{i} \times (v_i \cos \theta \mathbf{i} - v_i \sin \theta \mathbf{j}) \right] = -mRv_i \mathbf{k} \sin \theta = \frac{-mv_i^3 \sin 2\theta \sin \theta}{g} \mathbf{k} \qquad \lozenge$$

(d) The downward force of gravity exerts a torque in the $-z$ direction. $\qquad \lozenge$

---

31. A particle with a mass of 0.400 kg is attached to the 100-cm mark of a meter stick with a mass of 0.100 kg. The meter stick rotates on a horizontal, frictionless table with an angular speed of 4.00 rad/s. Calculate the angular momentum of the system when the stick is pivoted about an axis (a) perpendicular to the table through the 50.0-cm mark and (b) perpendicular to the table through the 0-cm mark.

Solution

G: Since the angular speed is constant, the angular momentum will be greater when the center of mass of the system is farther from the axis of rotation (larger I).

O: Use the equation $L = I\omega$ to find L for each I.

A: Defining the distance from the pivot to the center of mass as d, we first find the rotational inertia of the system for each case, from the information

$M = 0.100$ kg, $\qquad\qquad m = 0.400$ kg, $\qquad\qquad D = 1.00$ m

241

(a) For the meter stick rotated about its center, $I_m = MD^2/12$. For the weight,

$$I_w = md^2 = m(D/2)^2$$

$$I = I_m + I_w = (MD^2/12) - (mD^2/4)$$

$$I = \frac{(0.100 \text{ kg})(1.00 \text{ m})^2}{12} - \frac{(0.400 \text{ kg})(1.00 \text{ m})^2}{4} = 0.108 \text{ kg} \cdot \text{m}^2$$

$$L = I\omega = (0.108 \text{ kg} \cdot \text{m}^2)(4.00 \text{ rad} / \text{s}) = 0.433 \text{ kg} \cdot \text{m}^2/ \text{s} \qquad \lozenge$$

(b) For a stick rotated about a point at one end,

$$I_m = mD^2/3$$

$$I_m = (0.100 \text{ kg})(1.00 \text{ m})^2/3 = 0.0333 \text{ kg} \cdot \text{m}^2$$

For the point mass,

$$I_w = mD^2 = (0.400 \text{ kg})(1.00 \text{ m})^2 = 0.400 \text{ kg} \cdot \text{m}^2$$

$$I = I_m + I_w = 0.433 \text{ kg} \cdot \text{m}^2$$

Thus, $\qquad L = I\omega = (0.433 \text{ kg} \cdot \text{m}^2)(4.00 \text{ rad} / \text{s}) = 1.73 \text{ kg} \cdot \text{m}^2/ \text{s} \qquad \lozenge$

L: As we expected, the angular momentum is larger when the center of mass is further from the rotational axis. In fact, the angular momentum for part (b) is 4 times greater than for part (a). The units of $\text{kg} \cdot \text{m}^2/ \text{s}$ also make sense since $[L] = [r] \times [p]$.

37. A 60.0-kg woman stands at the rim of a horizontal turntable having a moment of inertia of 500 kg·m^2 and a radius of 2.00 m. The turntable is initially at rest and is free to rotate about a frictionless, vertical axle through its center. The woman then starts walking around the rim clockwise (as viewed from above the system) at a constant speed of 1.50 m/s relative to the Earth. (a) In what direction and with what angular speed does the turntable rotate? (b) How much work does the woman do to set herself and the turntable into motion?

Solution (a) The table rotates in a direction opposite to that in which the woman walks. There are no external torques acting on the system; therefore, from conservation of angular momentum, we have $L_f = L_i = 0$

so $\quad L_f = I_w \omega_w + I_t \omega_t = 0 \qquad$ and $\qquad \omega_t = -\dfrac{I_w}{I_t}\omega_w = -\left(\dfrac{m_w r^2}{I_t}\right)\left(\dfrac{v_w}{r}\right)$

$$\omega_t = -\frac{(60.0 \text{ kg})(2.00 \text{ m})(1.50 \text{ m / s})}{500 \text{ kg} \cdot \text{m}^2} = -0.360 \text{ rad / s} = 0.360 \text{ rad / s CCW} \quad \Diamond$$

(b) Work done $= \Delta K = K_f - 0 = \frac{1}{2} m_{\text{woman}} v_{\text{woman}}^2 + \frac{1}{2} I_{\text{table}} \omega_{\text{table}}^2$

$$W = \tfrac{1}{2}(60.0 \text{ kg})(1.50 \text{ m/s})^2 + \tfrac{1}{2}\left(500 \text{ kg} \cdot \text{m}^2\right)(0.360 \text{ rad/s})^2 = 99.9 \text{ J} \qquad \Diamond$$

Related Questions: (a) Why is angular momentum conserved? (b) Why is mechanical energy not conserved? (c) Is linear momentum conserved?

(a) The axle exerts no torque on the woman-plus-turntable system; only torques from outside the system can change the total angular momentum.

(b) The internal forces of the woman pushing backward on the turntable does positive work, converting chemical into kinetic energy.

(c) No. If the woman starts walking north, she pushes south on the turntable. Its axle holds it still against linear motion by pushing north on it, and this outside force delivers northward linear momentum into the system.

39. A wooden block of mass M resting on a frictionless horizontal surface is attached to a rigid rod of length ℓ and of negligible mass (Fig. P11.39). The rod is pivoted at the other end. A bullet of mass m traveling parallel to the horizontal surface and normal to the rod with speed v hits the block and becomes embedded in it. (a) What is the angular momentum of the bullet-block system? (b) What fraction of the original kinetic energy is lost in the collision?

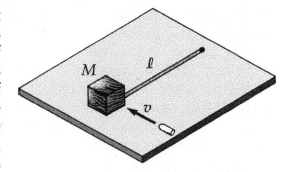

Figure P11.39

Solution

G: Since there are no external torques acting on the bullet-block system, the angular momentum of the system will be constant and will simply be that of the bullet before it hits the block.

We should expect there to be a significant, but not a total loss of kinetic energy in this perfectly inelastic "angular collision", since the block and bullet will move together after the collision with a velocity that can be found from conservation of angular momentum.

O: We have practically solved this problem already! We just need to work out the details for the loss of kinetic energy.

A: Taking the origin at the pivot point,

(a) Note that **r** is perpendicular to **v**, so $\sin(\theta) = 1$ and

$$L = rmv\sin(\theta) = lmv \qquad \Diamond$$

244

(b)   Taking v_f to be the speed of the bullet and the block together, we first apply conservation of angular momentum. $L_i = L_f$ becomes:

$$lmv = l(m + M)v_f \qquad \text{or} \qquad v_f = \left(\frac{m}{m + M}\right)v$$

The total kinetic energies before and after the collision are

$$K_i = \frac{1}{2}mv^2$$

$$K_f = \frac{1}{2}(m + M)v_f{}^2 = \frac{1}{2}(m + M)\left(\frac{m}{m + M}\right)^2 v^2 = \frac{1}{2}\left(\frac{m^2}{m + M}\right)v^2$$

So the fraction of the kinetic energy that is "lost" will be:

$$\text{Fraction} = \frac{-\Delta K}{K_i} = \frac{K_i - K_f}{K_i}$$

$$\text{Fraction} = \frac{\frac{1}{2}mv^2 - \frac{1}{2}\left(\frac{m^2}{m + M}\right)v^2}{\frac{1}{2}mv^2} = \frac{M}{m + M} \qquad \lozenge$$

L: We could have also used conservation of linear momentum to solve part (b), instead of conservation of angular moment; the answer would be the same. From the final equation, we can see that a larger fraction of kinetic energy is lost for a smaller bullet. This is consistent with reasoning that if the bullet were so small that the block barely moved after the collision, then nearly all the initial kinetic energy would disappear into internal energy.

47. A string is wound around a uniform disk of radius R and mass M. The disk is released from rest when the string is vertical and its top end tied to a fixed bar (Fig. P11.47). Show that (a) the tension in the string is one-third of the weight of the disk, (b) the magnitude of the acceleration of the center of mass is $2g/3$, and (c) the speed of the center of mass is $(4gh/3)^{1/2}$ as the disk descends. Verify your answer to part (c) using the energy approach.

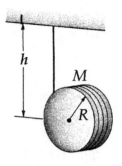

Figure P11.47

Solution

$\Sigma\mathbf{F}=m\mathbf{a}$ yields $\qquad \Sigma F = T - Mg = -Ma,\quad$ or $\qquad\qquad a = (Mg - T)/M$

$\Sigma\tau = m\alpha$ becomes $\quad \Sigma\tau = TR = I\alpha = \frac{1}{2}MR^2(a/R):\qquad a = 2T/M$

(a) Setting these two equations equal, $\qquad\qquad T = Mg/3 \qquad\qquad \Diamond$

(b) $\quad a = 2T/M = (2Mg/3M) \qquad$ or $\qquad\qquad a = \frac{2}{3}g \qquad\qquad \Diamond$

(c) Since $v_i = 0$ and $a = \frac{2}{3}g$, $\qquad\qquad v_f^2 = v_i^2 + 2ah$

gives us $\quad v_f^2 = 0 + 2\left(\frac{2}{3}g\right)h \qquad$ or $\qquad v_f = \sqrt{4gh/3} \qquad \Diamond$

Now we verify this answer. Requiring conservation of mechanical energy, we have $\Delta U + \Delta K_{\text{rot}} + \Delta K_{\text{trans}} = 0$:

$$mg\Delta h + \frac{1}{2}I\omega^2 + \frac{1}{2}mv^2 = 0$$

$$(0 - mgh) + \frac{1}{2}\left(\frac{1}{2}MR^2\right)\omega^2 - 0 + \left(\frac{1}{2}Mv^2 - 0\right) = 0$$

When there is no slipping, $\qquad \omega = \dfrac{v}{R} \qquad$ and $\qquad v = \sqrt{\dfrac{4gh}{3}} \qquad\qquad \Diamond$

55. A mass m is attached to a cord passing through a small hole in a frictionless, horizontal surface (Fig. P11.55). The mass is initially orbiting with speed v_i in a circle of radius r_i. The cord is then slowly pulled from below, and the radius of the circle decreases to r.   (a) What is the speed of the mass when the radius is r? (b) Find the tension in the cord as a function of r. (c) How much work W is done in moving m from r_i to r? (**Note**: The tension depends on r.) (d) Obtain numerical values for v, T, and W when $r = 1.00$ m, $m = 50.0$ g, $r_i = 0.300$ m,   and $v_i = 1.50$ m / s.

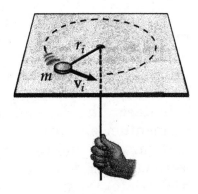

Figure P11.55

Solution

(a)   Although an external force (tension of rope) acts on the mass, no external torques act. Therefore L = constant, and at any time

$$mvr = mv_i r_i \qquad \text{gives us} \qquad v = \frac{v_i r_i}{r} \qquad \Diamond$$

(b) $T = \dfrac{mv^2}{r}$. Substituting for v from (a), we find that $T = m\dfrac{v_i^2 r_i^2}{r^3}$ $\Diamond$

(c) $W = \Delta K = \frac{1}{2}m\left(v^2 - v_i^2\right) = \dfrac{mv_i^2}{2}\left(\dfrac{r_i^2}{r^2} - 1\right)$ $\Diamond$

(d)   Substituting the given values into the previous equations, we find

$$v = 4.50 \text{ m / s} \qquad\qquad T = 10.1 \text{ N} \qquad\qquad W = 0.450 \text{ J} \quad \Diamond$$

59. Two astronauts (Fig. P11.59), each having a mass of 75.0 kg, are connected by a 10.0-m rope of negligible mass. They are isolated in space, orbiting their center of mass at speeds of 5.00 m/s. (a) Treating the astronauts as particles, calculate the magnitude of the angular momentum and (b) the rotational energy of the system. By pulling on the rope, one of the astronauts shortens the distance between them to 5.00 m. (c) What is the new angular momentum of the system? (d) What are the astronauts' new speeds? (e) What is the new rotational energy of the system? (f) How much work is done by the astronaut in shortening the rope?

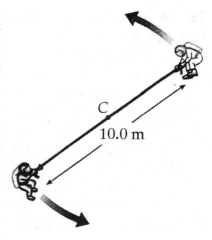

C

10.0 m

Figure P11.59

Solution

(a) $|\mathbf{L}| = |m\mathbf{r} \times \mathbf{v}|$

In this case, $\mathbf{r}$ and $\mathbf{v}$ are perpendicular, so the magnitude of $\mathbf{L}$ about C is

$$L = \sum mrv = 2(75.0 \text{ kg})(5.00 \text{ m})(5.00 \text{ m / s}) = 3.75 \times 10^3 \text{ kg} \cdot \text{m}^2 / \text{s} \qquad \lozenge$$

(b) $K = \frac{1}{2}mv^2 + \frac{1}{2}mv^2 = \frac{1}{2}(75.0 \text{ kg})(5.00 \text{ m / s})^2(2) = 1.88 \times 10^3 \text{ J} \qquad \lozenge$

(c) With a lever arm of zero, the rope tension generates no torque about C. Thus, the angular momentum is unchanged:

$$L = 3.75 \times 10^3 \text{ kg} \cdot \text{m}^2 / \text{s} \qquad \lozenge$$

(d) Again, $L = 2mrv$

$$3.75 \times 10^3 \text{ kg} \cdot \text{m}^2 / \text{s} = 2(75.0 \text{ kg})(2.50 \text{ m})(v \sin 90°)$$

$$v = 10.0 \text{ m/s} \qquad \lozenge$$

(e) $K = 2\left(\frac{1}{2}mv^2\right) = 2\left(\frac{1}{2}75\ \text{kg}\right)(10\ \text{m/s})^2 = 7.50 \times 10^3\ \text{J}$ ◊

(f) $W_{nc} = K_f - K_0 = 7.50 \times 10^3\ \text{J} - 1.88 \times 10^3\ \text{J}$

   $W_{nc} = 5.62 \times 10^3\ \text{J}$ ◊

63. A spool of wire of mass M and radius R is unwound under a constant force $\mathbf{F}$ (Fig. P11.63). Assuming that the spool is a uniform solid cylinder that does not slip, show that (a) the acceleration of the center of mass is $4\mathbf{F}/3M$ and that (b) the force of friction is to the **right** and is equal in magnitude to $F/3$. (c) If the cylinder starts from rest and rolls without slipping, what is the speed of its center of mass after it has rolled through a distance d?

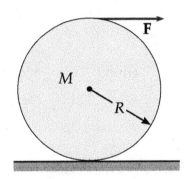

Figure P11.63

Solution

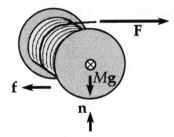

To keep the spool from slipping, there must be static friction acting at its contact point with the floor. Assume friction is directed to the left. Then the full set of Newton's second-law equations are:

$\sum F_x = ma_x:$ $\qquad F - f = Ma$

$\sum F_y = ma_y:$ $\qquad -Mg + n = 0$

$\sum \tau_{CM} = I_{CM}\alpha_{CM}:$ $\qquad -FR + Mg(0) + n(0) - fR = \frac{1}{2}MR^2\left(-\dfrac{a}{R}\right)$

We have written $\alpha_{CM} = \dfrac{-a}{R}$ to describe the clockwise rotation. Regarding F, M, and R as known, the first and third of these equations allow us to solve for f and a:

$$f = F - Ma \qquad \text{and} \qquad -F - f = -\frac{1}{2}Ma$$

(a) Substituting for f, $\qquad\qquad\qquad F + F - Ma = \dfrac{1}{2}Ma$

Thus, $2F = \dfrac{3}{2}Ma$ and $a = \dfrac{4F}{3M}$ ◊

(b) Then solving for f, $\qquad f = F - M\left(\dfrac{4F}{3M}\right) = F - \dfrac{4F}{3} = -\dfrac{F}{3}$ ◊

The negative sign means that the force of friction is opposite the direction we assumed. That is, **f** is to the right, so the spool does not spin like a car stuck in the snow.

(c) Since a is constant, we can use

$$v^2 = v_0{}^2 + 2a(x - x_0) = 0 + 2\left(\frac{4F}{3M}\right)d$$ ◊

$$v = \sqrt{\frac{8Fd}{3m}}$$ ◊

65. Suppose a solid disk of radius R is given an angular speed ω_i about an axis through its center and is then lowered to a horizontal surface and released, as shown in Problem 64 (see Fig. P11.64). Furthermore, assume that the coefficient of friction between the disk and the surface is μ. (a) Show that the time it takes for pure rolling motion to occur is $R\omega_i / 3\mu g$. (b) Show that the distance the disk travels before pure rolling occurs is $R^2\omega_i^2 / 18\mu g$.

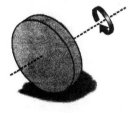

Figure P11.64

Solution

(a) If $v_0 = 0$, then at any time $v = at$

From Problem 64, once pure rolling occurs, $\omega = \tfrac{1}{3}\omega_i$, so that $v = \tfrac{1}{3}R\omega_i$.

Using these expressions for v and v_0 in the first equation, we find

$$\tfrac{1}{3}R\omega = at \qquad \text{where} \qquad a = \frac{F}{m} = \frac{-\mu mg}{m} = -\mu g$$

Therefore, $$t = \frac{R\omega_i}{3\mu g} \qquad \Diamond$$

(b) The distance of travel is $\Delta x = \tfrac{1}{2}at^2$. Using the result from part (a), we find

$$\Delta x = \tfrac{1}{2}(\mu g)\left(\frac{\tfrac{1}{3}R\omega_i}{\mu g}\right)^2 = \frac{R^2\omega_i^2}{18\mu g} \qquad \Diamond$$

Chapter 12

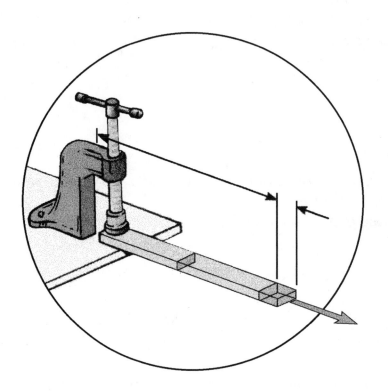

Static Equilibrium and Elasticity

STATIC EQUILIBRIUM AND ELASTICITY

INTRODUCTION

Part of this chapter is concerned with the conditions under which a rigid object is in equilibrium. The term **equilibrium** implies either that the object is at rest or that its center of mass moves with constant velocity. We deal here with the former, which are referred to as objects in **static equilibrium**.

In order for an extended object to be in static equilibrium, the net force on it must be zero **and** it must have no tendency to rotate. This **second** condition of equilibrium requires that **the net torque about any origin be zero**. In order to establish whether or not an object is in equilibrium, we must know its size and shape, the forces acting on different parts of it, and the points of application of the various forces.

The last section of this chapter deals with the realistic situation of objects that deform under load conditions. Such deformations are usually elastic in nature and do not affect the conditions of equilibrium. By **elastic** we mean that when the deforming forces are removed, the object returns to its original shape. Several elastic constants are defined, each corresponding to a different type of deformation.

EQUATIONS AND CONCEPTS

In general, **an object at rest or one moving with constant velocity will only do so if the resultant force on it is zero.** This is a statement of the **first condition of equilibrium,** and corresponds to the condition of translational equilibrium.

$$\Sigma F = 0 \qquad (12.1)$$

Equation 12.1 is a **vector sum** of all **external forces** acting on the body, so this necessarily implies that the sum of the x, y, and z components separately must be zero.

$$\Sigma F_x = 0$$
$$\Sigma F_y = 0$$
$$\Sigma F_z = 0$$

The **second condition of equilibrium** of a rigid body requires that the **vector sum of the torques relative to any origin must be zero.** This is the condition of rotational equilibrium.

$$\Sigma \tau = 0 \qquad (12.2)$$

If all the forces acting on a rigid body lie in a common plane, say the xy plane, then there is no z component of force. In this case, we only have to deal with three equations--two of which correspond to the first condition of equilibrium, the third coming from the second condition. In this case, the torque vector lies along the z axis. All problems in this chapter fall into this category.

$$\Sigma F_x = 0 \qquad (12.3)$$
$$\Sigma F_y = 0$$
$$\Sigma \tau_z = 0$$

In order to compute the torque due to the weight (gravitational force), all the weight can be considered to be concentrated at a single point called the center of gravity. When an object is in a uniform gravitational field, the center of gravity is located at the center of mass.

$$x_{CG} = \frac{m_1 x_1 + m_2 x_2 + m_3 x_3 + \cdots}{m_1 + m_2 + m_3 + \cdots} \qquad (12.4)$$

$$= \frac{\Sigma m_i x_i}{\Sigma m_i}$$

Young's modulus Y is a measure of the resistance of a body to elongation, and is equal to the ratio of the tensile stress (the force per unit area) to the tensile strain (the change in length over the original length).

$$Y = \frac{\text{tensile stress}}{\text{tensile strain}} = \frac{F/A}{\Delta L / L_i} \qquad (12.6)$$

The **Shear modulus** S is a measure of the deformation which occurs when a force is applied to one surface, while the opposite surface is held fixed. The shear modulus equals the ratio of the shearing stress to the shear strain.

$$S = \frac{\text{shear stress}}{\text{shear strain}} = \frac{F/A}{\Delta x / h} \qquad (12.7)$$

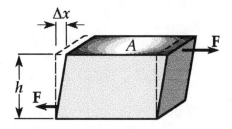

The **Bulk modulus** B is a parameter which characterizes the response of a body to uniform pressure on all sides. It is defined as the ratio of the volume stress (the pressure) to the volume strain ($\Delta V / V$).

$$B = \frac{\text{Volume stress}}{\text{Volume strain}} = -\frac{\Delta P}{\Delta V / V} \qquad (12.8)$$

SUGGESTIONS, SKILLS, AND STRATEGIES

Since this chapter represents the application of Newton's laws to a special situation, namely, rigid bodies in static equilibrium, it is important that you understand and follow the procedures for analyzing such problems. The following skills must be mastered in this regard:

- The need to recognize all **external** forces acting on the body, and the construction of an accurate free-body diagram.

- Resolving the external forces into their rectangular components, and applying the first condition of equilibrium $\Sigma F_x = 0$ and $\Sigma F_y = 0$.

- You must choose a convenient origin for calculating the net torque on the body. The choice of this origin is arbitrary. (The torque equation gives information which is not offered by applying $\Sigma \mathbf{F} = 0$.)

- Solving the set of simultaneous equations obtained from the two conditions of equilibrium.

The following procedure is recommended when analyzing a body in equilibrium under the action of several external forces:

- Make a sketch of the object under consideration.

- Draw a free-body diagram and label all external forces acting on the object. Try to guess the correct direction for each force. If you select an incorrect direction that leads to a negative sign in your solution for a force, do not be alarmed; this merely means that the direction of the force is the opposite of what you assumed.

- Resolve all forces into rectangular components, choosing a convenient coordinate system. Then apply the first condition for equilibrium, which balances forces. Remember to keep track of the signs of the various force components.

- Choose a convenient axis for calculating the net torque on the object. Remember that the choice of the origin for the torque equation is **arbitrary;** therefore, choose an origin that will simplify your calculation as much as possible. Becoming adept at this is a matter of practice.

- The first and second conditions of equilibrium give a set of linear equations with several unknowns. All that is left is to solve the simultaneous equations for the unknowns in terms of the known quantities.

REVIEW CHECKLIST

▷ Describe the two necessary conditions of equilibrium for a rigid body.

▷ Locate the center of gravity of a system of particles or a rigid body and understand the subtle difference between center of gravity and center of mass.

▷ Analyze problems of rigid bodies in static equilibrium using the procedures presented in Section 12.3 of the text.

ANSWERS TO SELECTED CONCEPTUAL QUESTIONS

12. A ladder is resting inclined against a wall. Would you feel safer climbing up the ladder if you were told that the ground is frictionless but the wall is rough or that the wall is frictionless but the ground is rough? Justify your answer.

Answer

The picture to the right shows the forces on the ladder if both the wall and floor exert friction. If the floor is perfectly smooth, it can exert no frictional force to the right, to counterbalance the wall's normal force. Therefore a ladder on a smooth floor cannot stand in equilibrium. On the other hand, a smooth wall can still exert a normal force to hold the ladder in equilibrium against horizontal motion. The counterclockwise torque of this force prevents rotation about the foot of the ladder. So you should choose a rough floor.

□   □   □   □

SOLUTIONS TO SELECTED END-OF-CHAPTER PROBLEMS

3. A uniform beam of mass m_b and length ℓ supports blocks of masses m_1 and m_2 at two positions, as shown in Figure P12.3. The beam rests on two points. For what value of x will the beam be balanced at P such that the normal force at O is zero?

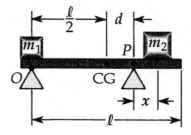

Figure P12.3

Solution

Refer to the free-body diagram, and take torques about point P.

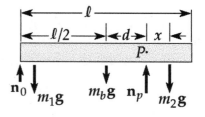

$$\Sigma \tau_p = -n_0 \left[\frac{\ell}{2} + d\right] + m_1 g \left[\frac{\ell}{2} + d\right] + m_b g d - m_2 g x = 0$$

We want to find x for which $n_0 = 0$. Let $n_0 = 0$ and solve for x.

$$m_1 \left[\frac{\ell}{2} + d\right] + m_b d - m_2 x = 0$$

so

$$x = \frac{m_1}{m_2}\left(\frac{\ell}{2} + d\right) + \frac{m_b}{m_2} d$$

◊

9.   Consider the following mass distribution: 5.00 kg at (0, 0) m, 3.00 kg at (0, 4.00) m, and 4.00 kg at (3.00, 0) m. Where should a fourth mass of 8.00 kg be placed so that the center of gravity of the four-mass arrangement will be at (0, 0)?

Solution

$$\mathbf{r}_{cg} = \frac{\Sigma m_i \mathbf{r}_i}{\Sigma m_i} \quad \text{or} \quad \mathbf{r}_{cg}(\Sigma m_i) = \Sigma m_i \mathbf{r}_i$$

We require the center of mass to be at the origin; this simplifies the situation, leaving

$$\Sigma m_i x_i = 0 \quad \text{and} \quad \Sigma m_i y_i = 0$$

To find the x-coordinate:

$$[5.00 \text{ kg}][0 \text{ m}] + [3.00 \text{ kg}][0 \text{ m}] + [4.00 \text{ kg}][3.00 \text{ m}] + [8.00 \text{ kg}]x = 0$$

and $x = -1.50 \text{ m}$

Likewise, to find the y-coordinate, we solve:

$$[5.00 \text{ kg}][0 \text{ m}] + [3.00 \text{ kg}][4.00 \text{ m}] + [4.00 \text{ kg}][3.00 \text{ m}] + [8.00 \text{ kg}]y = 0$$

and $y = -1.50 \text{ m}$

Therefore, a fourth mass of 8.00 kg should be located at

$$\mathbf{r}_4 = (-1.50\mathbf{i} - 1.50\mathbf{j}) \text{ m}$$

◊

13. A 15.0-m uniform ladder weighing 500 N rests against a frictionless wall. The ladder makes a 60.0° angle with the horizontal. (a) Find the horizontal and vertical forces that the ground exerts on the base of the ladder when an 800-N firefighter is 4.00 m from the bottom. (b) If the ladder is just on the verge of slipping when the firefighter is 9.00 m up, what is the coefficient of static friction between the ladder and the ground?

Solution

G: Refer to the free-body diagram at right, as needed. Since the wall is frictionless, only the ground exerts an upward force on the ladder to oppose the combined weight of the ladder and firefighter, so n_g = 1300 N. Based on the angle of the ladder, f < 1300 N. The coefficient of friction is probably somewhere between 0 and 1.

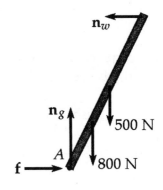

O: Draw a free-body diagram, apply Newton's second law, and sum torques to find the unknown forces. Since this is a statics problem (no motion), both the net force and net torque are zero.

A: (a) $\Sigma F_x = f - n_w = 0$

$\Sigma F_y = n_g - 800 \text{ N} - 500 \text{ N} = 0$ so that $n_g = 1300 \text{ N}$ (upwards)   ◊

Taking torques about an axis at the foot of the ladder, $\Sigma \tau_A = 0$

$-(8.00 \text{ N})(4.00 \text{ m})\sin 30° - (500 \text{ N})(7.50 \text{ m})\sin 30° + n_w(15.0 \text{ m})\cos 30° = 0$

Solving the torque equation for n_w,

$$n_w = \frac{\left[(4.00 \text{ m})(800 \text{ N}) + (7.50 \text{ m})(500 \text{ N})\right]\tan\ 30.0°}{15.0 \text{ m}} = 267.5 \text{ N}$$

Next substitute this value into the F_x equation to find

$f = n_w = 268 \text{ N}$ (**f** is directed toward the wall)   ◊

(b) When the firefighter is 9.00 m up the ladder, the torque equation $\Sigma\tau_A = 0$ gives

$$-(800 \text{ N})(9.00 \text{ m})\sin 30° - (500 \text{ N})(7.50 \text{ m})\sin 30° + n_w(15.0 \text{ m})\sin 60° = 0$$

Solving, $n_w = 421 \text{ N}$. Since $f = n_w = 421 \text{ N}$ and $f = f_{max} = \mu_s n_g$,

$$\mu_s = \frac{f_{max}}{n_g} = \frac{421 \text{ N}}{1300 \text{ N}} = 0.324 \qquad \lozenge$$

L: The calculated answers seem reasonable since they agree with our predictions. This problem would be more realistic if the wall were not frictionless, in which case an additional vertical force would be added. This more complicated problem could be solved if we knew at least one of the coefficients of friction.

---

17. A 1 500-kg automobile has a wheel base (the distance between the axles) of 3.00 m. The center of mass of the automobile is on the center line at a point 1.20 m behind the front axle. Find the force exerted by the ground on each wheel.

Solution

G: Since the center of mass lies in the front half of the car, there should be more force on the front wheels than the rear ones, and the sum of the wheel forces must equal the weight of the car.

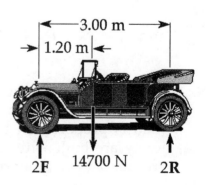

O: Draw a free-body diagram, apply Newton's second law, and sum torques to find the unknown forces for this statics problem.

A: The car's weight is $F_g = mg = (1500 \text{ kg})(9.80 \text{ m} / \text{s}^2) = 14\,700 \text{ N}$

Call **F** the force of the ground on each of the front wheels and **R** the normal force on each of the rear wheels.

If we take torques around the front axle, the equations are as follows:

$\Sigma F_x = 0$: $0 = 0$

$\Sigma F_y = 0$: $2R - 14\,700 \text{ N} + 2F = 0$

$\Sigma \tau = 0$: $-2R(3.00 \text{ m}) + (14\,700 \text{ N})(1.20 \text{ m}) + 2F(0) = 0$

The torque equation gives :

$$R = \frac{17\,640 \text{ N} \cdot \text{m}}{6.00 \text{ m}} = 2940 \text{ N} = 2.94 \text{ kN} \qquad \Diamond$$

Then, from the second force equation,

$$2(2.94 \text{ kN}) - 14.7 \text{ kN} + 2F = 0$$

and $F = 4.41 \text{ kN}$ $\qquad\qquad\qquad\qquad\qquad\qquad\qquad\qquad$ $\Diamond$

L: As expected, the front wheels experience a greater force (about 50% more) than the rear wheels. Since the frictional force between the tires and road is proportional to this normal force, it makes sense that most cars today are built with front wheel drive so that the wheels under power are the ones with more traction (friction).

25. A 200-kg load is hung on a wire with a length of 4.00 m, a cross-sectional area of 0.200×10^{-4} m^2, and a Young's modulus of 8.00×10^{10} N/m^2. What is its increase in length?

Solution

G: Since metal wire does not stretch very much, the length will probably not change by more than 1% (< 4 cm in this case) unless it is stretched beyond its elastic limit.

O: Apply the Young's Modulus strain equation to find the increase in length.

A: Young's Modulus is $\quad Y = \dfrac{F / A}{\Delta L / L_i}$

The load force is $\quad F = (200 \text{ kg})(9.80 \text{ m/s}^2) = 1960 \text{ N}$

So $\qquad \Delta L = \dfrac{FL_i}{AY} = \dfrac{(1960 \text{ N})(4.00 \text{ m})(1000 \text{ mm / m})}{\left(0.200 \times 10^{-4} \text{ m}^2\right)\left(8.00 \times 10^{10} \text{ N / m}^2\right)}$

$\Delta L = 4.90 \text{ mm}$ ◊

L: The wire only stretched about 0.1% of its length, so this seems like a reasonable amount.

33. If the shear stress exceeds about 4.00×10^8 N/m², steel ruptures. Determine the shearing force necessary (a) to shear a steel bolt 1.00 cm in diameter and (b) to punch a 1.00-cm-diameter hole in a steel plate 0.500 cm thick.

Solution

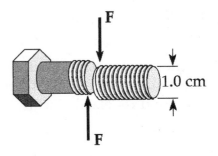

(a) We do not need the equation $S = \text{stress}/\text{strain}$. Rather, we use just the definition of stress $\sigma = F/A$, where A is the area of one of the layers sliding over each other:

$$F = \sigma A = \pi\left(4.00 \times 10^8 \text{ N}/\text{m}^2\right)\left(0.500 \times 10^{-2} \text{ m}\right)^2 = 31.4 \text{ kN} \qquad \Diamond$$

(b) Now the area of the molecular layers sliding over each other is the curved lateral surface area of the cylinder punched out, a cylinder of radius 0.50 cm and height 0.50 cm.

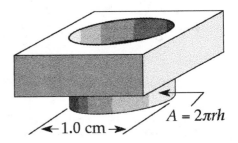

So, $F = \sigma A = \sigma(h)(2\pi r)$

$$= 2\pi\left(4.00 \times 10^8 \text{ N}/\text{m}^2\right)\left(0.500 \times 10^{-2} \text{ m}\right)\left(0.500 \times 10^{-2} \text{ m}\right)$$

$$= 62.8 \text{ kN} \qquad \Diamond$$

35. When water freezes, it expands by about 9.00%. What would be the pressure increase inside your automobile's engine block if the water in it froze? (The bulk modulus of ice is 2.00×10^9 N/m².)

Solution V represents the original volume; $0.0900V$ is the change in volume that happens if the block cracks open. Imagine squeezing the ice back down ($\Delta V = -0.0900V$) to its original volume, according to the definition of the bulk modulus (Eq. 12.8):

$$\Delta P = -\frac{B(\Delta V)}{V} = -\frac{\left(2.00\times10^9 \text{ N}/\text{m}^2\right)\left(-0.0900V\right)}{V} = 1.80\times10^8 \text{ N}/\text{m}^2 \ \lozenge$$

37.   A bridge with a length of 50.0 m and a mass of 8.00×10^4 kg is supported on a smooth pier at each end, as illustrated in Figure P12.37. A truck of mass 3.00×10^4 kg is located 15.0 m from one end. What are the forces on the bridge at the points of support?

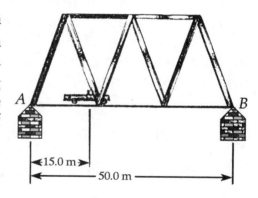

Figure P12.37

Solution   Let n_A and n_B be the normal forces at the points of support. Choosing the origin at point A, we find:

$$\Sigma F_y = 0: \quad n_A + n_B - \left(8.00\times10^4 \text{ kg}\right)g - \left(3.00\times10^4 \text{ kg}\right)g = 0$$

$$\Sigma \tau = 0: \quad -\left(3.00\times10^4 \text{ kg}\right)(15.0 \text{ m})g - \left(8.00\times10^4 \text{ kg}\right)(25.0 \text{ m})g + n_B(50.0 \text{ m}) = 0$$

The equations combine to give

$$n_A = 5.98\times10^5 \text{ N} \qquad \lozenge$$

$$n_B = 4.80\times10^5 \text{ N} \qquad \lozenge$$

45. A uniform sign of weight F_g and width 2L hangs from a light, horizontal beam hinged at the wall and supported by a cable (Fig. P12.45). Determine (a) the tension in the cable and (b) the components of the reaction force exerted by the wall on the beam in terms of F_g, d, L, and θ.

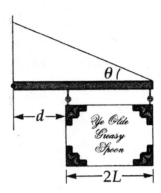

Figure P12.45

Solution Choose the beam for analysis, and draw a free-body diagram as shown. We know the direction of the tension force at the right end is along the cable, at an angle of θ above the horizontal.

Taking torques about the left end,

$$\Sigma F_x = 0: \qquad +R_x - T\cos\theta = 0$$

$$\Sigma F_y = 0: \qquad +R_y - F_g + T\sin\theta = 0$$

$$\Sigma \tau = 0: \qquad R_y(0) + R_x(0) - F_g(d+L) + (0)(T\cos\theta) + (d+2L)(T\sin\theta) = 0$$

(a) The torque equation gives $\qquad T = \dfrac{F_g(d+L)}{(d+2L)\sin\theta}$ ◊

(b) Now from the force equations,

$$R_x = \frac{F_g(d+L)}{(d+2L)\tan\theta} \qquad \text{and} \qquad R_y = F_g - \frac{F_g(d+L)}{d+2L} = \frac{F_gL}{d+2L}$$ ◊

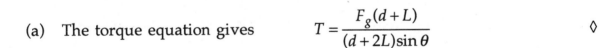

49. A 10 000-N shark is supported by a cable attached to a 4.00-m rod that can pivot at its base. Calculate the tension that the cable must have if it is to hold the system in the position shown in Figure P12.49. Find the horizontal and vertical forces exerted on the base of the rod. (Neglect the weight of the rod.)

Solution

G: Since the rod helps support the weight of the shark by exerting a vertical force, the tension in the upper portion of the cable must be less than 10 000 N. Likewise, the vertical force on the base of the rod should also be less than 10 kN.

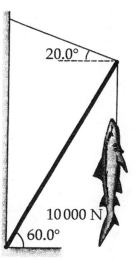

Figure P12.49

O: This is another statics problem where the sum of the forces and torques must be zero. To find the unknown forces, draw a free-body diagram, apply Newton's second law, and sum torques.

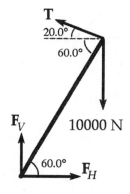

A: From the diagram, the angle **T** makes with the rod is

$$\theta = 60.0° + 20.0° = 80.0°$$

and the perpendicular component of **T** is $T\sin(80.0°)$.

Summing torques around the base of the rod, and applying Newton's second law in the horizontal and vertical directions,

$$\Sigma\tau = 0: \qquad -(4.00\text{ m})(10\,000\text{ N})\cos 60° + T(4.00\text{ m})\sin(80°) = 0$$

$$T = \frac{(10000\text{ N})\cos(60.0°)}{\sin(80.0°)} = 5.08 \times 10^3\text{ N} \qqu\qquad \Diamond$$

$$\Sigma F_x = 0: \qquad F_H - T\cos(20.0°) = 0$$

$$F_H = T\cos(20.0°) = 4.77 \times 10^3 \text{ N} \qquad \Diamond$$

$$\Sigma F_y = 0: \qquad F_V + T\sin(20.0°) - 10\,000 \text{ N} = 0$$

$$F_V = (10000 \text{ N}) - T\sin(20.0°) = 8.26 \times 10^3 \text{ N} \qquad \Diamond$$

L: The forces calculated are indeed less than 10 kN as predicted. That shark sure is a big catch; he weighs about a ton!

---

53. A force acts on a rectangular cabinet weighing 400 N, as illustrated in Figure P12.53. (a) If the cabinet slides with constant speed when $F = 200$ N and $h = 0.400$ m, find the coefficient of kinetic friction and the position of the resultant normal force. (b) If $F = 300$ N, find the value of h for which the cabinet just begins to tip.

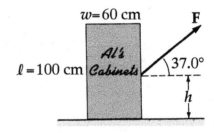

Figure P12.53

Solution

(a) Think of the normal force as acting at distance x from the lower left corner. Moving with constant speed, the cabinet is in equilibrium:

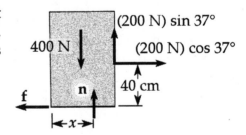

$$\Sigma F_x = 0: \qquad -f + (200 \text{ N})\cos(37.0°) = 0$$

$$\Sigma F_y = 0: \qquad -400 \text{ N} + n + (200 \text{ N})\sin(37.0°) = 0$$

We have already $f = 160$ N and $n = 280$ N, so $\mu_k = f/n = 0.571$ $\qquad \Diamond$

Take torques about the lower left corner; $\Sigma\tau = 0$ gives

$$-(400\text{ N})(30\text{ cm}) + nx + (200\text{ N})(60\text{ cm})\sin(37°) - (200\text{ N})(40\text{ cm})\cos(37°) = 0$$

Substituting $n = 280$ N (and converting cm to m, and back) gives

$$x = \frac{120\text{ N}\cdot\text{m} - 72.2\text{ N}\cdot\text{m} + 63.9\text{ N}\cdot\text{m}}{280\text{ N}} = 39.9\text{ cm} \qquad \lozenge$$

(b) When the cabinet is just about to tip, the normal force is located at the lower right corner, and $\Sigma\tau = 0$ is still true.

Because most of the forces are directed through the lower right corner, we choose to take torques about that point. This leaves only two forces to deal with.

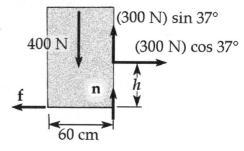

$$\Sigma\tau = 0: \qquad -(300\text{ N})(h)\cos(37.0°) + (400\text{ N})(30.0\text{ cm}) = 0$$

Solving for h, $\qquad h = \dfrac{120\text{ N}\cdot\text{m}}{240\text{ N}} = 50.1\text{ cm}$ $\qquad \lozenge$

---

57. A uniform beam of mass m is inclined at an angle θ to the horizontal. Its upper end produces a 90° bend in a very rough rope tied to a wall, and its lower end rests on a rough floor (Fig. P12.57). (a) If the coefficient of static friction between the beam and the floor is μ_s, determine an expression for the maximum mass M that can be suspended from the top before the beam slips. (b) Determine the magnitude of the reaction force at the floor and the magnitude of the force exerted by the beam on the rope at P in terms of m, M, and μ_s.

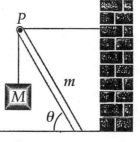

Figure P12.57

Solution

G: The solution to this problem is not as obvious as some other problems because there are three independent variables that affect the maximum mass M. We could at least expect that more mass can be supported for higher coefficients of friction (μ_s), larger angles (θ), and a more massive beam (m).

O: Draw a free-body diagram, apply Newton's second law, and sum torques to find the unknown forces for this statics problem.

A: (a)   Use $\Sigma F_x = \Sigma F_y = \Sigma \tau = 0$ and choose the origin at the point of contact on the floor to simplify the torque analysis. On the verge of slipping, the friction is $f = \mu_s n$, and

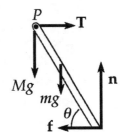

$\Sigma F_x = 0$: $T - \mu_s n = 0$

$\Sigma F_y = 0$: $n - Mg - mg = 0$

Solving these equations, $T = \mu_s g(M + m)$

From $\Sigma \tau = 0$, $Mg(\cos \theta)L + mg(\cos \theta)\dfrac{L}{2} - T(\sin \theta)L = 0$

Substituting for T, we get $M = \dfrac{m}{2}\left[\dfrac{2\mu_s \sin \theta - \cos \theta}{\cos \theta - \mu_s \sin \theta} \right]$ ◊

(b)   At the floor, we see that the normal force is in the y direction and frictional force is in the x direction. The reaction force then is

$$R = \sqrt{n^2 + (\mu_s n)^2} = g(M + m)\sqrt{1 + \mu_s^2}$$ ◊

At point P, the force of the beam on the rope is

$$F = \sqrt{T^2 + (Mg)^2} = g\sqrt{M^2 + \mu_s^2(M + m)^2}$$ ◊

L: In our answer to part (a), notice that this result does not depend on L, which is reasonable since the center of mass of the beam is proportional to the length of the beam.

The answer to this problem is certainly more complex than most problems. We can see that the maximum mass M that can be supported is proportional to m, but it is not clear from the solution that M increases proportional to μ_s and θ as predicted. To further examine the solution to part (a), we could graph or calculate the ratio M/m as a function of θ for several reasonable values of μ_s ranging from 0.5 to 1. Since the mass values must be positive , we find that only angles from about 40° to 60° are possible for this scenario, which explains why we don't encounter this precarious configuration very often.

59. A stepladder of negligible weight is constructed as shown in Figure P12.59. A painter with a mass of 70.0 kg stands on the ladder 3.00 m from the bottom. Assuming that the floor is frictionless, find (a) the tension in the horizontal bar connecting the two halves of the ladder, (b) the normal forces at A and B, and (c) the components of the reaction force at the single hinge C that the left half of the ladder exerts on the right half. (**Hint:** Treat each half of the ladder separately.)

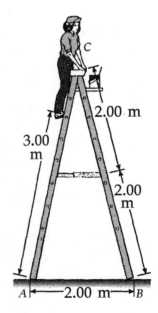

Figure P12.59

Solution If we think of the whole ladder, we can solve part (b).

(b) The painter is 3/4 of the way up the ladder, so the lever arm of her weight about A is

$$\frac{3}{4}(1.00 \text{ m}) = 0.750 \text{ m}$$

$\Sigma F_x = 0:\quad 0 = 0$

$\Sigma F_y = 0:\quad n_A - 686\text{ N} + n_B = 0$

$\Sigma \tau_A = 0:\quad n_A(0) - (686\text{ N})(0.750\text{ m}) + n_B(2.00\text{ m}) = 0$

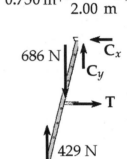

Thus, $n_B = 257\text{ N}$ and $n_A = 686\text{ N} - 257\text{ N} = 429\text{ N}$ ◊

Now consider the left half of the ladder. We know the direction of the bar tension, and we make guesses for the directions of the components of the hinge force. If a guess is wrong, the answer will be negative.

The side rails make an angle with the horizontal

$$\theta = \cos^{-1}(1/4) = 75.5°$$

Taking torques about the top of the ladder, we have

$\Sigma \tau_c = 0:\quad (-429\text{ N})(1.00\text{ m}) + T(2.00\text{ m})\sin 75.5° + (686\text{ N})(0.250\text{ m}) = 0$

$\Sigma F_x = 0:\quad T - C_x = 0$

$\Sigma \tau_A = 0:\quad 429\text{ N} - 686\text{ N} + C_y = 0$

(a)   From the torque equation, $T = \dfrac{257\text{ N}\cdot\text{m}}{1.94\text{ m}} = 133\text{ N}$ ◊

(c)   From the force equations, $C_x = T = 133\text{ N}$ to the left

$C_y = 686\text{ N} - 429\text{ N} = 257\text{ N}$ up

The force that the left half exerts on the right half has opposite components:

133 N to the right,   and 257 N down.   ◊

63. A wire of length L_i, Young's modulus Y, and cross-sectional area A is stretched elastically by an amount ΔL. According to Hooke's law, the restoring force is $-k\,\Delta L$. (a) Show that $k = YA/L_i$ (b) Show that the work done in stretching the wire by an amount ΔL is $W = YA(\Delta L)^2/2L_i$.

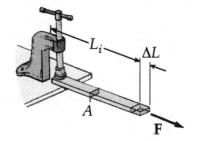

Figure 12.13

Solution

(a) According to Hooke's law $|\mathbf{F}| = k\,\Delta L$. Young's modulus is defined as

$$Y = \frac{F/A}{\Delta L/L_i} = k\frac{L_i}{A}$$

or
$$k = \frac{YA}{L_i} \qquad \Diamond$$

(b) Since we have determined that the wire stretches like a spring, we can determine the work done by integrating the force $F = -kx$ over the distance we stretched the wire.

$$W = -\int_0^{\Delta L} F\,dx = -\int_0^{\Delta L}(-kx)dx = \frac{YA}{L_i}\int_0^{\Delta L} x\,dx = \left[\frac{YA}{L_i}\left(\frac{1}{2}x^2\right)\right]_{x=0}^{x=\Delta L}$$

Therefore, $W = \dfrac{YA}{2L_i}\Delta L^2$ $\qquad \Diamond$

67. (a) Estimate the force with which a karate master strikes a board if the hand's speed at time of impact is 10.0 m/s and decreases to 1.00 m/s during a 0.00200-s time-of-contact with the board. The mass of coordinated hand and arm is 1.00 kg. (b) Estimate the shear stress if this force is exerted on a 1.00-cm-thick pine board that is 10.0 cm wide. (c) If the maximum shear stress a pine board can receive before breaking is 3.60×10^6 N/m², will the board break?

Solution

The impulse-momentum theorem describes the force of the board on his hand:

$$Ft = mv_f - mv_i$$

$$F = (0.00200 \text{ s}) = (1.00 \text{ kg})(1.00 \text{ m/s} - 10.0 \text{ m/s})$$

and $\qquad$ $F = -4500$ N

(a) Therefore, the force of his hand on board is 4500 N $\qquad$ ◊

(b) That force produces a shear stress on the area that is exposed when the board snaps:

$$\text{Stress} = \frac{F}{A} = \frac{4500 \text{ N}}{(10.0^{-1} \text{ m})(10.0^{-2} \text{ m})} = 4.50 \times 10^6 \text{ N/m}^2 \qquad ◊$$

(c) **Yes;** this suffices to break the board. $\qquad$ ◊

Chapter 13

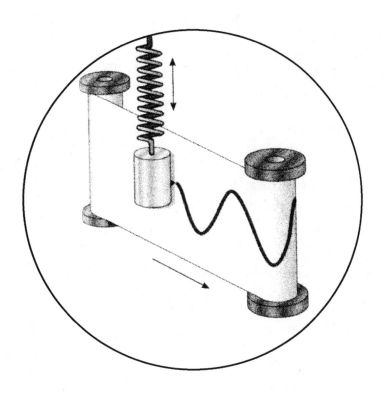

Oscillatory Motion

OSCILLATORY MOTION

INTRODUCTION

A very special kind of motion occurs when the force on a body is proportional to the displacement of the body from equilibrium. If this force always acts toward the equilibrium position of the body, there is a repetitive back-and-forth motion about this position. Such motion is an example of what is called **periodic** or **oscillatory** motion.

Most of the material in this chapter deals with **simple harmonic motion.** In this type of motion, an object oscillates between two spatial positions for an indefinite period of time with no loss in mechanical energy. In real mechanical systems, retarding (frictional) forces are always present and these forces are considered in an optional section at the end of the chapter.

EQUATIONS AND CONCEPTS

The force exerted by a spring on a mass attached to the spring and displaced a distance x from the unstretched position is given by Hooke's law. The force constant, k, is always positive and has a value which corresponds to the relative stiffness of the spring. The negative sign means that the force exerted on the mass is always directed opposite the displacement. The force is a restoring force, always directed toward the equilibrium position.

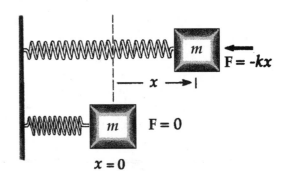

An object exhibits simple harmonic motion when the net force along the direction of motion is proportional to the displacement and in the opposite direction.

$$F = -kx \tag{13.1}$$

Applying **Newton's second law** to motion in the x direction gives $F = ma_x = -kx$. Since $a_x = d^2x/dt^2$, this is equivalent to Equation 13.17, where the ratio k/m is denoted ω^2.

$$\frac{d^2x}{dt^2} = -\omega^2 x \tag{13.17}$$

Equation (13.3) describes the displacement from equilibrium for a particle moving in simple harmonic motion along the x axis. In this expression, A represents the **amplitude** of the motion, $\omega t + \phi$ is the **phase**, ω is the **angular frequency** (rad/s), and ϕ is the **phase constant**.

$$x = A\cos(\omega t + \phi) \tag{13.3}$$

$$\text{with } \omega^2 = \frac{k}{m}$$

The **period of motion** T equals the time it takes the mass to complete **one** oscillation; that is, the time it takes the mass to return to its original position for the first time. The **frequency** of the motion, f, numerically equals the inverse of the period and represents the number of oscillations per unit time. T is measured in seconds, while f is measured in s^{-1} or Hertz (Hz).

$$T = \frac{2\pi}{\omega} \tag{13.4}$$

$$f = \frac{1}{T} = \frac{\omega}{2\pi} \tag{13.5}$$

Taking the first derivative of x with respect to time gives the **velocity of the oscillator as a function of time**.

$$v = \frac{dx}{dt} = -\omega A \sin(\omega t + \phi)$$ (13.7)

The acceleration as a function of time is equal to the time derivative of the velocity (or the second derivative of the displacement). Note from Equation 13.9 that the **acceleration** (and hence the force) **is always proportional to and opposite the displacement.**

$$a = \frac{dv}{dt} = -\omega^2 A \cos(\omega t + \phi)$$ (13.8)

or

$$a = -\omega^2 x$$ (13.9)

The kinetic energy of a simple harmonic oscillator is given by $\frac{1}{2}mv^2$, while the potential energy is equal to $\frac{1}{2}kx^2$. Using Equations 13.3 and 13.7, together with $\omega^2 = k/m$, gives the **total** energy E of the oscillator. Note that E remains constant since we have assumed there are no nonconservative forces acting on the system. The total energy of the simple harmonic oscillator is a constant of the motion and is proportional to the square of the amplitude.

$$E = \tfrac{1}{2}mv^2 + \tfrac{1}{2}kx^2$$

or

$$E = \tfrac{1}{2}kA^2$$ (13.22)

Energy conservation can be used to obtain an expression for velocity as a function of position. The speed of an object in simple harmonic motion is a maximum at $x = 0$; the speed is zero when the mass is at the points of maximum displacement $(x = \pm A)$.

$$v = \pm\sqrt{\frac{k}{m}\left(A^2 - x^2\right)} \qquad (13.23)$$

or

$$v = \pm\omega\sqrt{\left(A^2 - x^2\right)}$$

The equation of motion for the simple pendulum assumes a small displacement so that $\sin\theta \cong \theta$.

$$\frac{d^2\theta}{dt^2} = -\frac{g}{L}\theta \qquad (13.24)$$

The period and frequency of a simple pendulum depend only on the length of the supporting string and the value of the acceleration due to gravity.

$$\omega = \sqrt{\frac{g}{L}} \qquad (13.25)$$

$$T = \frac{2\pi}{\omega} = 2\pi\sqrt{\frac{L}{g}} \qquad (13.26)$$

SUGGESTIONS, SKILLS, AND STRATEGIES

Most of this chapter deals with simple harmonic motion, and the properties of the displacement expression

$$x(t) = A\cos(\omega t + \phi) \qquad (13.3)$$

In order to obtain the velocity $v(t)$ and acceleration $a(t)$ of the system, one must be familiar with the derivative operation as applied to trigonometric functions. In particular, note that

$$\frac{d}{dt}\cos(\omega t + \phi) = -\omega\sin(\omega t + \phi) \quad \text{and} \quad \frac{d}{dt}\sin(\omega t + \phi) = \omega\cos(\omega t + \phi)$$

Using these results, and $x(t)$ from Equation 13.3, we see that

$$v(t) = \frac{dx(t)}{dt} = -A\omega \sin(\omega t + \phi)$$ (13.7)

and

$$a(t) = \frac{dv(t)}{dt} = -A\omega^2 \cos(\omega t + \phi)$$ (13.8)

By direct substitution, you should be able to show that Equation 13.3 represents a general solution to the equation of motion for the mass-spring system (a second-order homogeneous differential equation) given by

$$\frac{d^2x}{dt^2} + \frac{k}{m}x = 0 \qquad \text{where} \qquad \omega = \sqrt{\frac{k}{m}}$$

In treating the motion of the simple pendulum, we made use of the small angle approximation $\sin\theta \cong \theta$. This approximation enables us to reduce the equation of motion to that of the simple harmonic oscillator. The small angle approximation for $\sin\theta$ follows from inspecting the series expansion for $\sin\theta$, where θ is in **radians**:

$$\sin\theta = \theta - \frac{\theta^3}{3!} + \frac{\theta^5}{5!} - \cdots$$

For small values of θ, the higher order terms in θ^3, θ^5 . . . are **small** compared to θ, so it follows that $\sin\theta \cong \theta$. The difference between $\sin\theta$ and θ is less than 1% for $0 < \theta < 15°$ (where $15° \cong 0.26$ rad).

REVIEW CHECKLIST

▷ Describe the general characteristics of simple harmonic motion, and the significance of the various parameters which appear in the expression for the displacement versus time, $x = A\cos(\omega t + \phi)$.

▷ Start with the expression for the displacement versus time for the simple harmonic oscillator, and obtain equations for the velocity and acceleration as functions of time.

▷ Understand the phase relations between displacement, velocity, and acceleration for simple harmonic motion, noting that acceleration is proportional to the displacement, but in the opposite direction.

▷ Describe and understand the conditions of simple harmonic motions executed by the mass-spring system (where the frequency depends on k and m) and the simple pendulum (where the frequency depends on L and g).

▷ Discuss the relationship between simple harmonic motion and the motion of a point on a circle moving with uniform angular velocity.

ANSWERS TO SELECTED CONCEPTUAL QUESTIONS

2. If the coordinate of a particle varies as $x = -A\cos\omega t$, what is the phase constant in Equation 13.3? At what position does the particle begin its motion?

Answer Equation 13.3 says $x = A\cos(\omega t + \phi)$. Since negating a cosine wave is equivalent to changing the phase by 180°, we can say that

$$\phi = 180°\left(\frac{\pi\,\text{rad}}{180°}\right) = \pi\,\text{rad}$$

At time $t = 0$, the particle is at a coordinate position of $x = -A$.

□   □   □   □

4.   Determine whether the following quantities can be in the same direction for a simple harmonic oscillator: (a) displacement and velocity, (b) velocity and acceleration, (c) displacement and acceleration.

□

Answer   In a simple harmonic oscillator, the velocity follows the displacement by 1/4 of a cycle, and the acceleration follows the displacement by 1/2 of a cycle. Referring to Figure 12.1 of this study guide, it can be noted that there exist times (a) when both the displacement and the velocity are positive, and therefore in the same direction. (b) There also exist times when both the velocity and the acceleration are positive, and therefore in the same direction. On the other hand, (c) when the displacement is positive, the acceleration is always negative, and therefore these two always act in opposite directions.

□   □   □   □

13.   Is it possible to have damped oscillations when a system is at resonance? Explain.

Answer   Yes. At resonance, the amplitude of a damped oscillator will remain constant. If the system were not damped, the amplitude would increase without limit at resonance.

□   □   □   □

SOLUTIONS TO SELECTED END-OF-CHAPTER PROBLEMS

1.   The displacement of a particle at $t = 0.250$ s is given by the expression $x = (4.00 \text{ m}) \cos(3.00\pi t + \pi)$, where x is in meters and t is in seconds. Determine (a) the frequency and period of the motion, (b) the amplitude of the motion, (c) the phase constant, and (d) the displacement of the particle at $t = 0.250$ s.

Solution   The particular displacement function $x = (4.00 \text{ m}) \cos(3.00\pi t + \pi)$ and the general one $x = A \cos(\omega t + \phi)$ have an especially powerful kind of equality called functional equality. They must give the same x value for all values of the variable t. This requires, then, that all parts be the same:

(a) $\omega = 3.00\pi = 2\pi f$ or $f = 1.50$ Hz $T = 1/f = 0.667$ s   ◊

(b) $A = 4.00$ m   ◊

(c) $\phi = \pi$ rad   ◊

(d) $x(t=0.250 \text{ s}) = (4.00 \text{ m}) \cos(1.75\pi \text{ rad}) = (4.00 \text{ m}) \cos(5.50 \text{ rad})$

Note that this is **not** 5.50°. Instead,

$x = (4.00 \text{ m}) \cos(5.50 \text{ rad}) = (4.00 \text{ m}) \cos 315° = 2.83$ m   ◊

---

5.   A particle moving along the x axis in simple harmonic motion starts from its equilibrium position, the origin, at $t = 0$ and moves to the right. The amplitude of its motion is 2.00 cm, and the frequency is 1.50 Hz. (a) Show that the displacement of the particle is given by $x = (2.00 \text{ cm}) \sin(3.00\pi t)$. Determine (b) the maximum speed and the earliest time $(t > 0)$ at which the particle has this speed, (c) the maximum acceleration and the earliest time $(t > 0)$ at which the particle has this acceleration, and (d) the total distance traveled between $t = 0$ and $t = 1.00$ s.

Solution

(a)   At $t = 0$, $x = 0$ and v is positive (to the right). The sine function is zero and the cosine is positive at $\theta = 0$, so this situation corresponds to $x = A \sin \omega t$ and $v = \omega A \cos \omega t = v_{max} \cos \omega t$.

Since $f = 1.50$ Hz, $\omega = (2\pi)f = 3.00\pi$ s^{-1}

Note that $T = 1/f = (2/3)$ s

Also, $A = 2.00$ cm, so $\qquad$ $x = (2.00 \text{ cm}) \sin\left((3.00\pi \text{ s}^{-1}) t\right)$ ◊

This is equivalent to $\qquad$ $x = A\cos(\omega t + \phi)$

with $\qquad A = 2.00$ cm, $\qquad \omega = 3.00\pi$ s^{-1} $\quad$ and $\quad \phi = -90° = -\pi/2$

(b) The velocity is $v = dx/dt$: $\quad v = (2.00 \text{ cm})(3.00\pi \text{ s}^{-1})\cos\left((3.00\pi \text{ s}^{-1})t\right)$

The maximum speed is $\qquad v_{max} = A\omega = (2.00 \text{ cm})(3.00\pi \text{ s}^{-1})$

$$v_{max} = 6.00\pi \text{ cm / s}$$ ◊

The particle first has this speed at $t = 0$, when $\cos(3\pi t) = +1$, and next has this speed at $\quad t = T/2 = 0.333$ s $\quad$ when $\cos\left((3.00\pi \text{ s}^{-1})(0.333 \text{ s})\right) = -1$ ◊

(c) Again, $a = dv/dt$ gives $\qquad a = (-2.00 \text{ cm})(3.00\pi \text{ s}^{-1})^2 \sin\left((3.00\pi \text{ s}^{-1})t\right)$

Its maximum value is $\qquad a_{max} = A\omega^2 = (2.00 \text{ cm})(3.00\pi \text{ s}^{-1})^2$

$$a_{max} = 18.0\pi^2 \text{ cm/s}^2$$ ◊

The acceleration first has this positive value at time $t = 3T/4 = 0.500$ s, when $\quad a = -(2.00 \text{ cm})(3.00\pi \text{ s}^{-1})^2 \sin\left((3.00\pi \text{ s}^{-1})(0.500 \text{ s})\right) = +18.0\pi^2$ ◊

(d) Since $T = 2/3$ s and $A = 2.00$ cm, the particle will travel 8.00 cm in this time. Hence, in $(1 \text{ s}) = (3T/2)$,

the particle will travel $\qquad$ 8.00 cm + 4.00 cm = 12.0 cm ◊

9. A 0.500-kg mass attached to a spring with a force constant of 8.00 N / m vibrates in simple harmonic motion with an amplitude of 10.0 cm. Calculate (a) the maximum value of its speed and acceleration, (b) the speed and acceleration when the mass is 6.00 cm from the equilibrium position, and (c) the time it takes the mass to move from $x = 0$ to $x = 8.00$ cm.

Solution $$\omega = \sqrt{\frac{k}{m}} = \sqrt{\frac{8.00 \text{ N} / \text{m}}{0.500 \text{ kg}}} = 4.00 \text{ s}^{-1}$$

Therefore, position can be represented by $x = (10.0 \text{ cm}) \sin\left(\left(4.00 \text{ s}^{-1}\right)t\right)$. From this we find

(a) $v = \dfrac{dx}{dt} = (40.0 \text{ cm} / \text{s}) \cos(4.00t)$ $\qquad$ $v_{max} = 40.0 \text{ cm} / \text{s}$ $\qquad$ ◊

$a = \dfrac{dv}{dt} = -\left(160 \text{ cm} / \text{s}^2\right) \sin(4.00t)$ $\qquad$ $a_{max} = 160 \text{ cm} / \text{s}^2$ $\qquad$ ◊

(b) $t = \dfrac{\sin^{-1}(x / 10.0 \text{ cm})}{4.00 \text{ s}^{-1}}$

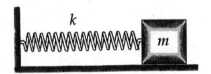

When $x = 6.00$ cm, $t = 0.161$ s and we find

$$v = (40.0 \text{ cm} / \text{s}) \cos\left[\left(4.00 \text{ s}^{-1}\right)(0.161 \text{ s})\right] = 32.0 \text{ cm} / \text{s} \qquad ◊$$

$$a = -\left(160 \text{ cm} / \text{s}^2\right) \sin\left[\left(4.00 \text{ s}^{-1}\right)(0.161 \text{ s})\right] = -96.1 \text{ cm} / \text{s}^2 \qquad ◊$$

(c) Using $t = \dfrac{\sin^{-1}(x / 10.0 \text{ cm})}{4.00 \text{ s}^{-1}}$,

When $x = 0$, $t = 0$ and when $x = 8.00$ cm, $t = 0.232$ s. Thus $\Delta t = 0.232$ s ◊

11. A 7.00-kg mass is hung from the bottom end of a vertical spring fastened to an overhead beam. The mass is set into vertical oscillations with a period of 2.60 s. Find the force constant of the spring.

Solution

A mass hanging from a vertical spring moves with simple harmonic motion just like a mass moving without friction attached to a horizontal spring.

We are given the period; by Equation 13.4, $\dfrac{2\pi}{\omega} = T = 2.60$ s

Solving for the angular frequency, $\omega = \dfrac{2\pi \text{ rad}}{2.60 \text{ s}} = 2.42$ rad / s

However, $\omega = \sqrt{\dfrac{k}{m}}$, so $k = \omega^2 m = (2.41 \text{ rad / s})^2 (7.00 \text{ kg})$

Thus we solve for the force constant: $k = 40.9 \text{ kg / s}^2 = 40.9$ N / m   ◊

---

17. An automobile having a mass of 1000 kg is driven into a brick wall in a safety test. The bumper behaves as a spring of constant 5.00×10^6 N / m and compresses 3.16 cm as the car is brought to rest. What was the speed of the car before impact, assuming that no energy is lost during impact with the wall?

Solution

G: If the bumper is only compressed 3 cm, the car is probably not permanently damaged, so v is most likely less than 10 mph ($\sim$ 5 m/s).

O: Assuming no energy is lost during impact with the wall, the initial energy (kinetic) equals the final energy (elastic potential).

A: Energy conservation gives $K_i = U_f$ or $\frac{1}{2}mv^2 = \frac{1}{2}kx^2$

Solving for the velocity, $v = x\sqrt{\dfrac{k}{m}} = \left(3.16 \times 10^{-2} \text{ m}\right)\sqrt{\dfrac{5.00 \times 10^6 \text{ N/m}}{1000 \text{ kg}}}$

Thus, $v = 2.23 \text{ m/s}$ ◊

L: The speed is less than 5 m/s as predicted, so the answer seems reasonable. If the speed of the car were sufficient to compress the bumper beyond its elastic limit, then some of the initial kinetic energy would be lost to deforming the front of the car. In that case, some other procedure would have to be used to estimate the car's initial speed.

---

23. A particle executes simple harmonic motion with an amplitude of 3.00 cm. At what displacement from the midpoint of its motion does its speed equal one half of its maximum speed?

Solution

G: If we consider the speed of the particle along its path as shown in the sketch, we can see that the particle is at rest momentarily at one endpoint while being accelerated toward the middle by an elastic force that decreases as the particle approaches the equilibrium position. When it reaches the midpoint, the direction of acceleration changes so that the particle slows down until it stops momentarily at the opposite endpoint. From this analysis, we can estimate that $v = v_{max}/2$ somewhere in the outer half of the travel (since this is the region where the speed is changing most rapidly):

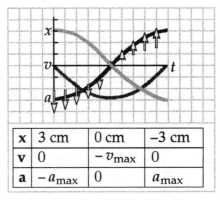

x	3 cm	0 cm	−3 cm
v	0	$-v_{max}$	0
a	$-a_{max}$	0	a_{max}

$1.50 < \pm x < 3.00$

O: One way to analyze this problem is to start with the equation for linear SHM, take the first derivative with respect to time to find $v(t)$, and solve this equation for x when $v = v_{max}/2$.

A: For linear SHM, $\qquad\qquad\qquad\qquad\qquad x = A\cos\omega t$

noting that the negative indicates direction, $\quad dx/dt = v = -A\omega\sin\omega t$

Since $A\omega$ is a constant, $v = \dfrac{v_{max}}{2}$ when $\qquad \sin\omega t = \pm\dfrac{1}{2}$

or $\qquad\qquad\qquad\qquad\qquad\qquad \omega t = \sin^{-1}\left(\pm\dfrac{1}{2}\right) = \pm\dfrac{\pi}{6}$

Thus, in our displacement equation, $\qquad \cos\omega t = \cos\left(\pm\dfrac{\pi}{6}\right) = \dfrac{\sqrt{3}}{2}$

Substituting $A = 3.00$ cm, $\qquad\qquad x = \pm\dfrac{A\sqrt{3}}{2} = \pm\dfrac{(3.00 \text{ cm})\sqrt{3}}{2}$

Solving, $\qquad\qquad\qquad\qquad\qquad x = \pm 2.60$ cm $\qquad\qquad\qquad \Diamond$

We could get the same answer from $\qquad v = \pm\omega\sqrt{A^2 - x^2}$

L: The calculated position is in the outer half of the travel as predicted, and is in fact very close to the endpoints. This means that the speed of the particle changes remarkably little until the particle reaches the ends of its travel, where it experiences the maximum restoring force of the spring, which is proportional to x.

29. A simple pendulum has a mass of 0.250 kg and a length of 1.00 m. It is displaced through an angle of 15.0° and then released. What are (a) the maximum speed, (b) the maximum angular acceleration, and (c) the maximum restoring force?

Solution We can solve this problem by either of two methods.

METHOD ONE

Since 15.0° is small enough that (in radians) $(\sin\theta = 0.259) = (\theta = 0.262)$ within 1%, we may treat the motion as simple harmonic motion. The constant angular frequency characterizing the motion is

$$\omega = \sqrt{\frac{g}{L}} = \sqrt{\frac{9.80 \text{ m} / \text{s}^2}{1.00 \text{ m}}} = 3.13 \text{ rad} / \text{s}$$

Rewriting the amplitude as a distance,

we find that $\qquad A = L\theta = (1.00 \text{ m})(0.262) = 0.262 \text{ m}$

(a) The maximum speed is $\quad v_{max} = \omega A = (3.13 \text{ s}^{-1})(0.262 \text{ m}) = 0.820 \text{ m} / \text{s}$ ◊

(b) Similarly, $\qquad a_{max} = \omega^2 A = (3.13 \text{ s}^{-1})^2(0.262 \text{ m}) = 2.57 \text{ m} / \text{s}^2$

This implies a maximum angular acceleration of

$$\alpha = \frac{a}{r} = \frac{2.57 \text{ m} / \text{s}^2}{1.00 \text{ m}} = 2.57 \text{ rad} / \text{s}^2 \qquad ◊$$

(c) $\qquad \Sigma F = ma = (0.250 \text{ kg})(2.57 \text{ m} / \text{s}^2) = 0.641 \text{ N}$ ◊

METHOD TWO

We may work out slightly more precise answers by using ideas we studied before. At release, the pendulum has height above its equilibrium position.

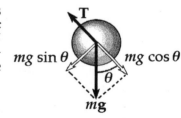

$$h = L - L\cos(15.0°)$$

$$h = 1.00 \text{ m} - (1.00 \text{ m})\cos(15.0°) = 0.0341 \text{ m}$$

(a)   In the case of this pendulum, the energy is constant as it swings down:

$$(K+U)_{\text{top}} = (K+U)_{\text{bottom}}$$

Applying this to the pendulum, $0 + mgh = \frac{1}{2}mv_{\text{max}}^2 + 0$

Solving, $v_{\text{max}} = \sqrt{2gh} = \sqrt{2(9.80 \text{ m / s}^2)(0.0341 \text{ m})}$

$$v_{\text{max}} = 0.817 \text{ m / s} \qquad \lozenge$$

(c)   We now draw a free body diagram of the forces on the   pendulum's mass, at the time of maximum displacement, when the pendulum is released. From the free body diagram, we find that the restoring force is $F = mg\sin(15.0°)$:

$$F = (0.250 \text{ kg})(9.80 \text{ m / s}^2)\sin(15.0°) = 0.634 \text{ N} \qquad \lozenge$$

(b)   This produces linear and angular accelerations of

$$a = \frac{\Sigma F}{m} = \frac{0.634\text{N}}{0.250 \text{ kg}} = 2.54 \text{ m / s}^2 \qquad \alpha = \frac{a}{r} = \frac{2.54 \text{ m / s}^2}{1.00 \text{ m}} = 2.54 \text{ rad / s}^2 \quad \lozenge$$

31. A particle of mass m slides without friction inside a hemispherical bowl of radius R. Show that, if it starts from rest with a small displacement from equilibrium, the particle moves in simple harmonic motion with an angular frequency equal to that of a simple pendulum of length R. That is, $\omega = \sqrt{g/R}$.

Solution   Locate the center of curvature C of the bowl. We can measure the excursion of the object from equilibrium by the angle θ between the radial line to C and the vertical. The distance the object moves from equilibrium is, $s = R\theta$.

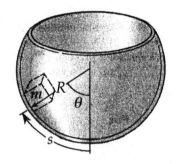

$\Sigma F_s = ma$ becomes $-mg\sin\theta = \dfrac{m\,d^2s}{dt^2}$

For small angles $\theta \cong \sin\theta$, so by substitution,

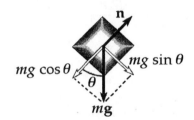

$$-mg\theta = \dfrac{m\,d^2s}{dt^2}$$

$\theta = s/R,$ so $-mg\,\dfrac{s}{R} = \dfrac{m\,d^2s}{dt^2}$ and $\dfrac{d^2s}{dt^2} = -\left(\dfrac{g}{R}\right)s$

The acceleration is proportional to the displacement and in the opposite direction, so we have simple harmonic motion.   ◊

We identify its angular frequency by comparing our equation to Eq. 13.17:

$$\dfrac{d^2x}{dt^2} = -\omega^2 x$$

Now x and s both measure displacement, so $\omega^2 = \dfrac{g}{R}$ and $\omega = \sqrt{\dfrac{g}{R}}$ ◊

33. A physical pendulum in the form of a planar body moves in simple harmonic motion with a frequency of 0.450 Hz. If the pendulum has a mass of 2.20 kg and the pivot is located 0.350 m from the center of mass, determine the moment of inertia of the pendulum.

Solution

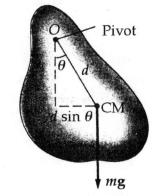

$$f = 0.450 \text{ Hz}, \quad d = 0.350 \text{ m}, \quad \text{and} \quad m = 2.20 \text{ kg}$$

Using equation 13.28,

$$T = 2\pi\sqrt{\frac{I}{mgd}} \qquad T^2 = \frac{4\pi^2 I}{mgd}$$

$$I = \frac{T^2 mgd}{4\pi^2} = \left(\frac{1}{f}\right)^2 \frac{mgd}{4\pi^2} = \frac{(2.20 \text{ kg})(9.80 \text{ m}/\text{s}^2)(0.350 \text{ m})}{(0.450 \text{ s}^{-1})^2(4\pi^2)} = 0.944 \text{ kg} \cdot \text{m}^2 \quad \Diamond$$

---

51. A compact mass M is attached to the end of a uniform rod, of equal mass M and length L, that is pivoted at the top (Fig. P13.51). (a) Determine the tensions in the rod at the pivot and at the point P when the system is stationary. (b) Calculate the period of oscillation for small displacements from equilibrium, and determine this period for $L = 2.00$ m. (**Hint:** Assume that the mass at the end of the rod is a point mass, and use Eq. 13.28.)

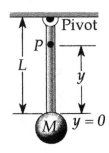

Figure P13.51

Solution

G: The tension in the rod at the pivot is the weight of the rod plus the weight of the mass M, so at the pivot point $T = 2Mg$. The tension at point P should be slightly less since the portion of the rod between P and the pivot does not contribute to the tension.

O: The tension can be found from applying Newton's Second Law. The period of this physical pendulum can be found by analyzing its moment of inertia and using Equation 13.28.

A: (a) When the pendulum is stationary, the tension at any point in the rod is simply the weight of everything below that point. This conclusion comes from applying $\Sigma F_y = ma_y = 0$ to everything below that point.

Thus, at the pivot the tension is $\quad F_T = F_{g,\text{ball}} + F_{g,\text{rod}} = Mg + Mg = 2Mg$ ◊

At point P, $\qquad\qquad\qquad\qquad\qquad F_T = F_{g,\text{ball}} + F_{g,\text{rod below }P}$

$$F_T = Mg + Mg\left(\frac{y}{L}\right) = Mg\left(1 + \frac{y}{L}\right) \qquad\qquad ◊$$

(b) For a physical pendulum where I is the moment of inertia about the pivot, $m = 2M$, and d is the distance from the pivot to the center of mass,

the period of oscillation is: $\qquad T = 2\pi\sqrt{I/(Mgd)}$

Relative to the pivot, $\qquad\qquad I_{\text{total}} = I_{rod} + I_{ball} = \frac{1}{3}ML^2 + ML^2 = \frac{4}{3}ML^2$

The center of mass distance is $\qquad d = \dfrac{\Sigma m_i x_i}{\Sigma m_i} = \dfrac{(ML/2 + ML)}{(M + M)} = \dfrac{3L}{4}$

so we have $\quad T = 2\pi\sqrt{\dfrac{I}{Mgd}}\quad$ or $\quad T = 2\pi\sqrt{\dfrac{(4ML^2/3)}{(2M)g(3L/4)}} = \dfrac{4\pi}{3}\sqrt{\dfrac{2L}{g}} \qquad ◊$

For $L = 2.00$ m, $\qquad\qquad\qquad T = \dfrac{4\pi}{3}\sqrt{\dfrac{2(2.00\text{ m})}{9.80\text{ m}/\text{s}^2}} = 2.68$ s $\qquad\qquad ◊$

L: In part (a), the tensions agree with the initial predictions. In part (b) we found that the period is slightly less (by about 6%) than a simple pendulum of length L. It is interesting to note that we were able to calculate a value for the period despite not knowing the mass value. This is because the period of any pendulum depends on the **location** of the center of mass and not on the **size** of the mass.

---

53. A large block P executes horizontal simple harmonic motion as it slides across a frictionless surface with a frequency of $f = 1.50$ Hz. Block B rests on it, as shown in Figure P13.53, and the coefficient of static friction between the two is $\mu_S = 0.600$. What maximum amplitude of oscillation can the system have if block B is not to slip?

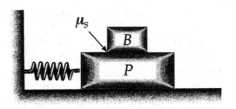

Figure P13.53

Solution If the block B does not slip, its motion is SHM with the same amplitude and frequency as those of P, and with its acceleration caused by the static friction force exerted on it by P. Think of the block when it is just ready to slip at a turning point in its motion:

$\Sigma F = ma$ becomes

$$f_{max} = \mu_s n = \mu_s mg = ma_{max} = mA\omega^2$$

Then $\qquad A = \dfrac{\mu_s g}{\omega^2} = \dfrac{0.600(9.80 \text{ m} / \text{s}^2)}{[2\pi(1.50 \text{ s}^{-1})]^2}$

$A = 6.62$ cm $\qquad\qquad\qquad\qquad\qquad\qquad \lozenge$

---

59. A pendulum of length L and mass M has a spring of force constant k connected to it at a distance h below its point of suspension (Fig. P13.59). Find the frequency of vibration of the system for small values of the amplitude (small θ). (Assume the vertical suspension of length L is rigid, but neglect its mass.)

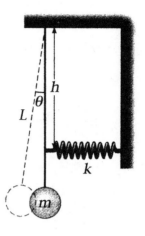

Solution

G: The frequency of vibration should be greater than that of a simple pendulum since the spring adds an additional restoring force:

Figure P13.59

$$f > \frac{1}{2\pi}\sqrt{\frac{g}{L}}$$

O: We can find the frequency of oscillation from the angular frequency, ω, which is found in the equation for angular SHM: $d^2\theta/dt^2 = -\omega^2\theta$. The angular acceleration can be found from analyzing the torques acting on the pendulum.

A: For the pendulum (see sketch),

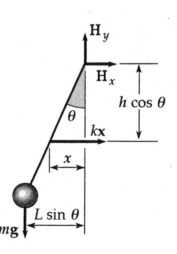

$$\Sigma\tau = I\alpha \qquad \text{and} \qquad d^2\theta/dt^2 = -\alpha$$

The negative sign appears because positive θ is measured clockwise in the picture. We take torque around the point of suspension:

$$\Sigma\tau = MgL\sin\theta + kxh\cos\theta = I\alpha$$

For small amplitude vibrations, use the approximations:

$$\sin\theta \approx \theta, \quad \cos\theta \approx 1, \quad \text{and} \quad x = h\tan\theta \cong h\theta$$

Therefore, with $I = mL^2$,
$$\frac{d^2\theta}{dt^2} = -\left[\frac{MgL + kh^2}{I}\right]\theta = -\left[\frac{MgL + kh^2}{ML^2}\right]\theta$$

This is of the SHM form
$$\frac{d^2\theta}{dt^2} = -\omega^2\theta$$

with angular frequency,
$$\omega = \sqrt{\frac{MgL + kh^2}{ML^2}} = 2\pi f$$

The ordinary frequency is
$$f = \frac{\omega}{2\pi} = \frac{1}{2\pi}\sqrt{\frac{MgL + kh^2}{ML^2}} \qquad \diamond$$

L: The frequency is greater than for a simple pendulum as we expected. In fact, the additional contribution inside the square root looks like the frequency of a mass on a spring scaled by h/L since the spring is connected to the rod and not directly to the mass. So we can think of the solution as:

$$f^2 = \frac{1}{4\pi^2}\left(\frac{MgL + kh^2}{ML^2}\right) = f^2_{pendulum} + \frac{h^2}{L^2}f^2_{spring}$$

---

63. A simple pendulum with a length of 2.23 m and a mass of 6.74 kg is given an initial speed of 2.06 m / s at its equilibrium position. Assume that it undergoes simple harmonic motion and determine its (a) period, (b) total energy and (c) maximum angular displacement.

Solution

(a) The period is $T = \dfrac{2\pi}{\omega}$:

$$T = 2\pi\sqrt{\dfrac{L}{g}} = 2\pi\sqrt{\dfrac{2.23 \text{ m}}{9.80 \text{ m}/\text{s}^2}} = 3.00 \text{ s} \quad \lozenge$$

(b) The total energy is $E = \tfrac{1}{2}mv_{max}^2$:

$$E = \tfrac{1}{2}(6.74 \text{ kg})(2.06 \text{ m}/\text{s})^2 = 14.3 \text{ J} \quad \lozenge$$

(c) At angular displacement θ_{max},

$$mgh = \tfrac{1}{2}mv_{max}^2 \quad \text{and} \quad h = \dfrac{v_{max}^2}{2g} = 0.217 \text{ m}$$

By geometry,

$$h = L - L\cos\theta_{max} = L(1 - \cos\theta_{max}),$$

Solving for θ_{max},

$$\cos\theta_{max} = 1 - \dfrac{h}{L} \quad \text{and} \quad \theta_{max} = 25.5° \quad \lozenge$$

Alternatively, we could write $v_{max} = \omega A$:

$$A = \dfrac{v_{max}}{\omega} = \dfrac{2.06 \text{ m}/\text{s}}{2.10/\text{s}} = 0.983 \text{ m} \quad \text{and} \quad \theta = \dfrac{A}{L} = \dfrac{0.983 \text{ m}}{2.23 \text{ m}} = 0.441 \text{ rad} = 25.2°$$

Our two answers are not precisely equal because the pendulum does not move with precisely simple harmonic motion.

---

67. A ball of mass m is connected to two rubber bands of length L, each under tension T, as in Figure P13.67. The ball is displaced by a small distance y perpendicular to the length of the rubber bands. Assuming that the tension does not change, show that (a) the restoring force is $-(2T/L)y$ and (b) the system exhibits simple harmonic motion with an angular frequency $\omega = \sqrt{2T/mL}$.

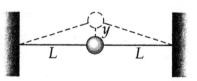

Figure P13.67

Solution

(a) Defining θ to be $\theta = \tan^{-1}(y\,/\,L)$, $\Sigma F = -2T \sin \theta \mathbf{j}$

Since for a small displacement, $\sin \theta \approx \tan \theta = \dfrac{y}{L}$

The resulting force is $F = \left(-\dfrac{2Ty}{L} \right)\mathbf{j}$ ◊

(b) For a spring system, $F = -kx$ becomes $F = -\left(\dfrac{2T}{L} \right)y$

Therefore, the angular frequency is $\omega = \sqrt{\dfrac{k}{m}} = \sqrt{\dfrac{2T}{mL}}$ ◊

---

69. A small, thin disk of radius r and mass m is attached rigidly to the face of a second thin disk of radius R and mass M as shown in Figure P13.69. The center of the small disk is located at the edge of the large disk. The large disk is mounted at its center on a frictionless axle. The assembly is rotated through an angle θ from its equilibrium position and released. (a) Show that the speed of the center of the small disk as it passes through the equilibrium position is

$$v = 2\sqrt{\dfrac{Rg(1 - \cos \theta)}{(M\,/\,m) + (r\,/\,R)^2 + 2}}$$

(b) Show that the period of the motion is

$$T = 2\pi\sqrt{\dfrac{(M + 2m)R^2 + mr^2}{2mgR}}$$

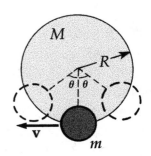

Figure P13.69

Solution

(a) $\Delta K + \Delta U = 0$, thus $K_{top} + U_{top} = K_{bot} + U_{bot}$ where $K_{top} = U_{bot} = 0$.

Therefore, $mgh = \frac{1}{2}I\omega^2$ where

$$h = R - R\cos\theta = R(1 - \cos\theta), \quad \omega = \frac{v}{R} \quad \text{and} \quad I = \frac{1}{2}MR^2 + \frac{1}{2}mr^2 + mR^2$$

Substituting, we find $mgR(1 - \cos\theta) = \frac{1}{2}\left(\frac{1}{2}MR^2 + \frac{1}{2}mr^2 + mR^2\right)\frac{v^2}{R^2}$

$$mgR(1 - \cos\theta) = \left(\frac{1}{4}M + \frac{1}{4}\frac{r^2}{R^2}m + \frac{1}{2}m\right)v^2$$

and $$v^2 = \frac{4gR(1 - \cos\theta)}{(M/m) + \left(r^2/R^2\right) + 2}$$

so $$v = 2\sqrt{\frac{Rg(1 - \cos\theta)}{(M/m) + (r/R)^2 + 2}} \qquad \Diamond$$

(b) $T = 2\pi\sqrt{\dfrac{I}{M_T g d}} \qquad M_T = m + M \qquad d = \dfrac{mR + M(0)}{m + M}$

$$T = 2\pi = \sqrt{\frac{\frac{1}{2}MR^2 + \frac{1}{2}mr^2 + mR^2}{mgR}} = 2\pi\sqrt{\frac{(M + 2m)R^2 + mr^2}{2mgR}} \qquad \Diamond$$

The Law of Gravity

Chapter 14

THE LAW OF GRAVITY

INTRODUCTION

In this chapter we study the law of gravity. Emphasis is placed on describing the motion of the planets, because astronomical data provide an important test of the validity of the law of gravity. We show that the laws of planetary motion developed by Johannes Kepler follow from the law of gravity and the concept of the conservation of angular momentum. A general expression for the gravitational potential energy is derived, and the energetics of planetary and satellite motion are treated. The law of gravity is also used to determine the force between a particle and an extended body.

EQUATIONS AND CONCEPTS

The **universal law of gravity** states that any two particles **attract** each other with a force that is proportional to the product of their masses and inversely proportional to the square of their separation.

$$F_g = G \frac{m_1 m_2}{r^2} \tag{14.1}$$

The constant G is called the **universal gravitational constant.**

$$G = 6.673 \times 10^{-11} \text{ N} \cdot \text{m}^2 / \text{kg}^2 \tag{14.2}$$

The acceleration due to gravity, g', decreases with increasing altitude, h, measured from the Earth's surface.

$$g' = \frac{GM_E}{r^2} = \frac{GM_E}{(R_E + h)^2} \qquad (14.6)$$

A **gravitational field g** exists at some point in space if a particle of mass m experiences a gravitational force $\mathbf{F}_g = m\mathbf{g}$ at that point. That is, the gravitational field represents the ratio of the gravitational force experienced by the mass divided by that mass.

$$\mathbf{g} \equiv \frac{\mathbf{F}_g}{m} \qquad (14.10)$$

The gravitational field at a distance r from the center of the Earth points radially inward toward the center of the Earth. Over a small region near the Earth's surface, **g** is an approximately uniform downward field.

$$\mathbf{g} = -\frac{GM_E}{r^2}\hat{\mathbf{r}} \qquad (14.11)$$

Since the gravitational force is conservative, we can define a gravitational energy function corresponding to that force. As a mass m moves from one position to another in the presence of the Earth's gravity, its potential energy changes by an amount given by Equation 14.13, where r_i and r_f are the initial and final distances of the mass from the center of the Earth.

$$U_f - U_i = -GM_E m\left(\frac{1}{r_f} - \frac{1}{r_i}\right) \qquad (14.13)$$

The **gravitational potential energy** associated with **any pair** of particles of masses m_1 and m_2 separated by a distance r is given by Equation 14.15. The negative sign in this expression corresponds to the attractive nature of the gravitational force. An external agent must do positive work to increase the separation of the particles.

$$U = -\frac{Gm_1m_2}{r}$$

(14.15)

In this expression it is assumed that $U_i = 0$ at $r_i = \infty$, and the equation is valid for an Earth-particle system when $r > R_E$.

As a body of mass m moves in an orbit around a very massive body of mass M (where $M \gg m$), the **total energy** of the system is the sum of the kinetic energy of m (taking the massive body to be at rest) and the potential energy of the system.

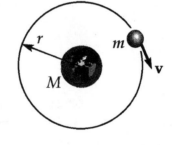

When the two contributions are evaluated for a circular orbit, one finds that the **total energy** E is negative, and given by Equation 14.19. This arises from the fact that the (positive) kinetic energy is equal to one half of the magnitude of the (negative) potential energy.

$$E = -\frac{GMm}{2r}$$

(14.19)

The **escape velocity** is defined as the **minimum** velocity a body must have, when projected from the Earth whose mass is M_E and radius is R_E, in order to escape the Earth's gravitational field (that is, to just reach $r = \infty$ with zero speed). Note that v_{esc} does not depend on the mass of the projected body.

$$v_{esc} = \sqrt{\frac{2GM_E}{R_E}} \qquad (14.22)$$

The **potential energy** associated with a particle of mass m and an **extended** body of mass M can be evaluated using Equation 14.23, where the extended body is divided into segments of mass dM, and r is the distance from dM to the particle.

$$U = -Gm \int \frac{dM}{r} \qquad (14.23)$$

If a particle of mass m is **outside** a **uniform** solid sphere or spherical shell of radius R, the sphere attracts the particle as though the mass of the sphere were concentrated at its center.

$$\mathbf{F}_g = -\frac{GMm}{r^2}\hat{\mathbf{r}} \qquad (r \geq R) \qquad (14.25a)$$

If a particle is located **inside a uniform spherical shell**, the force acting on the particle is **zero**.

$$\mathbf{F}_g = 0 \qquad (r < R) \qquad (14.25b)$$

If a particle of mass m is **inside** a **homogeneous solid sphere** of mass M and radius R, the gravitational force acting on the particle acts toward the center of the sphere and is proportional to the distance r from the center to the particle. This force is due only to that portion of the mass contained within the portion of the sphere of radius $r < R$.

$$\mathbf{F}_g = -\frac{GmM}{R^3} r \,\hat{\mathbf{r}} \quad (r < R) \tag{14.27}$$

SUGGESTIONS, SKILLS, AND STRATEGIES

In this chapter we made use of the definite integral in evaluating the potential energy function associated with the conservative gravitational force. You should be familiar with the following type of definite integral:

$$\int_{x_1}^{x_2} x^n dx = \frac{x^{n+1}}{n+1}\bigg]_{x_1}^{x_2} = \frac{x_2^{n+1} - x_1^{n+1}}{n+1} \qquad (n \neq -1)$$

For example, in deriving Equation 14.13, we used the above expression as follows:

$$\int_{r_i}^{r_f} \frac{dr}{r^2} = \int_{r_i}^{r_f} r^{-2} dr = \frac{r_f^{-1} - r_i^{-1}}{-2+1} = -\left(\frac{1}{r_f} - \frac{1}{r_i}\right)$$

REVIEW CHECKLIST

▷ State Kepler's three laws of planetary motion and recognize that the laws are empirical in nature; that is, they are based on astronomical data. Describe the nature of Newton's universal law of gravity, and the method of deriving Kepler's third law ($T^2 \propto r^3$) from this law for circular orbits. Recognize that Kepler's second law is a consequence of conservation of angular momentum and the central nature of the gravitational force.

▷ Understand the concepts of the gravitational field and the gravitational potential energy, and know how to derive the expression for the potential energy for a pair of particles separated by a distance r.

▷ Describe the total energy of a planet or Earth satellite moving in a circular orbit about a large body located at the center of motion. Note that the total energy is negative, as it must be for any closed orbit.

▷ Understand the meaning of escape velocity, and know how to obtain the expression for v_{esc} using the principle of conservation of energy.

▷ Learn the method for calculating the gravitational force between a particle and an extended object. In particular, you should be familiar with the force between a particle and a spherical body when the particle is located outside and inside the spherical body.

ANSWERS TO SELECTED CONCEPTUAL QUESTIONS

7.   Explain why it takes more fuel for a spacecraft to travel from the Earth to the Moon than for the return trip. Estimate the difference.

Answer The mass and radius of the Earth and Moon, and the distance between the Earth and the Moon are $M_E = 5.98 \times 10^{24}$ kg, $R_E = 6.37 \times 10^6$ m, $M_M = 7.36 \times 10^{22}$ kg, $R_M = 1.75 \times 10^6$ m, and $d = 3.84 \times 10^8$ m. To travel between the Earth and the Moon, a distance d, a motor must boost the spacecraft over the point of zero total gravitational field in between. Call x the distance of this point from Earth. To cancel, the Earth and Moon must here produce equal fields:

$$\frac{GM_E}{x^2} = \frac{GM_M}{(d-x)^2}$$

Isolating x, $\frac{M_E}{M_M}(d-x)^2 = x^2$ gives $x = \frac{d\sqrt{M_E}}{\sqrt{M_M}+\sqrt{M_E}} = 3.46 \times 10^8$ m

and $\sqrt{\frac{M_E}{M_M}}(d-x) = x$ gives $(d-x) = \frac{d\sqrt{M_M}}{\sqrt{M_M}+\sqrt{M_E}} = 3.83 \times 10^7$ m

In general, we can ignore the gravitational pull of the far planet when we're close to the other; our results will still be approximately correct. Thus the approximate energy difference between point x and the Earth's surface, is:

$$\frac{\Delta E_{E \to x}}{m} = \frac{-GM_E}{x} - \frac{-GM_E}{R_E}$$

Similarly, flying from the moon: $\frac{\Delta E_{M \to x}}{m} = \frac{-GM_M}{d-x} - \frac{-GM_M}{R_M}$

Taking a ratio of energy terms, $\frac{\Delta E_{E \to x}}{\Delta E_{M \to x}} = \left(\frac{M_E}{x} - \frac{M_E}{R_E}\right) \Big/ \left(\frac{M_M}{d-x} - \frac{M_M}{R_M}\right) \cong 23.0$

This would also be the minimum fuel ratio, if the spacecraft was driven by a method other than rocket exhaust. If rockets were used, Equation 8.38 would apply, and the total fuel ratio would be much larger.

□   □   □   □

8. Why don't we put a geosynchronous weather satellite in orbit around the 45th parallel? Wouldn't this be more useful for the United States than such a satellite in orbit around the equator?

Answer While a satellite in orbit around the 45th parallel might be more useful, it isn't possible. The center of a satellite orbit must be the center of the Earth, since that is the force center for the gravitational force.

□   □   □   □

13. At what position in its elliptical orbit is the speed of a planet a maximum? At what position is the speed a minimum?

Answer The speed is a maximum at the perihelion, p, and is a minimum at the   aphelion, a.

□   □   □   □

16. In his 1798 experiment, Cavendish was said to have "weighed the Earth." Explain this statement.

Answer The Earth creates a gravitational field at its surface according to $g = GM_E/R_E^2$. The factors g and R_E were known, so as soon as Cavendish measured G, he could compute the mass of the Earth.

□   □   □ .   □

SOLUTIONS TO SELECTED END-OF-CHAPTER PROBLEMS

9.   When a falling meteor is at a distance above the Earth's surface of 3.00 times the Earth's radius, what is its acceleration due to the Earth's gravity?

Solution   The acceleration of gravity, $g = GM_e / r^2$, follows an inverse-square law. At the surface, a distance of one Earth-radius (R_E) from the center, it is 9.80 m/s².

At an altitude $3.00R_E$ above the surface (at distance $4.00R_E$ from the center), the acceleration of gravity will be $4.00^2 = 16.0$ times smaller:

$$g = \frac{GM_E}{(4.00R_E)^2} = \frac{GM_E}{16.0R_E^2} = \frac{9.80 \text{ m}/\text{s}^2}{16.0} = 0.612 \text{ m}/\text{s}^2 \text{ down} \qquad \Diamond$$

---

15.   Plaskett's binary system consists of two stars that revolve in a circular orbit about a center of mass midway between them. This means that the masses of the two stars are equal (Figure P14.15). If the orbital velocity of each star is 220 km/s and the orbital period of each is 14.4 days, find the mass M of each star. (For comparison, the mass of our Sun is 1.99×10^{30} kg.)

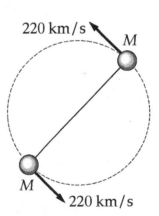

Figure P14.15

Solution

G: From the given data, it is difficult to estimate a reasonable answer to this problem without actually working through the details to actually solve it. A reasonable guess might be that each star has a mass equal to or larger than our Sun since our Sun happens to be less massive than most stars in the universe.

O: The only force acting on the two stars is the central gravitational force of attraction which results in a centripetal acceleration. When we solve Newton's 2nd law, we can find the unknown mass in terms of the variables given in the problem.

A: Applying Newton's 2nd Law, $\Sigma F = ma$ yields $F_g = ma_c$ for each star:

$$\frac{GMM}{(2r)^2} = \frac{Mv^2}{r} \qquad \text{so} \qquad M = \frac{4v^2r}{G}$$

We can write r in terms of the period, T, by considering the time and distance of one complete cycle. The distance traveled in one orbit is the circumference of the stars' common orbit , so $2\pi r = vT$. Therefore

$$M = \frac{4v^2r}{G} = \left(\frac{4v^2}{G}\right)\left(\frac{vT}{2\pi}\right) = \frac{2v^3T}{\pi G}$$

$$M = \frac{2(220 \times 10^3 \text{ m / s})^3 (14.4 \text{ d})(86400 \text{ s / d})}{\pi\left(6.67 \times 10^{-11} \text{ N} \cdot \text{m}^2 / \text{kg}^2\right)} = 1.26 \times 10^{32} \text{ kg} \qquad \lozenge$$

L: The mass of each star is about 63 solar masses, much more than our initial guess! A quick check in an astronomy book reveals that stars over 8 solar masses are considered to be **heavyweight** stars, and astronomers estimate that the maximum theoretical limit is about 100 solar masses before a star becomes unstable. So these 2 stars are exceptionally massive!

19. Io, a satellite of Jupiter, has an orbital period of 1.77 days and an orbital radius of 4.22×10^5 km. From these data, determine the mass of Jupiter.

Solution We start by converting the orbital period to SI units:

$$T = (1.77 \text{ d})(86\ 400 \text{ s} / \text{d}) = 1.53 \times 10^5 \text{ s}$$

The gravitational force of Jupiter is the central force on Io, so $\Sigma F_{Io} = M_{Io}a$:

$$\frac{GM_J M_{Io}}{r^2} = \frac{M_{Io}v^2}{r} = \frac{M_{Io}}{r}\left(\frac{2\pi r}{T}\right)^2 = \frac{4\pi^2 r M_{Io}}{T^2}$$

Thus, $\quad M_J = \frac{4\pi^2 r^3}{GT^2} = \left(\frac{(4\pi^2)(4.22 \times 10^8 \text{ m})^3}{(6.67 \times 10^{-11} \text{ N} \cdot \text{m}^2 / \text{kg}^2)(1.53 \times 10^5 \text{ s})^2}\right)\left(\frac{1 \text{ N}}{\text{kg} \cdot \text{m} / \text{s}^2}\right)$

and $\quad M_J = 1.90 \times 10^{27}$ kg (about 300 times the mass of the earth) ◊

21. A synchronous satellite, which always remains above the same point on a planet's equator, is put in orbit around Jupiter so that scientists can study the famous red spot. Jupiter rotates once every 9.84 h. Use the data of Table 14.2 to find the altitude of the satellite.

Solution Jupiter's rotational period, in seconds, is

$$T = (9.84 \text{ h})(3600 \text{ s} / \text{h}) = 35424 \text{ s}$$

Jupiter's gravitational force is the central force on the satellite:

$$\frac{GM_s M_J}{r^2} = \frac{M_s v^2}{r} = \left(\frac{M_s}{r}\right)\left(\frac{2\pi r}{T}\right)^2 \quad \text{so} \quad GM_J T^2 = 4\pi^2 r^3$$

$$r = \sqrt[3]{\frac{GM_J T^2}{4\pi^2}} = \left(\frac{(6.67\times10^{-11}\ \text{N}\cdot\text{m}^2/\text{kg}^2)(1.90\times10^{27}\ \text{kg})(3.54\times10^4\ \text{s})^2}{4\pi^2}\right)^{1/3}$$

$$r = 15.9\times10^7\ \text{m}$$

Thus, we can calculate the altitude of the synchronous satellite to be

$$\text{Altitude} = (15.98\times10^7\ \text{m}) - (6.99\times10^7\ \text{m}) = 8.92\times10^7\ \text{m} \qquad \Diamond$$

25. Compute the magnitude and direction of the gravitational field at a point P on the perpendicular bisector of two equal masses separated by a distance $2a$ as shown in Figure P14.25.

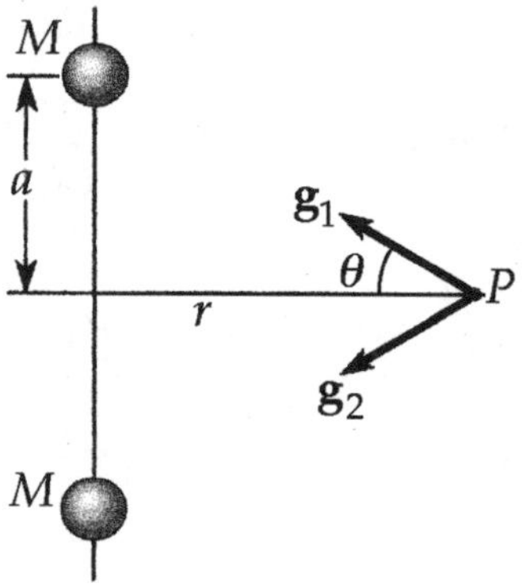

Solution   We must add the vector fields created by each mass. In equation form, $\mathbf{g} = \mathbf{g}_1 + \mathbf{g}_2$ where

$$\mathbf{g}_1 = \frac{GM}{r^2+a^2} \quad \text{to the left and upward at } \theta$$

and $\quad \mathbf{g}_2 = \frac{GM}{r^2+a^2} \quad$ to the left and downward at θ

Figure P14.25
(modified)

Therefore,

$$\mathbf{g} = \frac{GM}{r^2+a^2}\cos\theta(-\mathbf{i}) + \frac{GM}{r^2+a^2}\sin\theta(\mathbf{j}) + \frac{GM}{r^2+a^2}\cos\theta(-\mathbf{i}) + \frac{GM}{r^2+a^2}\sin\theta(-\mathbf{j})$$

$$\mathbf{g} = \frac{2GM}{r^2+a^2}\frac{r}{\sqrt{r^2+a^2}}(-\mathbf{i}) + 0\mathbf{j} = \frac{-2GMr}{(r^2+a^2)^{3/2}}\mathbf{i} \qquad \Diamond$$

29. After our Sun exhausts its nuclear fuel, its ultimate fate may be to collapse to a **white dwarf** state, in which it has approximately the same mass it has now but a radius equal to the radius of the Earth. Calculate (a) the average density of the white dwarf, (b) the acceleration due to gravity at its surface, and (c) the gravitational potential energy associated with a 1.00-kg object at its surface.

Solution

(a) $\rho = \dfrac{M_s}{V} = \dfrac{M_s}{\left(\frac{4}{3}\right)\pi R_E^{\,3}} = \dfrac{1.99\times10^{30}\text{ kg}}{\left(\frac{4}{3}\right)\pi(6.37\times10^6\text{ m})^3} = 1.84\times10^9\text{ kg / m}^3$ ◊

(This is on the order of 1 million times the density of concrete!)

(b) For an object of mass m on its surface, $mg = GM_s m/R_E^{\,2}$. Thus,

$g = \dfrac{GM_s}{R_E^{\,2}} = \dfrac{(6.67\times10^{-11}\text{ N}\cdot\text{m}^2/\text{kg}^2)(1.99\times10^{30}\text{ kg})}{(6.37\times10^6\text{ m})^2} = 3.27\times10^6\text{ m / s}^2$ ◊

(This acceleration is on the order of 1 million times more than g_{Earth})

(c) $U_g = \dfrac{-GM_s m}{R_E} = \dfrac{(-6.67\times10^{-11}\text{ N}\cdot\text{m}^2/\text{kg}^2)(1.99\times10^{30}\text{ kg})(1\text{ kg})}{(6.37\times10^6\text{ m})}$

$U_g = -2.08\times10^{13}\text{ J}$ ◊

(Such a large potential energy could yield a big gain in kinetic energy with even small changes in height. For example, dropping the 1.00-kg object from 1.00 m would result in a final velocity of 2 560 m/s.)

---

37. A spaceship is fired from the Earth's surface with an initial speed of 2.00×10^4 m / s. What will its speed be when it is very far from the Earth? (Neglect friction.)

Solution   Energy is conserved between the surface and the distant point:

$$(K + U_g)_i + W_{nc} = (K + U_g)_f$$

$$\tfrac{1}{2}mv_i^2 - \frac{GM_E m}{R_E} + 0 = \tfrac{1}{2}mv_f^2 - \frac{GM_E m}{\infty}$$

$$v_f^2 = v_i^2 - 2GM_E / R_E \qquad (\text{Note: } 2GM_E / R_E \text{ is simply } v_{escape}^2)$$

$$v_f^2 = \left(2.00 \times 10^4 \text{ m / s}\right)^2 - \frac{2(6.67 \times 10^{-11} \text{ N} \cdot \text{m}^2 / \text{ kg}^2)(5.98 \times 10^{24} \text{ kg})}{(6.37 \times 10^6 \text{ m})}$$

Thus, $v_f^2 = 2.75 \times 10^8 \text{ m}^2 / \text{s}^2$ and $v_f = 1.66 \times 10^4 \text{ m / s}$ ◊

---

39. A "treetop satellite" moves in a circular orbit just above the surface of a planet, which is assumed to offer no air resistance. Show that its orbital speed v and the escape speed from the planet are related by the expression $v_{esc} = \sqrt{2}v$.

Solution   Call M the mass of the planet and R its radius. For the orbiting "treetop satellite," $\Sigma F = ma$ becomes

$$\frac{GMm}{R^2} = \frac{mv^2}{R} \qquad \text{which then yields} \qquad v = \sqrt{\frac{GM}{R}}$$

Applying conservation of energy for an object launched with escape velocity gives

$$\tfrac{1}{2}mv_{esc}^2 - \frac{GMm}{R} = 0 \qquad \text{so that} \qquad v_{esc} = \sqrt{\frac{2GM}{R}} = \sqrt{2}\,v \qquad ◊$$

---

45. A uniform rod of mass M is in the shape of a semicircle of radius R (Fig. P14.45). Calculate the force on a point mass m placed at the center of the semicircle.

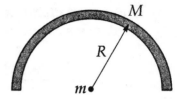

Figure P14.45

Solution

G: If the rod completely encircled the point mass, the net force on m would be zero (by symmetry), and if all the mass M would be concentrated at the middle of the rod, the force would be $F = GmM/R^2$ directed upwards. Since the given configuration is somewhere between these two extreme cases, we can expect the net force on m to be upwards and $F < GmM/R^2$

O: The net force on the point mass can be found by integrating the contributions from each small piece of the semicircular rod.

A: If we consider a segment of the curved rod, dM, to act like a point mass, we can apply Eq. 14.1 to find the force exerted on m due to dM as:

$$d\mathbf{F} = \frac{Gm(dM)}{R^2} \qquad \text{(directed toward } dM\text{)}$$

We must integrate this differential force, to find the total force on the point mass, m, but we first write the differential mass element in terms of a variable that we can integrate easily: the angle, θ, which ranges from $0°$ to $180°$ for this semicircular rod.

One segment of the arc, dM, subtends an angle $d\theta$ and has length $Rd\theta$. Since the whole rod has length πR and mass M, this incremental element has mass

$$dM = \left(\frac{M}{\pi R}\right)R\,d\theta = \frac{M\,d\theta}{\pi}$$

This mass element exerts a force $d\mathbf{F}$ on the point mass at the center.

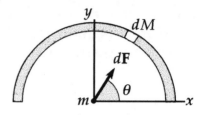

$$dF = \frac{Gm\,dM}{R^2}\hat{\mathbf{r}} = \frac{Gm}{R^2}\left(\frac{M}{\pi}\,d\theta\right)\hat{\mathbf{r}}$$

where $\hat{\mathbf{r}}$ is at an angle θ above the x-axis,

or $\qquad dF = \frac{GmM}{\pi R^2}\,d\theta\,(\cos\theta\,\mathbf{i} + \sin\theta\,\mathbf{j})$

To find the net force on the point mass, we integrate the contributions for all mass elements from $\theta = 0°$ to $\theta = 180°$:

$$\mathbf{F} = \int_{\text{all }m} d\mathbf{F} = \int_{\theta=0}^{180°}\left(\frac{GMm\,d\theta}{\pi R^2}\right)(\cos\theta\,\mathbf{i} + \sin\theta\,\mathbf{j})$$

$$\mathbf{F} = \left(\frac{GMm\,\mathbf{i}}{\pi R^2}\right)\int_0^{180°}\cos\theta\,d\theta + \left(\frac{GMm\,\mathbf{j}}{\pi R^2}\right)\int_0^{180°}\sin\theta\,d\theta$$

$$\mathbf{F} = \left(\frac{GMm\,\mathbf{i}}{\pi R^2}\right)\sin\theta\Big|_0^{180°} + \left(\frac{GMm\,\mathbf{j}}{\pi R^2}\right)[-\cos\theta]_0^{180°}$$

$$\mathbf{F} = \left(\frac{GMm\,\mathbf{i}}{\pi R^2}\right)(0-0) + \left(\frac{GMm\,\mathbf{j}}{\pi R^2}\right)[-(-1)+1]$$

$$\mathbf{F} = 0\mathbf{i} + \left(2G\frac{Mm}{\pi R^2}\right)\mathbf{j} \qquad\qquad \Diamond$$

L: As predicted, the direction of the force on the point mass at the center is vertically upward. Also, the net force has the same algebraic form as for two point masses, reduced by a factor of $2/\pi$, so this answer agrees with our prediction.

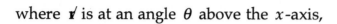

47. A 500-kg uniform solid sphere has a radius of 0.400 m. Find the magnitude of the gravitational force exerted by the sphere on a 50.0-g particle located (a) 1.50 m from the center of the sphere, (b) at the surface of the sphere, and (c) 0.200 m from the center of the sphere.

Solution $F = GmM/r^2$:

(a) $F = \dfrac{\left(6.67 \times 10^{-11} \text{ N} \cdot \text{m}^2 / \text{kg}^2\right)(0.0500 \text{ kg})(500 \text{ kg})}{(1.50 \text{ m})^2}$

Therefore, $\qquad F = 7.41 \times 10^{-10}$ N $\qquad\qquad\qquad \Diamond$

(b) $F = \dfrac{\left(6.67 \times 10^{-11} \text{ N} \cdot \text{m}^2 / \text{kg}^2\right)(0.0500 \text{ kg})(500 \text{ kg})}{(0.400 \text{ m})^2} = 1.04 \times 10^{-8}$ N $\qquad \Diamond$

(c) In this case the mass m is a distance r from a sphere of mass,

$$M = (500 \text{ kg})(0.200 \text{ m}/0.400 \text{ m})^3 = 62.5 \text{ kg}$$

$F = \dfrac{\left(6.67 \times 10^{-11} \text{ N} \cdot \text{m}^2 / \text{kg}^2\right)(0.0500 \text{ kg})(62.5 \text{ kg})}{(0.200 \text{ m})^2} = 5.21 \times 10^{-9}$ N $\qquad \Diamond$

53. A particle of mass m is located inside a uniform solid sphere of radius R and mass M, at a distance r from its center. (a) Show that the gravitational potential energy of the system is $U = (GmM / 2R^3)r^2 - 3GmM / 2R$. (b) Write an expression for the amount of work done by the gravitational force in bringing the particle from the surface of the sphere to its center.

Solution The potential energy at any point can be found from $U = \int F dr$.

(a)   Initially take the particle from ∞ and move it to the sphere's surface.

Then the potential energy is $$U = \int_{\infty}^{R} \left(\frac{GmM}{r^2} \right) dr = -\frac{GmM}{R}$$

Now move it to a position r from the center of the sphere. The force in this case is a function of the mass enclosed by r at any point.

Since the density $\rho = \dfrac{M}{\frac{4}{3}\pi R^3}$, $$U = \int_{R}^{r} \frac{Gm\left(4\pi r^3\right)\rho}{3r^2} dr = \frac{GmM}{R^3}\left(\frac{r^2 - R^2}{2} \right)$$

The gravitational energy is $$U = \frac{GmM}{2R^3}\left(r^2 - 3R^2\right) = \left(\frac{GmM}{2R^3}\right)r^2 - \frac{3GmM}{2R} \qquad ◊$$

(b)   At the surface and center, $U(R) = -\dfrac{GMm}{R}$ and $U(0) = -\dfrac{3GMm}{2R}$

so the work done is $$W_g = -\left[U(0) - U(R)\right] = \frac{GMm}{2R} \qquad ◊$$

---

57.   In introductory physics laboratories, a typical Cavendish balance for measuring the gravitational constant G uses lead spheres with masses of 1.50 kg and 15.0 g whose centers are separated by about 4.50 cm. Calculate the gravitational force between these spheres, treating each as a point mass located at the center of the sphere.

Solution   The force between the spheres is given by $F_g = GmM/r^2$. Substituting in values of $m = 0.0150$ kg, $M = 1.50$ kg, and $r = 0.0450$ m,

$$F_g = \frac{\left(6.673 \times 10^{-11}\ \text{N} \cdot \text{m}^2/\text{kg}^2\right)\left(0.0150\ \text{kg}\right)\left(1.50\ \text{kg}\right)}{\left(0.0450\ \text{m}\right)^2} = 7.41 \times 10^{-10}\ \text{N} \qquad ◊$$

---

61. Two hypothetical planets of masses m_1 and m_2 and radii r_1 and r_2, respectively, are nearly at rest when they are an infinite distance apart. Because of their gravitational attraction, they head toward each other on a collision course. (a) When their center-to-center separation is d, find expressions for the speed of each planet and their **relative**

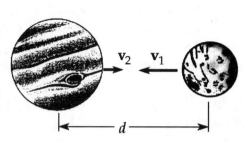

velocity. (b) Find the kinetic energy of each planet **just** before they collide, if $m_1 = 2.00 \times 10^{24}$ kg, $m_2 = 8.00 \times 10^{24}$ kg, $r_1 = 3.00 \times 10^6$ m, and $r_2 = 5.00 \times 10^6$ m. (**Hint:** Both energy and momentum are conserved.)

Solution (a) At infinite separation, $U = 0$; and at rest, $K = 0$. Since energy is conserved, we have

$$0 = \tfrac{1}{2}m_1 v_1^2 + \tfrac{1}{2}m_2 v_2^2 - \frac{Gm_1 m_2}{d} \qquad (1)$$

The initial momentum is zero and momentum is conserved. Therefore,

$$0 = m_1 v_1 - m_2 v_2 \qquad (2)$$

Combine (1) and (2) to find $\quad v_1 = m_2 \sqrt{\dfrac{2G}{d(m_1 + m_2)}} \quad$ and $\quad v_2 = m_1 \sqrt{\dfrac{2G}{d(m_1 + m_2)}}$

The relative velocity is $\quad v_r = v_1 - (-v_2) = \sqrt{\dfrac{2G(m_1 + m_2)}{d}} \qquad \Diamond$

(b) Substitute the given numerical values into the equation found for v_1 and v_2 in part (a) to find $v_1 = 1.03 \times 10^4$ m/s and $v_2 = 2.58 \times 10^3$ m/s.

Therefore, $\qquad K_1 = \tfrac{1}{2}m_1 v_1^2 = 1.07 \times 10^{32}$ J $\qquad \Diamond$

and $\qquad K_2 = \tfrac{1}{2}m_2 v_2^2 = 2.67 \times 10^{31}$ J $\qquad \Diamond$

63. A sphere of mass M and radius R has a nonuniform density that varies with r, the distance from its center, according to the expression $\rho = Ar$, for $0 \leq r \leq R$. (a) What is the constant A in terms of M and R? (b) Determine an expression for the force exerted on a particle of mass m placed outside the sphere. (c) Determine an expression for the force exerted on the particle if it is inside the sphere. (**Hint:** See Section 14.10 and note that the distribution is spherically symmetric.)

Solution

(a) If we consider a hollow shell in the sphere with radius r and thickness dr, then

$$dM = \rho \, dV = \rho \left(4\pi r^2 dr \right)$$

The total mass is thus
$$M = \int_0^R \rho \, dV = \int_0^R (Ar)\left(4\pi r^2 dr \right) = \pi A R^4$$

and the constant A is
$$A = \frac{M}{\pi R^4} \qquad\qquad \diamond$$

(b) The total mass of the sphere acts as if it were at the center of the sphere and $F = GmM / r^2$ directed toward the center of the sphere.

(c) Inside the sphere at a distance r from the center, where dM is just the mass of a shell enclosed within the radius r,

$$dF = \left(Gm / r^2 \right) dM$$

$$F = \frac{Gm}{r^2} \int_0^r dM = \frac{Gm}{r^2} \int_0^r Ar \left(4\pi r^2 \right) dr = \frac{Gm}{r^2} \frac{M 4\pi}{\pi R^4} \frac{r^4}{4} = \frac{GmMr^2}{R^4} \qquad \diamond$$

69. Two stars of masses M and m, separated by a distance d, revolve in circular orbits about their center of mass (Fig. P14.69). Show that each star has a period given by

$$T^2 = \frac{4\pi^2 d^3}{G(M+m)}$$

(**Hint:** Apply Newton's second law to each star, and note that the center-of-mass condition requires that $Mr_2 = mr_1$, where $r_1 + r_2 = d$)

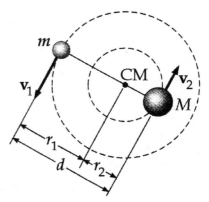

Figure P14.69

Solution   For the star of mass M and orbital radius r_2,

$\Sigma F = ma$ gives

$$\frac{GMm}{d^2} = \frac{Mv_2^2}{r_2} = \frac{M}{r_2}\left(\frac{2\pi r_2}{T}\right)^2$$

For the star of mass m, $\Sigma F = ma$ gives

$$\frac{GMm}{d^2} = \frac{mv_1^2}{r_1} = \frac{m}{r_1}\left(\frac{2\pi r_1}{T}\right)^2$$

Cross-multiplying, we then obtain

$$GmT^2 = 4\pi^2 d^2 r_2$$

and simultaneously,

$$GMT^2 = 4\pi^2 d^2 r_1$$

Adding the two equations, we find

$$G(M+m)T^2 = 4\pi^2 d^2(r_1 + r_2) = 4\pi^2 d^3$$

$$T^2 = \frac{4\pi^2 d^3}{G(M+m)}$$

In a visual binary star system T, d, r_1, and r_2 can be measured, so the mass of each component can be computed.

Chapter 15

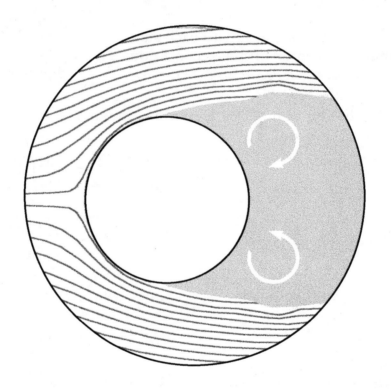

Fluid Mechanics

FLUID MECHANICS

INTRODUCTION

In this chapter we first consider a fluid at rest and derive an expression for the pressure exerted by the fluid as a function of its density and depth. We then treat fluids in motion, an area of study called **fluid dynamics**. We use a model based on simplified assumptions to analyze some situations of practical importance. An underlying principle known as the **Bernoulli principle** enables us to determine relationships among the pressure, density, and velocity at every point in a fluid. We conclude the chapter with a brief discussion of internal friction in a fluid and turbulent motion.

EQUATIONS AND CONCEPTS

The **density** of a homogeneous substance is defined as its ratio of mass per unit volume. The value of density is characteristic of a particular type of material and independent of the total quantity of material in the sample.

$$\rho \equiv \frac{m}{V}$$

The SI units of density are kg per cubic meter.

$$1\,\text{g/cm}^3 = 1000\,\text{kg/m}^3$$

The (average) pressure of a fluid is defined as the normal force per unit area acting on a surface immersed in the fluid.

$$P \equiv \frac{F}{A} \tag{15.1}$$

Atmospheric pressure is often expressed in other units: atmospheres, mm of mercury (Torr), or pounds per square inch.

$$1 \, atm = 1.013 \times 10^5 \, Pa$$

$$1 \, Torr = 133.3 \, Pa$$

$$1 \, lb \, / \, in^2 = 6895 \, Pa$$

The SI units of pressure are newtons per square meter, or Pascal (Pa).

$$1 \, Pa \equiv 1 \, N/m^2 \qquad (15.3)$$

The absolute pressure, P, at a depth, h, below the surface of a liquid which is open to the atmosphere is greater than atmospheric pressure, P_0, by an amount which depends on the depth below the surface.

$$P = P_0 + \rho g h \qquad (15.4)$$

$$P_0 = 1.013 \times 10^5 \, Pa = 1 \, atm$$

The quantity $P_g = \rho g h$ is called the gauge pressure and P is the absolute pressure. Therefore,

$$P = P_0 + P_g$$

Pascal's law states that a change in the pressure applied to an enclosed fluid (liquid or gas) is transmitted undiminished to every point within the fluid and over the walls of the vessel which contain the fluid.

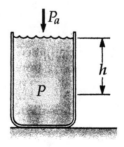

324

Archimedes' principle states that when an object is partially or fully submersed in a fluid, the fluid exerts an upward buoyant force on the object. The magnitude of the buoyant force depends on the density of the fluid and the volume of displaced fluid, V. In particular, note that B equals the weight of the displaced fluid.

Archimedes' principle.

$$B = \rho_{fluid}Vg \qquad (15.5)$$

Fluid dynamics, the treatment of fluids in motion, is greatly simplified under the assumption that the fluid and its flow are ideal, with the following four characteristics:

- **nonviscous** - internal friction between adjacent fluid layers is negligible

- **incompressible** - the density of the fluid is constant throughout the fluid

- **steady** - the velocity, density, and pressure at each point in the fluid are constant in time

- **irrotational** (nonturbulent) - there are no eddy currents within the fluid (each element of the fluid has zero angular velocity about its center)

The equation of continuity is really just a statement of conservation of mass: **the mass that flows into a tube equals the mass that flows out.**

$$\rho A_1 v_1 = \rho A_2 v_2$$

For an **incompressible fluid** ($\rho = $ constant), the equation of continuity can be written as Equation 15.7. The product Av is called the flow rate. Thus, the flow rate of an incompressible fluid at any point along a pipe is constant.

$$A_1 v_1 = A_2 v_2 = \text{constant} \qquad (15.7)$$

Bernoulli's equation is the most fundamental law in fluid mechanics. The equation is a statement of the law of conservation of mechanical energy as applied to a fluid. Bernoulli's equation states that the sum of pressure, kinetic energy per unit volume, and potential energy per unit volume remains constant along a streamline of an ideal fluid.

$$P + \tfrac{1}{2}\rho v^2 + \rho g y = \text{constant} \qquad (15.9)$$

REVIEW CHECKLIST

▷ Understand the concept of pressure at a point in a fluid, and the variation of pressure with depth. Understand the relationships among absolute, gauge, and atmospheric pressure values; and know the several different units commonly used to express pressure.

▷ Understand the origin of buoyant forces; and state and explain Archimedes' principle.

▷ State the simplifying assumptions of an ideal fluid moving with streamline flow.

▷ State and understand the physical significance of the **equation of continuity** (constant flow rate) and **Bernoulli's equation** for fluid flow (relating **flow velocity, pressure,** and **pipe elevation).**

ANSWERS TO SELECTED CONCEPTUAL QUESTIONS

1. Two drinking glasses of the same weight but of different shape and different cross-sectional area are filled to the same level with water. According to the expression $P = P_0 + \rho g h$, the pressure at the bottom of both glasses is the same. In view of this, why does one glass weigh more than the other?

Answer The hydrostatic pressure represents the force per unit area which would be exerted on any object placed at that depth. The weight of each tumbler depends on the density of the fluid, the volume of the fluid, and the weight of the empty tumbler.

□   □   ·   □   □

5. A fish rests on the bottom of a bucket of water while the bucket is being weighed. When the fish begins to swim around, does the weight change?

Answer In either case, the scale is supporting the container, the water, and the fish. Therefore the weight will remains the same. The reading on the scale, however, can change if the net center of mass accelerates in the vertical direction, as when the fish jumps out of the water. In that case, the scale, which registers force, will also show an additional force caused by Newton's law, $F = ma$.

□   □   □   □

8. The water supply for a city is often provided by reservoirs built on high ground. Water flows from the reservoir, through pipes, and into your home when you turn the tap on your faucet. Why is the flow of water more rapid out of a faucet on the first floor of a building than it is in an apartment on a higher floor?

Answer   The water supplied to the building flows through a pipe connected to the water tower. Near the earth, the water pressure is greater because the pressure increases with increasing depth beneath the surface of the water. The penthouse apartment is not as far below the water surface, hence the water flow will not be as rapid as in a lower floor.

☐   ☐   ☐   ☐

23.   An unopened can of diet cola floats when placed in a tank of water, whereas a can of regular cola of the same brand sinks in the tank. What do you suppose could explain this phenomenon?

Answer   Ultimately, only Archimedes' Principle can explain this; we can be sure that the total density of the diet cola is less than the density of water, while the total density of the regular cola is greater than the density of water. This could be due to several things, such as   the presence of more water, more carbonation, a lighter sweetener, or less cola in the diet version.   It could be due to the diet cola having a different internal pressure, causing the cans' sides to bow out more. It could even be due to the existence of an air bubble **underneath** the floating diet can. If you look, most cans have ridged bottoms, and could hold an air bubble there.

☐   ☐   ☐   ☐

SOLUTIONS TO SELECTED END-OF-CHAPTER PROBLEMS

1.   Calculate the mass of a solid iron sphere that has a diameter of 3.00 cm.

Solution   The definition of density $\rho = m/V$ is often written as $m = \rho V$.

Here $V = \frac{4}{3} \pi r^3$,   so $m = \rho V = \rho \left(\frac{4}{3} \pi (d/2)^3 \right)$

Thus, $m = \left(7.86 \times 10^3 \text{ kg / m}^3\right)\left(\frac{4}{3}\right)\left((1.50 \text{ cm})^3\right)\left(10^{-6} \text{ m}^3 / \text{ cm}^3\right) = 111 \text{ g}$ ◊

3. A 50.0-kg woman balances on one heel of a pair of high-heel shoes. If the heel is circular with radius 0.500 cm, what pressure does she exert on the floor?

Solution

The area of the circular base of the heel is

$$\pi r^2 = \pi(0.500 \text{ cm})^2\left(\frac{1 \text{ m}^2}{10\ 000 \text{ cm}^2}\right) = 7.85 \times 10^{-5} \text{ m}^2$$

The force she exerts is her weight, $mg = (50.0 \text{ kg})(9.80 \text{ m / s}^2) = 490 \text{ N}$.

Then $\qquad P = \dfrac{F}{A} = \dfrac{490 \text{ N}}{7.85 \times 10^{-5} \text{ m}^2} = 6.24 \times 10^6 \text{ Pa}$ ◊

7. The spring of the pressure gauge shown in Figure 15.2 has a force constant of 1000 N/m, and the piston has a diameter of 2.00 cm. When the gauge is lowered into the water, at what depth does the piston move in by 0.500 cm?.

Solution

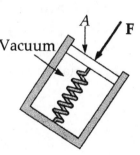

$$F_{\text{spring}} = F_{\text{fluid}} \quad \text{or} \quad kx = \rho g h A$$

Figure 15.2

and $\qquad h = \dfrac{kx}{\rho g A} = \dfrac{(1000 \text{ N / m}^2)(0.00500 \text{ m})}{(10^3 \text{ kg / m}^3)(9.80 \text{ m / s}^2)\pi(0.0100 \text{ m})^2}$

yields $\qquad h = 1.62 \text{ m}$ ◊

9. What must be the contact area between a suction cup (completely exhausted) and a ceiling if the cup is to support the weight of an 80.0-kg student?

Solution

G: The suction cups used by burglars seen in movies are about 10 cm in diameter, and it seems reasonable that one of these might be able to support the weight of an 80-kg student. The face area of a 10-cm cup is approximately:

$$A = \pi r^2 \approx 3(0.05 \text{ m})^2 \approx 0.008 \text{ m}^2$$

O: "Suction" is not a new kind of force. Familiar forces hold the cup in equilibrium, one of which is the atmospheric pressure acting over the area of the cup. This problem is simply another application of Newton's 2nd law.

A: The vacuum between cup and ceiling exerts no force on either. The atmospheric pressure of the air below the cup pushes up on it with a force $(P_{atm})(A)$. If the cup barely supports the student's weight, then the normal force of the ceiling is approximately zero, and

$$\Sigma F_y = 0 + (P_{atm})(A) - mg = 0$$

$$A = \frac{mg}{P_{atm}} = \frac{784 \text{ N}}{1.013 \times 10^5 \text{ N / m}^2} = 7.74 \times 10^{-3} \text{ m}^2 \qquad \Diamond$$

L: This calculated area agrees with our prediction and corresponds to a suction cup that is 9.93 cm in diameter (Our 10 cm estimate was right on – a lucky guess considering that a burglar would probably use at least two suction cups, not one.) As an aside, the suction cup we have drawn, above, appears to be about 30 cm in diameter, plenty big enough to support the weight of the student.

17. Blaise Pascal duplicated Torricelli's barometer, using a red Bordeaux wine, of density 984 kg/m³, as the working liquid (Fig. P15.17). What was the height h of the wine column for normal atmospheric pressure? Would you expect the vacuum above the column to be as good as that for mercury?

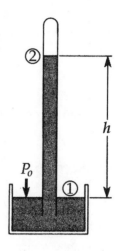

Solution

In Bernoulli's equation,

$$P_1 + \tfrac{1}{2}\rho v_1^2 + \rho g y_1 = P_2 + \tfrac{1}{2}\rho v_2^2 + \rho g y_2$$

Take point 1 at the wine surface in the pan, where $P_1 = P_0$, and point 2 at the wine surface up in the tube. Here we approximate $P_2 = 0$, although some alcohol and water will evaporate.

Figure P15.17
(modified)

Thus, the vacuum is not as good as with mercury. Unless you are careful, a lot of dissolved oxygen or carbon dioxide may come bubbling out. ◊

Now $\qquad P_1 = P_2 + \rho g(y_2 - y_1).$

$$1 \text{ atm} = 0 + (984 \text{ kg/m}^3)(9.80 \text{ m/s}^2)(y_2 - y_1)$$

$$y_2 - y_1 = \frac{1.013 \times 10^5 \text{ N/m}^2}{9640 \text{ N/m}^3} = 10.5 \text{ m} \qquad\qquad ◊$$

A water barometer in a stairway of a three-story building is a nice display. Red wine makes the fluid level easier to see. On television, Mister Wizard uses grape juice.

25. A cube of wood having side dimensions of 20.0 cm and a density of $650 \text{ kg}/\text{m}^3$ floats on water. (a) What is the distance from the horizontal top surface of the cube to the water level? (b) How much lead weight must be placed on top of the cube so that its top is just level with the water?

Solution Set h equal to the distance from the top of the cube to the water level.

(a) According to Archimedes' principle,

$$B = \rho_w V g = \left(1.00 \text{ g}/\text{cm}^3\right)\left[(20.0 \text{ cm})^2(20.0 \text{ cm} - h)\right]g$$

But B is also equal to the weight of the block, so

$$B = Mg = \rho_{wood}V_{wood}g = \left(0.650 \text{ g}/\text{cm}^3\right)(20.0 \text{ cm})^3 g$$

Setting these two equations equal,

$$\left(0.650 \text{ g}/\text{cm}^3\right)(20.0 \text{ cm})^3 g = \left(1.00 \text{ g}/\text{cm}^3\right)(20.0 \text{ cm})^2(20.0 \text{ cm} - h)g$$

$$h = (20.0 \text{ cm})\left(1 - \frac{0.650 \text{ g}/\text{cm}^3}{1.00 \text{ g}/\text{cm}^3}\right) = 7.00 \text{ cm} \qquad \Diamond$$

(b) $B = w + Mg$ where $M = $ mass of lead

$$\left(1 \text{ g}/\text{cm}^3\right)(20.0 \text{ cm})^3 g = \left(0.650 \text{ g}/\text{cm}^3\right)(20.0 \text{ cm})^3 g + Mg$$

$$M = (20.0 \text{ cm})^3\left(1 \text{ g}/\text{cm}^3 - 0.650 \text{ g}/\text{cm}^3\right) = (20.0 \text{ cm})\left(0.350 \text{ g}/\text{cm}^3\right)$$

$$M = 2800 \text{ g} = 2.80 \text{ kg} \qquad \Diamond$$

27. A plastic sphere floats in water with 50.0% of its volume submerged. This   same sphere floats in oil with 40.0% of its volume submerged. Determine the densities of the oil and the sphere.

Solution

The forces on the ball are its weight

$$F_g = mg = \rho_{plastic}V_{ball}g$$

and the buoyant force of the liquid

$$B = \rho_{fluid}V_{immersed}g$$

When floating in water, $\Sigma F_y = 0$:

$$-\rho_{plastic}V_{ball}g + \rho_{water}(0.500V_{ball})g = 0$$

$$\rho_{plastic} = 0.500\rho_{water} = 500 \text{ kg / m}^3 \qquad \lozenge$$

When floating in oil, $\Sigma F_y = 0$:

$$-\rho_{plastic}V_{ball}g + \rho_{oil}(0.400V_{ball})g = 0$$

$$\rho_{plastic} = 0.400\rho_{oil}$$

$$\rho_{oil} = \frac{500 \text{ kg / m}^3}{0.400} = 1250 \text{ kg / m}^3 \qquad \lozenge$$

This oil would sink in water.

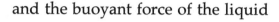

29. How many cubic meters of helium are required to lift a balloon with a 400-kg payload to a height of 8000 m? (Take $\rho_{He} = 0.180 \text{ kg} / \text{m}^3$.) Assume that the balloon maintains a constant volume and that the density of air decreases with altitude z according to the expression $\rho_{air} = \rho_0 e^{-z/8000}$, where z is in meters, and $\rho_0 = 1.25 \text{ kg}/\text{m}^3$ is the density of air at sea level.

Solution

At $z = 8000$ m, the density of air is

$$\rho_{air} = \rho_0 e^{-z/8000} = \left(1.25 \text{ kg} / \text{m}^3\right) e^{-1} = \left(1.25 \text{ kg} / \text{m}^3\right)(0.368)$$

$$\rho_{air} = 0.460 \text{ kg} / \text{m}^3$$

Think of the balloon reaching equilibrium at this height. The weight of the helium in the balloon, and the balloon's payload, respectively, are

$$F_{g,He} = \rho_{He}Vg \quad \text{and} \quad F_{g,\,load} = (400 \text{ kg})(9.80 \text{ m} / \text{s}^2) = 3920 \text{ N}$$

$\Sigma F_y = 0$ gives $\qquad +\rho_{air}Vg - 3920 \text{ N} - \rho_{He}Vg = 0$

Solving, $\qquad (\rho_{air} - \rho_{He})Vg = 3920 \text{ N}$

and $\qquad V = \dfrac{400 \text{ kg}}{\rho_{air} - \rho_{He}} = \dfrac{400 \text{ kg}}{(0.460 - 0.180) \text{ kg} / \text{m}^3}$

Solving, $\qquad V = 1.43 \times 10^3 \text{m}^3$ $\hfill \Diamond$

35. A large storage tank, open at the top and filled with water, develops a small hole in its side at a point 16.0 m below the water level. If the rate of flow from the leak is 2.50×10^{-3} m^3 / min, determine (a) the speed at which the water leaves the hole and (b) the diameter of the hole.

Solution

Assuming the top is open to the atmosphere, then $P_1 = P_0$.

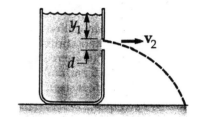

(a) $P_1 + \frac{1}{2}\rho v_1^2 + \rho g y_1 = P_2 + \frac{1}{2}\rho v_2^2 + \rho g y_2$

   $A_1 >> A_2$, so $v_1 << v_2$

   Assuming $v_1 \approx 0$ and $P_1 = P_2 = P_0$,

   $v_2 = \sqrt{2 g y_1} = \sqrt{2(9.80 \text{ m / s}^2)(16 \text{ m})} = 17.7 \text{ m / s}$ ◊

(b) From the flow rate, we find that

$$A_2 v_2 = 2.50 \times 10^{-3} \text{ m}^3 / \text{ min}$$

$$\left(\frac{\pi d^2}{4}\right)(17.7 \text{ m / s})(60 \text{ s / min}) = 2.50 \times 10^3 \text{ m}^3 / \text{ min}$$

Thus, $d = 1.73 \times 10^{-3} \text{ m} = 1.73 \text{ mm}$ ◊

37. Water flows through a fire hose of diameter 6.35 cm at a rate of 0.0120 m^3/s. The fire hose ends in a nozzle with an inner diameter of 2.20 cm. What is the speed at which the water exits the nozzle?

Solution   Take point 1 inside the hose and point 2 at the surface of the water stream leaving the nozzle. The volume flow rate is constant:

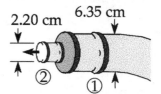

$$0.0120 \text{ m}^3/\text{s} = A_1 v_1 = A_2 v_2$$

$$v_1 = \frac{0.0120 \text{ m}^3/\text{s}}{\pi (6.35 \times 10^{-2} \text{ m}/2)^2} = 3.79 \text{ m/s}$$

and

$$v_2 = \frac{0.0120 \text{ m}^3/\text{s}}{\pi (1.10 \times 10^{-2} \text{ m})^2} = 31.6 \text{ m/s} \qquad ◊$$

Related Calculation   Find the pressure at both points.

At point 2, the water pressure is 101.3 kPa. The air exerts this much force per unit area on the water. By Newton's third law, the water exerts the same force on the air. If we have a nonviscous incompressible fluid in steady nonturbulent flow, and the height at both points is the same,

$$P_1 + \tfrac{1}{2}\rho v_1^{\,2} + \rho g h_1 = P_2 + \tfrac{1}{2}\rho v_2^{\,2} + \rho g h_2$$

$$P_1 = P_2 + \tfrac{1}{2}\rho\left(v_2^{\,2} - v_1^{\,2}\right)$$

$$P_1 = 1.01 \times 10^5 \ \frac{\text{N}}{\text{m}^2} + \tfrac{1}{2}\left(1000 \ \frac{\text{kg}}{\text{m}^3}\right)\left(\left(31.6 \ \frac{\text{m}}{\text{s}}\right)^2 - \left(3.79 \ \frac{\text{m}}{\text{s}}\right)^2\right)$$

$$P_1 = 101 \text{ kPa} + 492 \text{ kPa} = 593 \text{ kPa} \qquad ◊$$

45. A large storage tank is filled to a height h_0. The tank is punctured at a height h above the bottom of the tank (Fig. P15.45). Find an expression for how far from the tank the exiting stream lands.

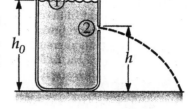

Figure P15.45
(modified)

Solution Take point 1 at the top liquid surface. Since the tank is large, the fluid level falls only slowly and $v_1 \cong 0$. Take point 2 at the surface of the water stream leaving the hole. At both points the pressure is one atmosphere because the water can push no more or less strongly than the air pushes on it, as described by Newton's third law.

$$P_1 + \tfrac{1}{2}\rho v_1{}^2 + \rho g y_1 = P_2 + \tfrac{1}{2}\rho v_2{}^2 + \rho g y_2$$

$$P_0 + 0 + \rho g h_0 = P_0 + \tfrac{1}{2}\rho v_2{}^2 + \rho g h$$

$$v_2 = \sqrt{2g(h_0 - h)}$$

Now each drop of water moves as a projectile. Its velocity v_2, which we now call its original velocity, has zero vertical component. Its time of fall is given by $y = v_{y0}t + \tfrac{1}{2}a_y t^2$:

$$-h = 0 - \tfrac{1}{2}gt^2 \qquad \text{and} \qquad t = \sqrt{\frac{2h}{g}}$$

and its horizontal displacement is

$$x = v_{x0}t + \tfrac{1}{2}a_x t^2 = \sqrt{2g(h_0 - h)}\sqrt{2h/g} + 0 = \sqrt{4h(h_0 - h)} \qquad \lozenge$$

47. A Ping-Pong ball has a diameter of 3.80 cm and average density of 0.0840 g/cm^3. What force would be required to hold it completely submerged under water?

Solution

G: According to Archimedes' Principle, the buoyant force acting on the submerged ball will be equal to the weight of the water the ball will displace. The ball has a volume of about 30 cm^3, so the weight of this water is approximately:

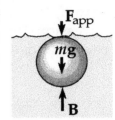

$$B = F_g = \rho V g \approx \left(1\text{ g / cm}^3\right)\left(30\text{ cm}^3\right)\left(10\text{ m / s}^2\right) = 0.3\text{ N}$$

Since the ball is much less dense than the water, the applied force will approximately equal this buoyant force.

O: Apply Newton's 2nd law to find the applied force.

A: At equilibrium, $\Sigma F = 0$ or $-F_{app} - mg + B = 0$

Where the buoyant force is $B = \rho_w V g$ and $\rho_w = 1000\text{ kg / m}^3$

The applied force is then $F_{app} = \rho_w V g - mg$

Using $m = \rho_{ball} V$ to eliminate the unknown mass of the ball, this becomes

$$F_{app} = Vg\left(\rho_w - \rho_{ball}\right) = \frac{4}{3}\pi r^3 g\left(\rho_w - \rho_{ball}\right)$$

$$F_{app} = \frac{4}{3}\pi\left(1.90\times10^{-2}\text{ m}\right)^3\left(9.80\text{ m / s}^2\right)\left(1000\text{ kg / m}^3 - 84\text{ kg / m}^3\right)$$

$$F_{app} = 0.258\text{ N} \qquad\qquad ◊$$

L: The force is approximately what we expected, so our result is reasonable. If a force greater than 0.258 N were to be applied, the ball would accelerate down until it hit the bottom (which would then provide a normal force directed upwards).

51. The true weight of an object is measured in a vacuum, where buoyant forces are absent. A body of volume V is weighed in air on a balance with the use of weights of density ρ. If the density of air is ρ_{air} and the balance reads F'_g, show that the true weight F_g is

$$F_g = F'_g + \left(V - \frac{F'_g}{\rho g} \right) \rho_{air} g$$

Solution The "balanced" condition is one in which the net torque on the balance is zero. Since the balance has lever arms of equal length, the total force on each pan is equal. Applying $\Sigma\tau = 0$ around the pivot gives us this in equation form:

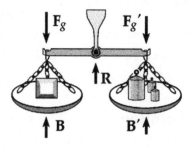

$$F_g - B = F'_g - B'$$

where B and B' are the buoyant forces on the body and weights respectively. The buoyant force experienced by an object of volume V in air is $B = V\rho_{air}g$. So for the test mass and for the weights, respectively,

$$B = V\rho_{air}g \qquad \text{and} \qquad B' = V'\rho_{air}g$$

Since the volume of the weights is not given explicitly, we must use the density equation to eliminate it:

$$V' = \frac{m'}{\rho} = \frac{m'g}{\rho g} = \frac{F'_g}{\rho g}$$

With this substitution, the buoyant force on the weights is $B' = \left(F'_g / \rho g \right) \rho_{air} g$

Therefore, $\qquad\qquad F_g = F'_g + \left(V - \frac{F'_g}{\rho g} \right) \rho_{air} g \qquad\qquad\qquad \lozenge$

Related   We can now answer the popular riddle: Which weighs more, a
Comment: pound of feathers or a pound of bricks? Like in the problem
above, the feathers have a greater buoyant force than the bricks,
so if they "weigh" the same on a scale as a pound of bricks, then
the feathers must have more mass and therefore a greater "true
weight."

57. With reference to Figure 15.7, show that the total torque exerted by the water behind the dam about an axis through O is $\frac{1}{6}\rho gwH^3$. Show that the effective line of action of the total force exerted by the water is at a distance $\frac{1}{3}H$ above O.

Solution   The torque is $\tau = \int d\tau = \int r\,dF$.

From Figure 15.5, we have

$$\tau = \int_0^H y\big[\rho g(H-y)w\big]dy = \frac{1}{6}\rho gwH^3$$

To calculate the total force, $dF = P\,dA$

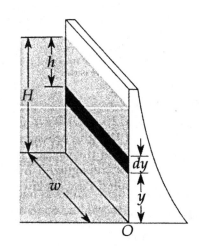

Figure 15.7

where $P = \rho gy$ and $dA = wdy$

$$F = \int_0^H dF = \int_0^H (\rho gy)w\,dy$$

$$F = \rho gw\int_0^H y\,dy = \frac{1}{2}\rho gwy^2\Big]_0^H = \frac{1}{2}\rho gwH^2$$

If this were applied at a height y_{eff} such that the torque remains unchanged,

$$\frac{1}{6}\rho gwH^3 = y_{eff}\Big[\frac{1}{2}\rho gwH^2\Big] \quad \text{and} \quad y_{eff} = \frac{1}{3}H \qquad \diamond$$

340

59. In 1983 the United States began coining the cent piece out of copper-clad zinc rather than pure copper. The mass of the old copper cent is 3.083 g, whereas that of the new cent is 2.517 g. Calculate the percent of zinc (by volume) in the new cent. The density of copper is 8.960 g/cm³ and that of zinc is 7.133 g/cm³. The new and old coins have the same volume.

Solution

Let f represent the fraction of the volume V occupied by zinc in the new coin. We have $m = \rho V$ for both coins:

$$3.083 \text{ g} = (8.960 \text{ g/cm}^3)V$$

$$2.517 \text{ g} = (7.133 \text{ g/cm}^3)(fV) + (8.960 \text{ g/cm}^3)(1-f)V$$

By substitution,

$$2.517 \text{ g} = (7.133 \text{ g/cm}^3)fV + 3.083 \text{ g} - (8.960 \text{ g/cm}^3)fV$$

$$fV = \frac{3.083 \text{ g} - 2.517 \text{ g}}{8.960 \text{ g/cm}^3 - 7.133 \text{ g/cm}^3}$$

$$f = \frac{0.566 \text{ g}}{1.827 \text{ g/cm}^3}\left(\frac{8.960 \text{ g/cm}^3}{3.083 \text{ g}}\right) = 0.9004 = 90.04\% \qquad \lozenge$$

Wave Motion

WAVE MOTION

INTRODUCTION

In this chapter we are going to study the properties of mechanical waves. In the case of mechanical waves, what we interpret as a wave corresponds to the disturbance of a body or medium through which the wave travels. Therefore, we can consider a wave to be the **motion of a disturbance.**

The mathematics used to describe wave phenomena is common to all waves. In general, we shall find that mechanical wave motion is described by specifying the positions of all points of the disturbed medium as a function of time.

The mechanical waves discussed in this chapter require (1) some source of disturbance, (2) a medium that can be disturbed, and (3) some physical connection through which adjacent portions of the medium can influence each other. We shall find that all waves carry energy. The amount of energy transmitted through a medium and the mechanism responsible for that transport of energy differ from case to case.

EQUATIONS AND CONCEPTS

The **wave speed** or **phase velocity** is in general the rate at which the profile of the disturbance moves along the direction of travel (e.g. the x axis).

$$v = \frac{dx}{dt} \qquad (16.3)$$

In the special case of a transverse pulse moving along a stretched string, the wave speed depends on the tension in the string T and the linear density μ of the string (mass per unit length).

$$v = \sqrt{\frac{T}{\mu}} \qquad (16.4)$$

For a sinusoidal wave, the displacement repeats itself when x is increased by an integral multiple of λ and the wave moves to the right a distance of vt in a time t.

$$y = A\sin\left[\frac{2\pi}{\lambda}(x - vt)\right] \qquad (16.6)$$

It is convenient to define three additional characteristic wave quantities: **the wave number k**, the **angular frequency** ω, and the **harmonic frequency** f.

$$k \equiv \frac{2\pi}{\lambda} \qquad (16.9)$$

$$\omega \equiv \frac{2\pi}{T} \qquad (16.10)$$

$$f = \frac{1}{T} \qquad (16.12)$$

The expression for the **w a v e function** can be written in a more compact form in terms of the parameters defined above. If the transverse displacement is not zero at $x = 0$ and $t = 0$, it is necessary to include a phase constant, ϕ.

$$y = A\sin(kx - \omega t) \qquad (16.11)$$

$$y = A\sin(kx - \omega t + \phi) \qquad (16.15)$$

The **wave speed** v or phase velocity can also be expressed in alternative forms.

$$v = \frac{\omega}{k} \qquad (16.13)$$

$$v = \lambda f \qquad (16.14)$$

The **transverse velocity** v_y of a point on a harmonic wave is out of phase with the **transverse acceleration** a_y of that point by $\pi / 2$ radians.

$$v_y = -\omega A \cos(kx - \omega t) \qquad (16.16)$$

$$a_y = -\omega^2 A \sin(kx - \omega t) \qquad (16.17)$$

The **power** transmitted by any harmonic wave is proportional to the square of the frequency and the square of the amplitude, where μ is the mass per unit length of the string.

$$\mathcal{P} = \tfrac{1}{2}\mu\omega^2 A^2 v \qquad (16.21)$$

Any wave function having the form $y = f(x \pm vt)$ satisfies the **linear wave equation**.

$$\frac{\partial^2 y}{\partial x^2} = \frac{1}{v^2}\frac{\partial^2 y}{\partial t^2} \qquad (16.26)$$

REVIEW CHECKLIST

▷   Recognize whether or not a given function is a possible description of a traveling wave.

▷   Express a given harmonic wave function in several alternative forms involving different combinations of the wave parameters: wavelength, period, phase velocity, wave number, angular frequency, and harmonic frequency.

▷   Given a specific wave function for a harmonic wave, obtain values for the characteristic wave parameters: A, ω, k, λ, f, and ϕ.

▷   Calculate the rate at which energy is transported by harmonic waves in a string.

ANSWERS TO SELECTED CONCEPTUAL QUESTIONS

2. How would you set up a longitudinal wave in a stretched spring? Would it be possible to set up a transverse wave in a spring?

Answer A longitudinal wave can be set up in a stretched spring by compressing the coils in a small region, and releasing the compressed region. The disturbance will proceed to propagate as a longitudinal pulse. It is quite possible to set up a transverse wave in a spring, simply by displacing the spring in a direction perpendicular to its length.

□　　□　　□　　□

8. A vibrating source generates a sinusoidal wave on a string under constant tension. If the power delivered to the string is doubled, by what factor does the amplitude change? Does the wave speed change under these circumstances?

Answer Power is always proportional to the square of the amplitude, so if power doubles, amplitude increases by $\sqrt{2}$ times, a factor of 1.41. The wave speed does not depend on amplitude, but stays constant.

□　　□　　□　　□

15. If you stretch a rubber hose and pluck it, you can observe a pulse traveling up and down the hose. What happens to the speed of the pulse if you stretch the hose more tightly? What happens to the speed if you fill the hose with water?

Answer If you stretch the hose tighter, you increase the tension, and increase the speed of the wave. If you fill it with water, you increase the linear density of the hose, and decrease the speed of the wave.

□　　□　　□　　□

17. When two waves interfere, can the amplitude of the resultant wave be greater than either of the two original waves? Under what conditions?

Answer They can, wherever the two waves are enough in phase that their displacements will add to create a total displacement greater than the amplitude of either of the two original waves.

When two one-dimensional waves of the same amplitude interfere, these conditions are satisfied whenever the absolute value of the phase difference between the two waves is less than 120°.

□   □   □   □

SOLUTIONS TO SELECTED END-OF-CHAPTER PROBLEMS

1. At $t = 0$, a transverse wave pulse in a wire is described by the function

$$y = \frac{6}{x^2 + 3}$$

where x and y are in meters. Write the function $y(x, t)$ that describes this wave if it is traveling in the positive x direction with a speed of 4.50 m/s.

Solution Recall from calculus that in order to shift any function $y(x)$ in the positive direction by an amount C, it is necessary to replace x with $(x - C)$. Therefore, if at any time t we want the function shifted right by a distance of $(4.50 \text{ m}/\text{s})(t \text{ s})$, then we should replace x with $(x - 4.50t)$:

$$y(x, t) = \frac{6}{(x - 4.50t)^2 + 3} \qquad \Diamond$$

We can cause any waveform to move along the x axis at a velocity v_x by substituting $(x - v_x t)$ for x. The same principle applies to other directions.

7. Two sinusoidal waves in a string are defined by the functions

$$y_1 = (2.00 \text{ cm}) \sin(20.0x - 32.0t) \quad \text{and} \quad y_2 = (2.00 \text{ cm}) \sin(25.0x - 40.0t)$$

where y and x are in centimeters and t is in seconds. (a) What is the phase difference between these two waves at the point $x = 5.00$ cm at $t = 2.00$ s? (b) What is the positive x value closest to the origin for which the two phases differ by $\pm\pi$ at $t = 2.00$ s? (This is where the sum of the two waves is zero.)

Solution   At any time and place, the phase shift between the waves is found by subtracting the phase of the two waves, $\Delta\phi = \phi_1 - \phi_2$:

$$\Delta\phi = (20.0 \text{ rad / cm})x - (32.0 \text{ rad / s})t - \left[(25.0 \text{ rad / cm})x - (40.0 \text{ rad / s})t\right]$$

Collecting terms, $\Delta\phi = -(5.00 \text{ rad / cm})x + (8.00 \text{ rad / s})t$

(a) At $x = 5.00$ cm and $t = 2.00$ s,

$$\Delta\phi = (-5.00 \text{ rad / cm})(5.00 \text{ cm}) + (8.00 \text{ rad / s})(2.00 \text{ s})$$

$$\Delta\phi = 9.00 \text{ rad} = 516° = 156° \qquad \Diamond$$

(b) The phase shift equals $\pm\pi$ whenever $\Delta\phi = \pi + 2n\pi$, for all integer n. Substituting this into the phase equation,

$$\pi + 2n\pi = -(5.00 \text{ rad / cm})x + (8.00 \text{ rad / s})t$$

At $t = 2.00$ s, $\pi + 2n\pi = -(5.00 \text{ rad / cm})x + (8.00 \text{ rad / s})(2.00 \text{ s})$

or $(5.00 \text{ rad / cm})x = (16.0 - \pi - 2n\pi) \text{ rad}$

The smallest positive value of x is found when $n = 2$:

$$x = \frac{(16.0 - 5\pi) \text{ rad}}{5.00 \text{ rad / cm}} = 0.0584 \text{ cm} \qquad \Diamond$$

9. Two pulses traveling on the same string are described by the functions

$$y_1 = \frac{5}{(3x-4t)^2+2} \quad \text{and} \quad y_2 = \frac{-5}{(3x+4t-6)^2+2}$$

(a) In which direction does each pulse travel? (b) At what time do the two cancel? (c) At what point do the two waves always cancel?

Solution

(a) At constant phase, $\phi = 3x - 4t$, or $\quad x = \dfrac{\phi + 4t}{3}$

As t increases, x increases, so the first wave moves to the right. ◊

In the same way, in the second case $\quad x = \dfrac{\phi - 4t + 6}{3}$

As t increases, x must decrease, so the second wave moves to the left. ◊

(b) We require that $y_1 + y_2 = 0$: $\qquad \dfrac{5}{(3x-4t)^2+2} + \dfrac{-5}{(3x+4t-6)^2+2} = 0$

This can be written as $\qquad (3x-4t)^2 = (3x+4t-6)^2$

Solving for the positive root, $\qquad t = 0.750 \text{ s}$ ◊

(c) The negative root yields $\qquad (3x-4t) = -(3x+4t-6)$

The time terms cancel, leaving $\qquad x = 1.00 \text{ m}$ ◊

At this point, the waves **always** cancel.

15. Transverse waves travel with a speed of 20.0 m/s in a string under a tension of 6.00 N. What tension is required to produce a wave speed of 30.0 m/s in the same string?

Solution

G: Since $v \propto \sqrt{F}$, the new tension must be about twice as much as the original to achieve a 50% increase in the wave speed.

O: The equation for a transverse wave on a string under tension can be used if we assume that the linear density of the string is constant. Then the ratio of the two wave speeds can be used to find the new tension.

A: The two wave speeds can be written as:

$$v_1 = \sqrt{F_1/\mu} \qquad \text{and} \qquad v_2 = \sqrt{F_2/\mu}$$

Dividing, $\qquad \dfrac{v_2}{v_1} = \sqrt{\dfrac{F_2}{F_1}}$

so $\qquad F_2 = \left(\dfrac{v_2}{v_1}\right)^2 F_1 = \left(\dfrac{30.0 \text{ m/s}}{20.0 \text{ m/s}}\right)^2 (6.00 \text{ N}) = 13.5 \text{ N}$ $\qquad\qquad \lozenge$

L: The new tension is slightly more than twice the original, so the result agrees with our initial prediction and is therefore reasonable.

21. A 30.0-m steel wire and a 20.0-m copper wire, both with 1.00-mm diameters, are connected end to end and are stretched to a tension of 150 N. How long does it take a transverse wave to travel the entire length of the two wires?

Solution   The total time of travel is the sum of the two times.

In each wire, $t = \dfrac{L}{v} = L\sqrt{\dfrac{\mu}{T}}$ where $\mu = \rho A = \dfrac{\pi \rho d^2}{4}$,   so $t = L\sqrt{\dfrac{\pi \rho d^2}{4T}}$

For copper, $t_1 = (20.0 \text{ m})\sqrt{\dfrac{\pi(8920 \text{ kg / m}^3)(0.00100 \text{ m})^2}{4(150 \text{ kg} \cdot \text{m / s}^2)}} = 0.1366 \text{ s}$

For steel, $t_2 = (30.0 \text{ m})\sqrt{\dfrac{\pi(7860 \text{ kg / m}^3)(0.001 \text{ m})^2}{4(150 \text{ kg} \cdot \text{m / s}^2)}} = 0.1925 \text{ s}$

The total time is $(0.137 \text{ s}) + (0.192 \text{ s}) = 0.329 \text{ s}$ ◊

---

25. A sinusoidal wave is traveling along a rope. The oscillator that generates the wave completes 40.0 vibrations in 30.0 s. Also, a given maximum travels 425 cm along the rope in 10.0 s. What is the wavelength?

Solution $f = \dfrac{40.0 \text{ waves}}{30.0 \text{ s}} = 1.33 \text{ s}^{-1}$ and $v = \dfrac{425 \text{ cm}}{10.0 \text{ s}} = 42.5 \text{ cm / s}$

Since $v = \lambda f$, $\lambda = \dfrac{v}{f} = \dfrac{42.5 \text{ cm / s}}{1.33 \text{ s}^{-1}} = 0.319 \text{ m}$ ◊

---

31. (a) Write the expression for y as a function of x and t for a sinusoidal wave traveling along a rope in the **negative** x direction with the following characteristics: $A = 8.00$ cm, $\lambda = 80.0$ cm, $f = 3.00$ Hz, and $y(0,t) = 0$ at $t = 0$. (b) Write the expression for y as a function of x and t for the wave in (a), assuming that $y(x,0) = 0$ at the point $x = 10.0$ cm.

Solution

(a)   The characteristic variables are $A = y_{max} = 8.00$ cm $= 0.0800$ m

$$k = \frac{2\pi}{\lambda} = \frac{2\pi}{0.800 \text{ m}} = 7.85 \text{ m}^{-1}$$

$$\omega = 2\pi f = 2\pi(3.00 \text{ s}^{-1}) = 6.00\pi \text{ rad / s}$$

Since $\phi = 0$, $y = A\sin(kx + \omega t)$,   and

$$y = (0.800 \text{ m}) \sin(7.85x + 6.00\pi t) \qquad \lozenge$$

(b)   In general, $y = (0.800 \text{ m}) \sin(7.85x + 6.00\pi t + \phi)$

If $y(x,0) = 0$ at $x = 0.100$ m $0 = (0.0800 \text{ m}) \sin(0.785 + \phi)$

and $\phi = -0.785$ rad

Therefore, $y = (0.800 \text{ m}) \sin(7.85x + 6.00\pi t - 0.785) \qquad \lozenge$

35. A wave is described by $y = (2.00 \text{ cm}) \sin(kx - \omega t)$, where $k = 2.11 \text{ rad} / \text{m}$, $\omega = 3.62 \text{ rad} / \text{s}$, x is in meters, and t is in seconds. Determine the amplitude, wavelength, frequency, and speed of the wave.

Solution

Given the general sinusoidal wave equation, $\quad y = A\sin(kx - \omega t + \phi)$

By comparison, the amplitude is $\qquad\qquad\qquad A = 2.00 \text{ cm}$ ◊

The angular wave number $k = 2.11 \text{ rad} / \text{m}$: $\qquad \lambda = \dfrac{2\pi}{k} = 2.98 \text{ m}$ ◊

The angular frequency $\omega = 3.62 \text{ rad} / \text{s}$: $\qquad f = \dfrac{\omega}{2\pi} = 0.576 \text{ Hz}$ ◊

The speed is $v = f\lambda$, so $\quad v = (0.576 \text{ s}^{-1})(2.98 \text{ m}) = 1.72 \text{ m} / \text{s}$ ◊

---

41. Sinusoidal waves 5.00 cm in amplitude are to be transmitted along a string that has a linear density of $4.00 \times 10^{-2} \text{ kg/m}$. If the source can deliver a maximum power of 300 W and the string is under a tension of 100 N, what is the highest vibrational frequency at which the source can operate?

Solution The wave speed is $\quad v = \sqrt{\dfrac{F}{\mu}} = \sqrt{\dfrac{100 \text{ N}}{4.00 \times 10^{-2} \text{ kg} / \text{m}}} = 50.0 \text{ m} / \text{s}$

From $\quad P = \dfrac{1}{2}\mu\omega^2 A^2 v$, $\qquad\qquad \omega^2 = \dfrac{2P}{\mu A^2 v}$

Solving for the angular frequency,

$$\omega^2 = \frac{2(300 \text{ N} \cdot \text{m} / \text{s})}{\left(4.00 \times 10^{-2} \text{ kg} / \text{m}\right)\left(5.00 \times 10^{-2} \text{ m}\right)^2 (50.0 \text{ m} / \text{s})}$$

and $\omega = 346.4 \text{ rad} / \text{s}$

The frequency is then $f = \dfrac{\omega}{2\pi} = 55.1 \text{ Hz}$ ◊

---

43. A sinusoidal wave on a string is described by the equation

$$y = (0.15 \text{ m}) \sin (0.80x - 50t)$$

where x and y are in meters and t is in seconds. If the mass per unit length of this string is 12.0 g/m, determine (a) the speed of the wave, (b) the wavelength, (c) the frequency, and (d) the power transmitted to the wave.

Solution Compare the given wave function, $y = (0.15 \text{ m}) \sin (0.80x - 50t)$

with the general wave equation, $y = A \sin(kx - \omega t)$

(a) $v = f\lambda = \dfrac{\omega}{k} = \dfrac{50 \text{ rad} / \text{s}}{0.80 \text{ rad} / \text{m}} = 62.5 \text{ m} / \text{s}$ ◊

(b) $\lambda = \dfrac{2\pi}{k} = \dfrac{2\pi \text{ rad}}{0.80 \text{ rad} / \text{m}} = 7.85 \text{ m}$ ◊

(c) $f = \dfrac{\omega}{2\pi} = \dfrac{50 \text{ rad} / \text{s}}{2\pi \text{ rad}} = 7.96 \text{ Hz}$ ◊

(d) $P = \frac{1}{2}\mu\omega^2 A^2 v = \frac{1}{2}(0.0120 \text{ kg} / \text{m})(50.0 \text{ s}^{-1})^2 (0.150 \text{ m})^2 (62.5 \text{ m} / \text{s}) = 21.1 \text{ W}$ ◊

---

47. Show that the wave function $y = \ln[b(x - vt)]$ is a solution to Equation 16.26, where b is a constant.

Solution

The important thing to remember with partial derivatives is that **you treat all variables as constants, except the single variable of interest.** Keeping this in mind, we must apply two common rules of derivation to the function $y = \ln[b(x - vt)]$:

$$\frac{\partial}{\partial x}[\ln f(x)] = \left(\frac{1}{f(x)}\right)\frac{\partial(f(x))}{\partial x} \tag{1}$$

$$\frac{\partial}{\partial x}\left[\frac{1}{f(x)}\right] = \frac{\partial}{\partial x}[f(x)]^{-1} = (-1)[f(x)]^{-2}\frac{\partial(f(x))}{\partial x} = -\left(\frac{1}{[f(x)]^2}\right)\frac{\partial(f(x))}{\partial x} \tag{2}$$

Applying (1), $\quad\quad \dfrac{\partial y}{\partial x} = \left(\dfrac{1}{b(x - vt)}\right)\dfrac{\partial(bx - bvt)}{\partial x} = \left(\dfrac{1}{b(x - vt)}\right)(b) = \dfrac{1}{x - vt}$

Applying (2), $\quad\quad \dfrac{\partial^2 y}{\partial x^2} = -\dfrac{1}{(x - vt)^2}$

In a similar way, $\quad \dfrac{\partial y}{\partial t} = \dfrac{-v}{(x - vt)} \quad$ and $\quad \dfrac{\partial^2 y}{\partial t^2} = -\dfrac{v^2}{(x - vt)^2}$

From the second-order partial derivatives, we see that $\quad\quad \dfrac{\partial^2 y}{\partial x^2} = \dfrac{1}{v^2}\dfrac{\partial^2 y}{\partial t^2} \quad \lozenge$

51. The wave function for a traveling wave on a taut string is (in SI units)

$$y(x, t) = (0.350 \text{ m}) \sin (10\pi t - 3\pi x + \pi / 4)$$

(a) What are the speed and direction of travel of the wave? (b) What is the vertical displacement of the string at $t = 0$, $x = 0.100$ m? (c) What are the wavelength and frequency of the wave? (d) What is the maximum magnitude of the transverse speed of the string?

Solution We compare the given equation with $y = A\sin(kx - \omega t + \phi)$; and find

$$k = 3\pi \text{ rad} / \text{m} \quad \text{and} \quad \omega = 10\pi \text{ rad} / \text{s}$$

(a) The speed is $\quad v = f\lambda = \dfrac{\omega}{k} = \dfrac{10\pi \text{ rad} / \text{s}}{3\pi \text{ rad} / \text{m}} = 3.33 \text{ m} / \text{s}$ ◊

(b) Substituting $t = 0$ and $x = 0.100$ m,

$$y = (0.350 \text{ m}) \sin (-0.300\pi + 0.250\pi) = -0.0548 \text{ m} = -5.48 \text{ cm}$$ ◊

Note that when you take the sine of a quantity with no units, it is not in degrees, but in radians.

(c) $\quad \lambda = \dfrac{2\pi \text{ rad}}{k} = \dfrac{2\pi \text{ rad}}{3\pi \text{ rad} / \text{m}} = 0.667 \text{ m}$ ◊

$$f = \dfrac{\omega}{2\pi \text{ rad}} = \dfrac{10\pi \text{ rad} / \text{s}}{2\pi \text{ rad}} = 5.00 \text{ Hz}$$ ◊

(d) $\quad v_y = \dfrac{\partial y}{\partial t} = (0.350 \text{ m})(10\pi \text{ rad} / \text{s}) \cos \left(10\pi t - 3\pi x + \dfrac{\pi}{4} \right)$

The maximum occurs when the cosine term is 1:

$$v_{y,\text{max}} = (10\pi \text{ rad} / \text{s})(0.350 \text{ m} / \text{s}) = 11.0 \text{ m} / \text{s}$$ ◊

Note how different the maximum particle speed is from the wave speed found in part (a).

59. A rope of total mass m and length L is suspended vertically. Show that a transverse wave pulse travels the length of the rope in a time $t = 2\sqrt{L/g}$. (**Hint:** First find an expression for the wave speed at any point a distance x from the lower end by considering the tension in the rope as resulting from the weight of the segment below that point.)

Solution   The tension in the rope at any point is the weight of the rope below that point. Therefore, the weight of the portion of the rope below point x is

$$F = \mu x g$$

The equation for the phase velocity is

$$v = \sqrt{F/\mu}$$

Substituting for the tension F,

$$v = \sqrt{gx}$$

But at each point x from 0 to L, the wave phase travels at

$$v = \frac{dx}{dt}$$

So,

$$\frac{dx}{dt} = \sqrt{gx} \quad \text{or} \quad dt = \frac{dx}{\sqrt{gx}}$$

Integrating both sides,

$$t = \frac{1}{\sqrt{g}} \int_0^L \frac{dx}{\sqrt{x}} = \frac{\left[2\sqrt{x}\right]_0^L}{\sqrt{g}} = 2\sqrt{\frac{L}{g}} \qquad \Diamond$$

---

61. It is stated in Problem 59 that a wave pulse travels from the bottom to the top of a rope of length L in a time $t = 2\sqrt{L/g}$. Use this result to answer the following questions. (It is **not** necessary to set up any new integrations.) (a) How long does it take for a wave pulse to travel halfway up the rope? (Give your answer as a fraction of the quantity $2\sqrt{L/g}$.) (b) A pulse starts traveling up the rope. How far has it traveled after a time $\sqrt{L/g}$?

Solution

G: The wave pulse travels faster as it goes up the rope because the tension higher in the rope is greater (to support the weight of the rope below it). Therefore it should take more than half the total time t for the wave to travel halfway up the rope. Likewise, the pulse should travel less than halfway up the rope in time $t/2$.

O: By using the time relationship given in the problem and making suitable substitutions, we can find the required time and distance.

A: (a) From the equation given, the time for a pulse to travel any distance, d, up from the bottom of a rope is $t_d = 2\sqrt{d/g}$. So the time for a pulse to travel a distance $L/2$ from the bottom is

$$t_{L/2} = 2\sqrt{\frac{L}{2g}} = 0.707\left(2\sqrt{\frac{L}{g}}\right) \qquad \lozenge$$

(b) Likewise, the distance a pulse travels from the bottom of a rope in a time t_d is $d = gt_d^2/4$. So the distance traveled by a pulse after a time $t_d = \sqrt{L/g}$ is

$$d = \frac{g(L/g)}{4} = \frac{L}{4} \qquad \lozenge$$

L: As expected, it takes the pulse more than 70% of the total time to cover 50% of the distance. In half the total trip time, the pulse has climbed only $1/4$ of the total length.

63. An aluminum wire under zero tension at room temperature is clamped at each end. The tension in the wire is increased by reducing the temperature, which results in a decrease in the wire's equilibrium length. What strain $(\Delta L / L)$ results in a transverse wave speed of 100 m/s? Take the cross-sectional area of the wire to be 5.00×10^{-6} m^2, the density to be 2.70×10^3 kg/m^3, and Young's modulus to be 7.00×10^{10} N/m^2.

Solution

The expression for the elastic modulus,
$$Y = \frac{F/A}{\Delta L / L}$$

becomes an equation for strain
$$\frac{\Delta L}{L} = \frac{F/A}{Y} \qquad (1)$$

Substitute into the equation for the wave speed
$$v = \sqrt{F/\mu}$$

the definition for μ
$$\mu = \frac{m}{L} = \frac{\rho(AL)}{L} = \rho A$$

and solve,
$$v^2 = \frac{F}{\mu} = \frac{1}{\rho}\left(\frac{F}{A}\right) \qquad \text{or} \qquad \left(\frac{F}{A}\right) = \rho v^2 \qquad (2)$$

Substituting (2) into (1),
$$\frac{\Delta L}{L} = \frac{\rho v^2}{Y} = \frac{(2.70 \times 10^3 \text{ kg}/\text{m}^3)(100 \text{ m}/\text{s})^2}{7.00 \times 10^{10} \text{ N}/\text{m}^2}$$

$$\left(\frac{\Delta L}{L}\right) = 3.86 \times 10^{-4} \qquad \Diamond$$

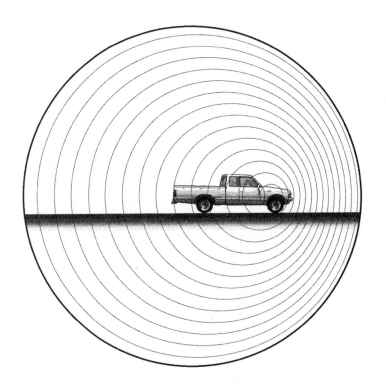

Sound Waves

SOUND WAVES

INTRODUCTION

Sound waves are the most important example of longitudinal waves. They can travel through any material medium with a speed that depends on the properties of the medium. As the waves travel, the particles in the medium vibrate to produce density and pressure changes along the direction of motion of the wave. These changes result in a series of high- and low-pressure regions called **condensations** and **rarefactions,** respectively. If the source of the sound waves vibrates sinusoidally, the pressure variations are also sinusoidal. We shall find that the mathematical description of harmonic sound waves is identical to that of harmonic string waves discussed in the previous chapter.

EQUATIONS AND CONCEPTS

A sound wave propagates as a **compressional wave** with a **speed** v which depends on the bulk modulus B and equilibrium density ρ of the medium in which the wave is traveling. The **Bulk Modulus** is equal to the negative ratio of the pressure variation to the fractional change in volume of the medium.

$$v = \sqrt{\frac{B}{\rho}} \qquad (17.1)$$

$$B = -\frac{\Delta P}{\Delta V / V}$$

The **speed of sound in a gas** depends on the characteristic parameter γ (the ratio of the specific heat at constant pressure to the specific heat at constant volume), and can be expressed either in terms of the pressure and density or the absolute temperature T and molar mass M.

$$v = \sqrt{\frac{\gamma P}{\rho}}$$

$$v = \sqrt{\frac{\gamma R T}{M}}$$

The speed of sound in a solid rod depends on the value of Young's modulus Y and the density of the material. Young's modulus is equal to the ratio of the tensile stress to the tensile strain (see Chapter 12).

$$v = \sqrt{\frac{Y}{\rho}}$$

A harmonic sound wave is produced in a gas when the source is a body vibrating in simple harmonic motion (such as a vibrating guitar string). Under these conditions, both the **displacement** of the medium s and the **pressure variations** of the gas ΔP vary harmonically in time. Note that the displacement and the pressure variation are out of phase by $\pi / 2$.

$$s(x,\ t) = s_{max} \cos(kx - \omega t) \tag{17.2}$$

$$\Delta P = \Delta P_{max} \sin(kx - \omega t) \tag{17.3}$$

A sound wave may be considered as either a displacement wave or a pressure wave. The **pressure amplitude** is **proportional** to the **displacement amplitude.**

$$\Delta P_{max} = \rho v \omega s_{max} \tag{17.4}$$

The **intensity** of a harmonic sound wave is proportional to the square of the source frequency and to the square of the displacement (or pressure) amplitude.

$$I = \frac{\mathcal{P}}{A} = \tfrac{1}{2}\rho v (\omega s_{max})^2 \qquad (17.5)$$

$$I = \frac{\Delta P^2_{max}}{2\rho v} \qquad (17.6)$$

The human ear is sensitive to a wide **range of intensities**. For this reason, a **logarithmic intensity level scale** is defined using a reference intensity of 10^{-12} W / m^2. On this scale, the unit of intensity level is the decibel (dB).

$$\beta \equiv 10 \, \log\!\left(\frac{I}{I_0}\right) \qquad (17.7)$$

$$I_0 = 10^{-12} \text{ W / m}^2$$

The **amplitude** of a periodic spherical wave varies inversely with the distance r from the source.

$$\psi(r, \, t) = \frac{s_0}{r}\sin(kr - \omega t) \qquad (17.9)$$

(Spherical wave)

At large distances from the source ($r \gg \lambda$), a spherical wave can be approximated as a **plane wave**. In this case, the **amplitude** of the wave is approximately **constant**.

$$\psi(x, \, t) = A\sin(kx - \omega t) \qquad (17.10)$$

(Plane wave)

The **intensity** I of a spherical wave decreases as the square of the distance from the source. This is so because the intensity is proportional to the **square of the amplitude**, which in turn varies inversely with distance from the source.

$$I = \frac{\mathcal{P}_{av}}{4\pi r^2} \qquad (17.8)$$

The change in frequency heard by an observer whenever there is **relative motion between the source and the observer** is called the **Doppler effect**. The upper signs refer to motion of one toward the other and the lower signs apply to motion of one away from the other. In Equation 17.17, v_0 and v_s are measured **relative to the medium in which the sound travels**.

$$f' = \left(\frac{v \pm v_0}{v \mp v_s} \right) f \qquad (17.17)$$

Shock waves are produced when a sound source moves through a medium with a speed v_s which is greater than the wave speed in that medium. The shock wave front has a conical shape with a half angle which depends on the **Mach number** of the source, defined as the ratio v_s / v.

$$\sin \theta = \frac{v}{v_s}$$

SUGGESTIONS, SKILLS, AND STRATEGIES

When making calculations using Equation 17.7 which defines the intensity of a sound wave on the decibel scale, the properties of logarithms must be kept clearly in mind.

In order to determine the decibel level corresponding to two sources sounded simultaneously, you must first find the intensity, I, of each source in W/m²; add these values, and then convert the resulting intensity to the decibel scale. As an illustration of this technique, note that if two sounds of intensity 40 dB and 45 dB are sounded together, the intensity level of the combined sources **is 46.2 dB (not 85 dB)**.

Your most likely error in using Equation 17.17 to calculate the Doppler frequency shift due to relative motion between a sound source and an observer is using the incorrect algebraic sign for the velocity of either the observer or the source. These sign conventions are illustrated in the chart below in which the directions of motion of the observer O and source S are indicated by arrows; and the correct choice of signs is used in Equation 17.17,

$$f' = \left(\frac{v \pm v_o}{v \mp v_s} \right) f$$

where f is the true frequency of the source and f' is the apparent frequency as measured by the observer.

Observer	Source	Equation	Remark
O→	S	$f' = \left(\dfrac{v + v_o}{v \mp v_s} \right) f$	Observer moving toward source, v_o is **positive**.
←O	S	$f' = \left(\dfrac{v - v_o}{v \mp v_s} \right) f$	Observer moving away from source, v_o is **negative**.
O	←S	$f' = \left(\dfrac{v \pm v_o}{v - v_s} \right) f$	Source moving toward observer, v_s is **negative**
O	S→	$f' = \left(\dfrac{v \pm v_o}{v + v_s} \right) f$	Source moving away from observer, v_s is **positive**

REVIEW CHECKLIST

▷ Calculate the speed of sound in various media in terms of the appropriate elastic properties of the medium (including bulk modulus, Young's modulus, and the pressure-volume relationships of an ideal gas), and the corresponding inertial properties (usually mass density).

▷ Describe the harmonic displacement and pressure variation as functions of time and position for a harmonic sound wave. Relate the displacement amplitude to the pressure amplitude for a harmonic sound wave and calculate the wave intensity from each of these parameters.

▷ Understand the basis of the logarithmic intensity level scale (decibel scale). Determine the intensity ratio for two sound sources whose decibel levels are known. Calculate the decibel level for some combination of sources whose individual sound levels are known.

▷ Describe the wave function for spherical and planar harmonic waves. Understand the amplitude dependence on distance from the source for spherical and plane waves.

▷ Describe the various situations under which a Doppler shifted frequency is produced. Note that a Doppler shift is observed as long as there is a **relative** motion between the observer and the source.

ANSWERS TO SELECTED CONCEPTUAL QUESTIONS

2. If an alarm clock is placed in a good vacuum and then activated, no sound is heard. Explain.

Answer We assume that a perfect vacuum surrounds the clock. The sound waves require a medium for them to travel to your ear. The hammer on the alarm will strike the bell, and the vibration will spread as sound waves through the body of the clock. If a bone of your skull were in contact with the clock, you would hear the bell. However, in the absence of a surrounding medium like air or water, no sound can be radiated away.

What happens to the sound energy? Here is the answer: As the sound wave travels through the steel and plastic, traversing joints and going around corners, its energy is converted into additional internal energy, raising the temperature of the materials. After the sound has died away, the clock will glow very slightly brighter in the infrared spectrum. For a larger-scale example of the same effect: Colossal storms raging on the Sun are deathly still for us.

□   □   □   □

5. If the distance from a point source is tripled, by what factor does the intensity decrease?

Answer We suppose that a point source has no structure, and radiates sound equally in all directions (isotropically). The sound wavefronts are expanding spheres, so the area over which the sound energy spreads increases according to $A = 4\pi r^2$. Thus, if the distance is tripled, the area increases by a factor of nine, and the new intensity will be one ninth of the old intensity. This answer according to the inverse-square law applies if the medium is uniform and unbounded.

For contrast, suppose that the sound is confined to move in a horizontal layer. (Thermal stratification in an ocean can have this effect on sonar "pings,") Then the area over which the sound energy is dispersed will only increase according to the circumference of an expanding circle: $A = 2\pi rh$, and so three times the distance will result in one third the intensity.

In the case of an entirely enclosed speaking tube (such as a ship's telephone), the area perpendicular to the energy flow stays the same, and increasing the distance will not change the intensity appreciably.

□   □   □   □

11. How can an object move with respect to an observer so that the sound from it is not shifted in frequency?

Answer   For the sound from a source not to shift in frequency, the radial velocity of the source relative to the observer must be zero; that is, the source must not be moving toward or away from the observer.

This can happen if the source and observer are not moving at all; if they have equal velocities relative to the medium; or, it can happen if the source moves around the observer in a circular pattern of constant radius. Even if the source accelerates along the circle, decelerates, or stops, the frequency heard will equal the frequency emitted by the source.

□   □   □   □

SOLUTIONS TO SELECTED END-OF-CHAPTER PROBLEMS

1. Suppose that you hear a clap of thunder 16.2 s after seeing the associated lightning stroke. The speed of sound waves in air is 343 m/s, and the speed of light in air is 3.00×10^8 m/s. How far are you from the lightning stroke?

Solution

G: There is a common rule of thumb that lightning is about a mile away for every 5 seconds of delay between the flash and thunder (or ~3 s/km). Therefore, this lightning strike is about 3 miles (~5 km) away.

O: The distance can be found from the speed of sound and the elapsed time. The time for the light to travel to the observer will be much less than the sound delay, so the speed of light can be taken as ∞.

A: Assuming that the speed of sound is constant through the air between the lightning strike and the observer,

$$v_s = \frac{d}{\Delta t} \quad \text{or} \quad d = v_s \Delta t = (343 \text{ m / s})(16.2 \text{ s}) = 5.56 \text{ km} \qquad \lozenge$$

L: Our calculated answer is consistent with our initial estimate, but we should check the validity of our assumption that the speed of light could be ignored. The time delay for the light is

$$t_{light} = \frac{d}{c_{air}} = \frac{5560 \text{ m}}{3.00 \times 10^8 \text{ m / s}} = 1.85 \times 10^{-5} \text{ s}$$

and $\Delta t = t_{sound} - t_{light} = 16.2 \text{ s} - 1.85 \times 10^{-5} \text{ s} \approx 16.2 \text{ s}$ (when rounded)

Since the travel time for the light is much smaller than the uncertainty in the time of 16.2 s, t_{light} can be ignored without affecting the distance calculation. However, our assumption of a constant speed of sound in air is probably not valid due to local variations in air temperature during a storm. We must assume that the given speed of sound in air is an accurate **average** value for the conditions described.

11. Write an expression that describes the pressure variation as a function of position and time for a sinusoidal sound wave in air if $\lambda = 0.100$ m and $\Delta P_{max} = 0.200$ Pa.

Solution We write the pressure variation as $\Delta P = \Delta P_{max} \sin(kx - \omega t)$.

Noting that $k = \dfrac{2\pi}{\lambda}$, $\qquad\qquad k = \dfrac{2\pi \text{ rad}}{0.100 \text{ m}} = 62.8 \text{ rad} / \text{m}$

Likewise, $\omega = \dfrac{2\pi v}{\lambda}$, so $\qquad\qquad \omega = \dfrac{(2\pi \text{ rad})(343 \text{ m} / \text{s})}{0.100 \text{ m}} = 2.16 \times 10^4 \text{ rad} / \text{s}$

We now can create our equation: $\quad \Delta P = (0.200 \text{ Pa})\sin(62.8x - 21600t)$ ◊

---

15. An experimenter wishes to generate in air a sound wave that has a displacement amplitude of 5.50×10^{-6} m. The pressure amplitude is to be limited to 8.40×10^{-1} Pa. What is the minimum wavelength the sound wave can have?

Solution We are given $s_{max} = 5.50 \times 10^{-6}$ m and $\Delta P_{max} = 0.840$ Pa

The pressure amplitude is $\qquad\qquad \Delta P_{max} = \rho v \omega s_{max} = \rho v \left(\dfrac{2\pi v}{\lambda}\right) s_{max}$

Rearranging terms, we find that $\qquad \lambda = \dfrac{2\pi \rho v^2 s_{max}}{\Delta P_{max}}$

Thus $\quad \lambda_{min} = \dfrac{2\pi \left(1.20 \text{ kg} / \text{m}^3\right)(343 \text{ m} / \text{s})^2 \left(5.50 \times 10^{-6} \text{ m}\right)}{0.840 \text{ Pa}} = 5.81 \text{ m}$ ◊

---

17. Calculate the sound level, in decibels, of a sound wave that has an intensity of $4.00 \ \mu W / m^2$.

Solution We use the equation $\beta = 10 \ \log(I/I_0)$, where $I_0 = 10^{-12} \ W / m^2$.

$$\beta = 10 \ \log\left(\frac{4.00 \times 10^{-6} \ W / m^2}{10^{-12} \ W / m^2}\right) = 66.0 \ dB \qquad \Diamond$$

21. A family ice show is held in an enclosed arena. The skaters perform to music with a sound level of 80.0 dB. This is too loud for your baby, who constantly yells at a level of 75.0 dB. (a) What total sound intensity engulfs you? (b) What is the combined sound level?

Solution $\beta = 10 \ \log\left(\dfrac{I}{10^{-12} \ W / m^2}\right)$ so $I = \left[10^{\beta/10}\right]10^{-12} \ W / m^2$

(a) For your baby, $I_b = \left(10^{75.0/10}\right)\left(10^{-12} \ W / m^2\right) = 3.16 \times 10^{-5} \ W / m^2$

For the music, $I_m = \left(10^{80.0/10}\right)\left(10^{-12} \ W / m^2\right) = 10.0 \times 10^{-5} \ W / m^2$

The combined intensity is $I_{total} = I_m + I_b$

$$I_{total} = 10.0 \times 10^{-5} \ W / m^2 + 3.16 \times 10^{-5} \ W / m^2$$

$$I_{total} = 13.2 \times 10^{-5} \ W / m^2 \qquad \Diamond$$

(b) The combined sound level is then

$$\beta_{total} = 10 \ \log\left(\frac{I_{total}}{10^{-12} \ W / m^2}\right) = 10 \ \log\left(\frac{1.32 \times 10^{-4} \ W / m^2}{10^{-12} \ W / m^2}\right) = 81.2 \ dB \qquad \Diamond$$

371

23. A fireworks charge is detonated many meters above the ground. At a distance of 400 m from the explosion, the acoustic pressure reaches a maximum of 10.0 N/m². Assume that the speed of sound is constant at 343 m / s throughout the atmosphere over the region considered, that the ground absorbs all the sound falling on it, and that the air absorbs sound energy as described by the rate 7.00 dB/km. What is the sound level (in decibels) at 4.00 km from the explosion?

Solution

G: At a distance of 4 km, an explosion should be audible, but probably not extremely loud. So based on the data in Table 17.2, we might expect the sound level to be somewhere between 40 and 80 dB.

O: From the sound pressure data given in the problem, we can find the intensity, which is used to find the sound level in dB. The sound intensity will decrease with increased distance from the source and from the absorption of the sound by the air.

A: At a distance of 400 m from the explosion, $\Delta P_{max} = 10.0$ Pa. At this point,

$$I_{max} = \frac{\Delta P_{max}{}^2}{2\rho v} = \frac{\left(10.0 \text{ N / m}^2\right)^2}{2\left(1.20 \text{ kg / m}^3\right)\left(343 \text{ m / s}\right)} = 0.121 \text{ W / m}^2$$

Therefore, the maximum sound level is

$$\beta_{max} = 10\log\left(\frac{I_{max}}{I_0}\right) = 10\log\left(\frac{0.121 \text{ W / m}^2}{1.00 \times 10^{-12} \text{ W / m}^2}\right) = 111 \text{ dB}$$

From equations 17.8 and 17.7, we can calculate the intensity and decibel level (due to distance alone) 4 km away:

$$I' = I(400 \text{ m})^2 / (4000 \text{ m})^2 = 1.21 \times 10^{-3} \text{ W / m}^2$$

and

$$\beta = 10\log\left(\frac{I'}{I_0}\right) = \left(\frac{1.21 \times 10^{-3} \text{ W / m}^2}{1.00 \times 10^{-12} \text{ W / m}^2}\right) = 90.8 \text{ dB}$$

At a distance of 4 km from the explosion, absorption from the air will have decreased the sound level by an additional

$$\Delta\beta = (7.00 \text{ dB / km})(3.60 \text{ km}) = 25.2 \text{ dB}$$

So at 4 km, the sound level will be

$$\beta_f = \beta - \Delta\beta = 90.8 \text{ dB} - 25.2 \text{ dB} = 65.6 \text{ dB} \qquad \Diamond$$

L: This sound level falls within our expected range. Evidently, this explosion is rather loud (about the same as a vacuum cleaner) even at a distance of 4 km from the source. It is interesting to note that the distance and absorption effects each reduce the sound level by about the same amount (~20 dB). If the explosion were at ground level, the sound level would be further reduced by reflection and absorption from obstacles between the source and observer, and the calculation would be much more complicated (if not impossible).

---

29. The sound level at a distance of 3.00 m from a source is 120 dB. At what distance is the sound level (a) 100 dB and (b) 10.0 dB?

Solution $\quad \beta = 10 \, \log\left(\dfrac{I}{10^{-12} \text{ W / m}^2}\right) \quad$ so $\quad I = \left[10^{\beta/10}\right]10^{-12} \text{ W / m}^2$

$$I_{120} = 1 \text{ W / m}^2 \qquad I_{100} = 10^{-2} \text{ W / m}^2 \qquad I_{10} = 10^{-11} \text{ W / m}^2$$

(a) The power passing through any sphere around the source is $\quad \mathcal{P} = 4\pi r^2 I$, so conservation of energy requires that $r_{120}{}^2 I_{120} = r_{100}{}^2 I_{100} = r_{10}{}^2 I_{10}$.

and $\quad r_{100} = r_{120}\sqrt{\dfrac{I_{120}}{I_{100}}} = (3.00 \text{ m})\sqrt{\dfrac{1 \text{ W / m}^2}{10^{-2} \text{ W / m}^2}} = 30.0 \text{ m} \qquad \Diamond$

(b) $\quad r_{10} = r_{120}\sqrt{\dfrac{I_{120}}{I_{10}}} = (3.00 \text{ m})\sqrt{\dfrac{1 \text{ W / m}^2}{10^{-11} \text{ W / m}^2}} = 9.49\times10^5 \text{ m} \qquad \Diamond$

---

35. Standing at a crosswalk, you hear a frequency of 560 Hz from the siren of an approaching police car. After the police car passes, the observed frequency of the siren is 480 Hz. Determine the car's speed from these observations.

Solution

G: We can assume that a police car with its siren on is in a hurry to get somewhere, and is probably traveling between 20 and 100 mph ($\sim$10 m/s to 50 m/s), depending on the driving conditions.

O: We can use the equation for the Doppler effect to find the speed of the car.

A: Approaching car:

$$f' = \frac{f}{1 - v_s/v} \qquad \text{(Eq. 17.14)}$$

Departing car:

$$f'' = \frac{f}{1 + v_s/v} \qquad \text{(Eq. 17.15)}$$

Where $f' = 560$ Hz and $f'' = 480$ Hz. Solving the two equations above for f and setting them equal gives:

$$f'(1 - v_s/v) = f''(1 + v_s/v) \qquad \text{or} \qquad f' - f'' = (f' + f'')(v_s/v)$$

so the speed of the source is

$$v_s = \frac{v(f' - f'')}{f' + f''} = \frac{(343 \text{ m / s})(560 \text{ Hz} - 480 \text{ Hz})}{560 \text{ Hz} + 480 \text{ Hz}} = 26.4 \text{ m / s} \qquad \lozenge$$

L: This seems like a reasonable speed (about 50 mph) for a police car, unless the street is crowded or the car is traveling on an open highway. Of course, this problem is only valid if the siren emits a single tone. A warble could make it difficult or impossible to solve this problem.

37. A tuning fork vibrating at 512 Hz falls from rest and accelerates at 9.80 m/s². How far below the point of release is the tuning fork when waves with a frequency of 485 Hz reach the release point? Take the speed of sound in air to be 340 m/s.

Solution   In order to solve this problem, we must first determine how fast the tuning fork is falling when its frequency is 485 Hz.

The tuning fork (source) is moving **away** from a stationary listener.

Therefore, we use the equation $f' = \left(\dfrac{v}{v + v_s} \right) f$:

$$485 \text{ Hz} = (512 \text{ Hz}) \frac{340 \text{ m / s}}{340 \text{ m / s} + v_{\text{fall}}}$$

Solving, $v_{\text{fall}} = (340 \text{ m / s}) \left(\dfrac{512 \text{ Hz}}{485 \text{ Hz}} - 1 \right) = 18.93 \text{ m / s}$

Now, using the kinematic equation $v_2{}^2 = v_1{}^2 + 2as$, we calculate that

$$s = \frac{v_2{}^2}{2a} = \frac{(18.93 \text{ m / s})^2}{2 \left(9.80 \text{ m / s}^2 \right)} = 18.28 \text{ m}$$

Since $v = at$, $t = \dfrac{v_{\text{fall}}}{a} = \dfrac{18.93 \text{ m / s}}{9.80 \text{ m / s}^2} = 1.932 \text{ s}$

At this moment, the fork would appear to ring at 485 Hz **just below the fork**. However, it takes some additional time for the waves to reach the point of release.

$$\Delta t = \frac{s}{v} = \frac{18.28 \text{ m}}{340 \text{ m / s}} = 0.0540 \text{ s}$$

Over the total time $t + \Delta t$, the fork falls a distance

$$s_{\text{total}} = \tfrac{1}{2}a(t + \Delta t)^2 = \tfrac{1}{2}\left(9.80 \text{ m / s}^2\right)(1.986 \text{ s})^2 = 19.3 \text{ m} \qquad \lozenge$$

---

43. A supersonic jet traveling at Mach 3.00 at an altitude of 20000 m is directly over a person at time $t = 0$, as in Figure P17.43. (a) How long will it be before the person encounters the shock wave? (b) Where will the plane be when it is finally heard? (Assume that the speed of sound in air is 335 m/s.)

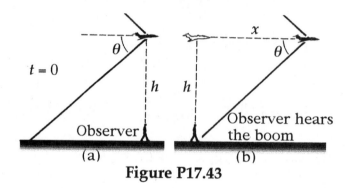

Figure P17.43

Solution Because the shock wave proceeds a set angle θ from the plane, we solve part (b) first.

(b) We use $\quad \sin\theta = \dfrac{v}{v_s} = \dfrac{\text{Mach 1}}{\text{Mach 3}} = \dfrac{1}{3},\quad$ and solve for $\quad \theta = 19.47°$

$$x = \frac{h}{\tan\theta} = \frac{20\,000 \text{ m}}{\tan 19.47°} = 56\,570 \text{ m} = 56.6 \text{ km} \qquad \lozenge$$

(a) It takes the plane $\quad t = \dfrac{x}{v_s} = \dfrac{56\,570 \text{ m}}{3(335 \text{ m / s})} = 56.3 \text{ s}$ to travel this distance $\qquad \lozenge$

---

53. To determine her own speed, a sky diver carries a buzzer that emits a steady tone at 1800 Hz. A friend at the landing site on the ground directly below the sky diver listens to the amplified sound he receives from the buzzer. Assume that the air is calm and that the speed of sound is 343 m/s, independent of altitude. While the sky diver is falling at terminal speed, her friend on the ground receives waves with a frequency of 2150 Hz. (a) What is the sky diver's speed of descent? (b) Suppose the sky diver is also carrying sound-receiving equipment that is sensitive enough to detect waves reflected from the ground. What frequency does she receive?

Solution

G: Sky divers typically reach a terminal speed of about 150 mph (~75 m/s), so this sky diver should also fall near this rate. Since her friend receives a higher frequency as a result of the Doppler shift, the sky diver should detect a frequency with twice the Doppler shift, at approximately $f' = 1800 \text{ Hz} + 2(2150 \text{ Hz} - 1800 \text{ Hz}) = 2500 \text{ Hz}$.

O: We can use the equation for the Doppler effect to answer both (a) and (b).

A: Call $f_e = 1800 \text{ Hz}$ the emitted frequency; v_e, the speed of the sky diver; and $f_g = 2150 \text{ Hz}$, the frequency of the wave crests reaching the ground.
(a) The sky diver source is moving toward the stationary ground, so we rearrange the equation

$$f_g = f_e \left(\frac{v}{v - v_e} \right)$$

to give $\quad v_e = v \left(1 - \frac{f_e}{f_g} \right) = (343 \text{ m / s}) \left(1 - \frac{1800 \text{ Hz}}{2500 \text{ Hz}} \right) = 55.8 \text{ m / s}$ ◊

(b) The ground now becomes a stationary source, reflecting crests with the 2150-Hz frequency at which they reach the ground, and sending them to a moving observer:

$$f_{e2} = f_g \left(\frac{v + v_e}{v} \right) = (2150 \text{ Hz}) \left(\frac{343 \text{ m / s} + 55.8 \text{ m / s}}{343 \text{ m / s}} \right) = 2500 \text{ Hz} \quad ◊$$

L: The answers appear to be consistent with our predictions, although the sky diver is falling somewhat slower than expected. The Doppler effect can be used to find the speed of many different types of moving objects, like raindrops (Doppler radar) and cars (police radar).

61. Two ships are moving along a line due east. The trailing vessel has a speed of 64.0 km/h relative to a land-based observation point, and the leading ship has a speed of 45.0 km/h relative to that point. The two ships are in a region of the ocean where the current is moving uniformly due west at 10.0 km/h. The trailing ship transmits a sonar signal at a frequency of 1200.0 Hz. What frequency is monitored by the leading ship? (Use 1520 m/s as the speed of sound in ocean water.)

Solution When the observer is moving in front of and in the same direction as the source,

$$f' = \left(\frac{v - v_0}{v - v_s}\right) f$$

where v_0 and v_s are measured relative to the **medium** in which the sound is propagated. In this case the ocean current is opposite the direction of travel of the ships and

$$v_0 = 45.0 \text{ km} / \text{h} - (-10.0 \text{ km} / \text{h}) = 55.0 \text{ km} / \text{h} = 15.3 \text{ m} / \text{s}$$

$$v_s = 64.0 \text{ km} / \text{h} - (-10.0 \text{ km} / \text{h}) = 74.0 \text{ km} / \text{h} = 20.6 \text{ m} / \text{s}$$

Therefore,

$$f' = (1200 \text{ Hz})\left(\frac{1520 \text{ m} / \text{s} - 15.3 \text{ m} / \text{s}}{1520 \text{ m} / \text{s} - 20.6 \text{ m} / \text{s}}\right) = 1204 \text{ Hz} \qquad \Diamond$$

63. A meteoroid the size of a truck enters the Earth's atmosphere at a speed of 20.0 km/s and is not significantly slowed before entering the ocean. (a) What is the Mach angle of the shock wave from the meteoroid in the atmosphere? (Use 331 m/s as the sound speed.) (b) Assuming that the meteoroid survives the impact with the ocean surface, what is the (initial) Mach angle of the shock wave that the meteoroid produces in the water? (Use the wave speed for sea water given in Table 17.1.)

Solution

(a)   The Mach angle in the air is

$$\theta_{atm} = \sin^{-1}\left(\frac{v}{v_s}\right) = \sin^{-1}\left(\frac{331 \text{ m}/\text{s}}{2.00 \times 10^4 \text{ m}/\text{s}}\right) = 0.948° \qquad \Diamond$$

(b)   At impact with the ocean,

$$\theta_{ocean} = \sin^{-1}\left(\frac{v}{v_s}\right) = \sin^{-1}\left(\frac{1533 \text{ m}/\text{s}}{2.00 \times 10^4 \text{ m}/\text{s}}\right) = 4.40° \qquad \Diamond$$

65. By proper excitation, it is possible to produce both longitudinal and transverse waves in a long metal rod. A particular metal rod is 150 cm long and has a radius of 0.200 cm and a mass of 50.9 g. Young's modulus for the material is $6.80 \times 10^{10} \text{ N}/\text{m}^2$. What must the tension in the rod be if the ratio of the speed of longitudinal waves to the speed of transverse waves is 8.00?

Solution

The longitudinal wave and the transverse wave have respective phase speeds of

$$v_L = \sqrt{\frac{Y}{\rho}} \qquad \text{where} \qquad \rho = \frac{\text{mass}}{\text{volume}} = \frac{m}{\pi r^2 L}$$

and $\qquad v_T = \sqrt{\frac{F}{\mu}} \qquad \text{where} \qquad \mu = \frac{m}{L}$

Since $v_L = 8.00 v_T$, $\qquad F = \frac{\mu Y}{64.0\rho} = \frac{\pi r^2 Y}{64.0}$

Solving, $\qquad F = \frac{\pi (0.00200 \text{ m})^2 (6.80 \times 10^{10} \text{ N / m}^2)}{64.0} = 1.34 \times 10^4 \text{ N} \qquad \Diamond$

---

71. Three metal rods are located relative to each other as shown in Figure P17.71, where $L_1 + L_2 = L_3$. The density and Young's modulus for the three materials are $\rho_1 = 2.70 \times 10^3 \text{ kg / m}^3$, $Y_1 = 7.00 \times 10^{10} \text{ N / m}^2$, $\rho_2 = 11.3 \times 10^3 \text{ kg / m}^3$, $Y_2 = 1.60 \times 10^{10} \text{ N / m}^2$, $\rho_3 = 8.80 \times 10^3 \text{ kg / m}^3$, and $Y_3 = 11.0 \times 10^{10} \text{ N / m}^2$. (a) If $L_3 = 1.50$ m,

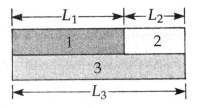

Figure P17.71

what must the ratio L_1 / L_2 be if a sound wave is to travel the combined length of rods 1 and 2 in the same time it takes to travel the length of rod 3? (b) If the frequency of the source is 4.00 kHz, determine the phase difference between the wave traveling along rods 1 and 2 and the one traveling along rod 3.

Solution

(a) The time required for a sound pulse to travel a distance L at a speed v is given by

$$t = \frac{L}{v} = \frac{L}{\sqrt{Y/\rho}}$$

Using this expression, we find

$$t_1 = L_1\sqrt{\frac{\rho_1}{Y_1}} = L_1\sqrt{\frac{2.70\times10^3 \text{ kg/m}^3}{7.00\times10^{10} \text{ N/m}^2}} = L_1\left(1.96\times10^{-4} \text{ s/m}\right)$$

$$t_2 = (1.50 - L_1)\sqrt{\frac{11.3\times10^3 \text{ kg/m}^3}{1.60\times10^{10} \text{ N/m}^2}} = 1.26\times10^{-3} \text{ s} - \left(8.40\times10^{-4} \text{ s/m}\right)L_1$$

$$t_3 = (1.50 \text{ m})\sqrt{\frac{8.80\times10^3 \text{ kg/m}^3}{11.0\times10^{10} \text{ N/m}^2}} = 4.24\times10^{-4} \text{ s}$$

We require $t_1 + t_2 = t_3$, or

$$\left(1.96\times10^{-4} \text{ s/m}\right)L_1 + \left(1.26\times10^{-3} \text{ s}\right) - \left(8.40\times10^{-4} \text{ s/m}\right)L_1 = 4.24\times10^{-4} \text{ s}$$

This gives $L_1 = 1.30 \text{ m}$ and $L_2 = (1.50 \text{ m}) - (1.30 \text{ m}) = 0.201 \text{ m}$

The ratio of lengths is $L_1/L_2 = 6.45$ ◊

(b) The ratio of lengths L_1/L_2 is adjusted in part (a) so that $t_1 + t_2 = t_3$. Therefore, sound travels the two paths in equal time intervals and the phase difference, $\Delta\phi = 0$. ◊

Chapter 18

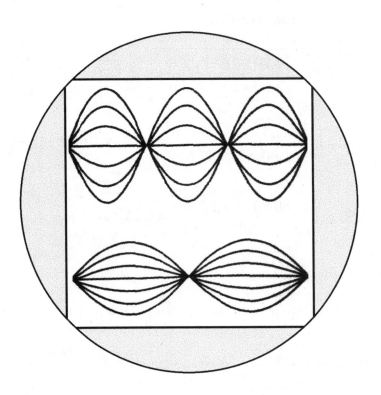

Superposition and Standing Waves

SUPERPOSITION AND STANDING WAVES

INTRODUCTION

An important aspect of waves is the combined effect of two or more of them traveling in the same medium.

In a linear medium, that is, one in which the restoring force of the medium is proportional to the displacement of the medium, the principle of superposition can be applied to obtain the resultant disturbance.

This chapter is concerned with the superposition principle as it applies to harmonic waves. If the harmonic waves that combine in a given medium have the same frequency and wavelength, and are traveling in opposite directions, one finds that a stationary pattern, called a **standing wave,** can be produced at certain frequencies.

EQUATIONS AND CONCEPTS

The wave function which is the resultant of two traveling harmonic waves having the same direction, frequency, and amplitude is also harmonic and has the same frequency and wavelength as the individual waves.

$$y = 2A\cos\left(\frac{\phi}{2}\right)\sin\left(kx - \omega t + \frac{\phi}{2}\right)$$

The amplitude of the resultant wave depends on the phase difference between the two individual waves.

$$\text{amplitude} = y_{max} = 2A\cos\left(\frac{\phi}{2}\right)$$

A **phase difference** can arise between two waves generated by the same source and arriving at a common point after having traveled along paths of **unequal path length.** This is also true for waves from different sources that are coherent, as in Young's double slit experiment. In this equation, ϕ is the phase difference and Δr is the **path difference** between the two waves.

$$\Delta r = \frac{\phi}{2\pi}\lambda \qquad\qquad (18.1)$$

A **standing wave** produced in a string, air column, or other medium is due to the interference of two sinusoidal waves with equal amplitude and frequency traveling in opposite directions.

$$y = (2A\sin kx)\cos\omega t \qquad\qquad (18.3)$$

The amplitude of a standing wave is a function of the position x along the string. **Antinodes,** A, are points of maximum displacement. **Nodes,** N, are points of zero displacement.

distance N to N $= \lambda/2$

A to A $= \lambda/2$

N to A $= \lambda/4$

A series of natural patterns of vibration called **normal modes** can be excited in a string fixed at both ends. Each mode corresponds to a characteristic frequency and wavelength.

$$\lambda_n = \frac{2L}{n} \qquad n = 1, 2, 3, \ldots \qquad (18.6)$$

$$f_n = n\frac{v}{2L} \qquad n = 1, 2, 3, \ldots \qquad (18.7)$$

$$f_n = \frac{n}{2L}\sqrt{\frac{T}{\mu}} \qquad n = 1, 2, 3, \ldots \qquad (18.8)$$

In an "open" pipe (open at both ends), all integral multiples of the fundamental frequency can be excited. In a "closed" pipe (closed at one end), only the odd multiples of the fundamental frequency--the odd harmonics--are possible.

$$f_n = n\frac{v}{2L} \qquad n = 1, 2, 3, \ldots \qquad (18.11)$$

$$f_n = n\frac{v}{4L} \qquad n = 1, 3, 5, \ldots \qquad (18.12)$$

Beats are formed by the combination of two waves of equal amplitude but slightly different frequencies traveling in the **same** direction.

$$(18.13)$$

$$y = \left[2A\cos 2\pi\left(\frac{f_1 - f_2}{2}\right)t\right]\cos 2\pi\left(\frac{f_1 + f_2}{2}\right)t$$

The amplitude of the wave described by Equation 18.13 is time dependent. Each occurrence of maximum amplitude results in a "beat"; the **beat frequency** f_b equals the difference in the frequencies of the individual waves.

$$A_{resultant} = 2A\cos 2\pi\left(\frac{f_1 - f_2}{2}\right)t \qquad (18.14)$$

$$f_b = |f_1 - f_2| \qquad (18.15)$$

Any **non-sinusoidal periodic wave form** can be represented by the combination of sinusoidal waves which form a harmonic series (combination of fundamental and various harmonics). Such a sum of sine and cosine terms is called a **Fourier series**.

$$(18.16)$$

$$y(t) = \sum_n \left(A_n \sin 2\pi f_n t + B_n \cos 2\pi f_n t\right)$$

REVIEW CHECKLIST

▷ Write out the wave function which represents the superposition of the two sinusoidal waves of equal amplitude and frequency traveling in opposite directions in the same medium.

▷ Identify the angular frequency, maximum amplitude, and determine the values of x which correspond to nodal and antinodal points of a standing wave, given an expression for the wave function.

▷ Calculate the normal mode frequencies for a string under tension, and for open and closed air columns.

▷ Describe the time dependent amplitude and determine the effective frequency of vibration when two waves of slightly different frequency interfere. Also, calculate the expected beat frequency for this situation.

ANSWERS TO SELECTED CONCEPTUAL QUESTIONS

2. Does the phenomenon of wave interference apply only to sinusoidal waves?

Answer No. Any waves moving in the same medium can interfere with each other. For example, two pulses moving in opposite directions on a stretched string interfere when they meet each other.

□   □   □   □

3. When two waves interfere constructively or destructively, is there any gain or loss in energy? Explain.

Answer No. The energy may be transformed into other forms of energy. For example, when two pulses traveling on a stretched string in opposite directions overlap, and one is inverted, some potential energy is transferred to kinetic energy when they overlap. In fact, if they have equal and opposite amplitudes, they completely cancel each other at one point. In this case, all of the energy is transverse kinetic energy when the resultant amplitude is zero.

□   □   □   □

12. An airplane mechanic notices that the sound from a twin-engine aircraft rapidly varies in loudness when both engines are running. What could be causing this variation from loudness to softness?

Answer Apparently the two engines are emitting sounds having frequencies which differ only by a very small amount from each other. This results in a beat frequency, causing the variation from loud to soft, and back again.

□   □   □   □

13. Why does a vibrating guitar string sound louder when placed on the instrument than it would if it were allowed to vibrate in the air while off the instrument?

Answer A vibrating string is not able to set very much air into motion when vibrated alone. Thus it will not be very loud. If it is placed on the instrument, however, the string's vibration sets the sounding board of the guitar into vibration. A vibrating piece of wood is able to move a lot of air, and the note is louder.

□   □   □   □

SOLUTIONS TO SELECTED END-OF-CHAPTER PROBLEMS

1. Two sinusoidal waves are described by the equations

$$y_1 = (5.00 \text{ m})\sin[\pi(4.00x - 1200t)]$$

and $$y_2 = (5.00 \text{ m})\sin[\pi(4.00x - 1200t - 0.250)]$$

where x, y_1, and y_2 are in meters and t is in seconds. (a) What is the amplitude of the resultant wave? (b) What is the frequency of the resultant wave?

Solution

We can represent the waves symbolically as

$$y_1 = A\sin(kx - \omega t)$$

and $$y_2 = A\sin(kx - \omega t - \phi)$$

with $$A = 5.00 \text{ m}, \quad \omega = 1200\pi \text{ s}^{-1}, \quad \text{and} \quad \phi = 0.250\pi$$

When we add the two waves, the resultant wave function has the form

$$y = y_1 + y_2 = 2A\cos(\phi/2)\sin(kx - \omega t - \phi/2)$$

(a) with amplitude $\quad A = 2A\cos(\phi/2) = 2(5.00)\cos(\pi/8) = 9.24 \text{ m}$ ◊

(b) and frequency $\quad f = \omega/2\pi = 1200\pi/2\pi = 600 \text{ Hz}$ ◊

388

11. Two speakers are driven by a common oscillator at 800 Hz and face each other at a distance of 1.25 m. Locate the points along a line joining the two speakers where relative minima would be expected. (Use $v = 343$ m / s.)

Solution

The wavelength is

$$\lambda = \frac{v}{f} = \frac{343 \text{ m / s}}{800 \text{ Hz}} = 0.429 \text{ m}$$

For minima, the difference in path length must be an odd number of half wavelengths. So, if x is the distance from the listener to the speaker on the left, then the path difference will be

$$(1.25 \text{ m} - x) - x = (2n+1)\frac{\lambda}{2}$$

where n is any integer. Use $\lambda = 0.429$ m and substitute allowed values of n to calculate corresponding values of x. Solving, gives the following results:

$n = 0,\quad x = 0.518$ m	$n = -1,\quad x = 0.732$ m
$n = 1,\quad x = 0.303$ m	$n = -2,\quad x = 0.947$ m
$n = 2,\quad x = 0.0891$ m	$n = -3,\quad x = 1.16$ m

The nodes are at distances of 0.0891 m, 0.304 m, 0.518 m, 0.732 m, 0.947 m, and 1.16 m from either speaker. ◊

13. Two sinusoidal waves are described by the equations

$$y_1 = (3.0 \text{ cm})\sin \pi(x + 0.60t) \qquad y_2 = (3.0 \text{ cm})\sin \pi(x - 0.60t)$$

where x is in centimeters and t is in seconds. Determine the **maximum** displacement of the motion at (a) $x = 0.250$ cm, (b) $x = 0.500$ cm, and (c) $x = 1.50$ cm. (d) Find the three smallest values of x corresponding to antinodes.

Solution   Adding y_1, and y_2, and using the trigonometry identity

$$\sin(\alpha + \beta) = \sin \alpha \cos \beta + \cos \alpha \sin \beta$$

We get $\qquad y = y_1 + y_2 = (6.0 \text{ cm})\sin(\pi x)\cos(0.60\pi t)$

Since $\cos(0) = 1$, we can find the maximum value of y by setting $t = 0$:

$$y_{max}(x) = y_1 + y_2 = (6.0 \text{ cm})\sin(\pi x)$$

(a)   At $x = 0.250$ cm, $y = (6.0 \text{ cm})\sin(0.250\pi) = 4.24$ cm ◊

(b)   At $x = 0.500$ cm, $y = (6.0 \text{ cm})\sin(0.500\pi) = 6.00$ cm ◊

(c)   At $x = 1.50$ cm, $y = (6.0 \text{ cm})\sin(1.50\pi) = -6.00$ cm ◊

(d)   The antinodes occur when $x = n\lambda/4$ $\quad (n = 1, 3, 5, \ldots)$

But $\quad k = 2\pi/\lambda = \pi,\quad$ so $\quad \lambda = 2.00$ cm, and

$$x_1 = \lambda/4 = (2.00 \text{ cm})/4 = 0.500 \text{ cm} \qquad ◊$$

$$x_2 = 3\lambda/4 = 3(2.00 \text{ cm})/4 = 1.50 \text{ cm} \qquad ◊$$

$$x_3 = 5\lambda/4 = 5(2.00 \text{ cm})/4 = 2.50 \text{ cm} \qquad ◊$$

17. Find the fundamental frequency and the next three frequencies that could cause a standing wave pattern on a string that is 30.0 m long, has a mass per length of 9.00×10^{-3} kg / m, and is stretched to a tension of 20.0 N.

Solution

G: The string described in the problem is very long, loose, and somewhat heavy, so it should have a very low fundamental frequency, maybe only a few vibrations per second.

O: The tension and linear density of the string can be used to find the wave speed, which can then be used along with the required wavelength to find the fundamental frequency.

A: The wave speed is

$$v = \sqrt{\frac{T}{\mu}} = \sqrt{\frac{20.0 \text{ N}}{9.00 \times 10^{-3} \text{ kg / m}}} = 47.1 \text{ m / s}$$

For a vibrating string of length L fixed at both ends, the wavelength of the fundamental is $\lambda = 2L = 60.0$ m; and the frequency is

$$f_1 = \frac{v}{\lambda} = \frac{v}{2L} = \frac{47.1 \text{ m / s}}{60.0 \text{ m}} = 0.786 \text{ Hz} \qquad \lozenge$$

The next three harmonics are

$$f_2 = 2f_1 = 1.57 \text{ Hz}, \quad f_3 = 3f_1 = 2.36 \text{ Hz}, \quad f_4 = 4f_1 = 3.14 \text{ Hz} \qquad \lozenge$$

L: The fundamental frequency is even lower than expected, less than 1 Hz. You could watch the string vibrating. It would weakly broadcast sound into the surrounding air, but all four of the lowest resonant frequencies are below the normal human hearing range (20 to 17 000 Hz), so the sounds are not even audible.

19. A cello A-string vibrates in its first normal mode with a frequency of 220 vibrations/s. The vibrating segment is 70.0 cm long and has a mass of 1.20 g. (a) Find the tension in the string. (b) Determine the frequency of vibration when the string vibrates in three segments.

Solution

G: The tension should be less than 500 N (~100 lb) since excessive force on the four cello strings would break the neck of the instrument. If a string vibrates in three segments, there will be three antinodes (instead of one for the fundamental mode), so the frequency should be three times greater than the fundamental.

O: From the string's length, we can find the wavelength,. We then can use the wavelength with the fundamental frequency to find the wave speed. Finally, we can find the tension from the wave speed and the linear mass density of the string.

A: When the string vibrates in the lowest frequency mode, the length of string forms a standing wave where $L = \lambda / 2$ so the fundamental harmonic wavelength is

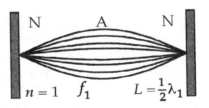

$$\lambda = 2L = 2(0.700 \text{ m}) = 1.40 \text{ m}$$

and the velocity is

$$v = f\lambda = (220 \text{ s}^{-1})(1.40 \text{ m}) = 308 \text{ m / s}$$

From the tension equation $\quad v = \sqrt{\dfrac{T}{\mu}} = \sqrt{\dfrac{T}{m/L}}$

(a) We get $T = v^2 m/L$, or $\quad T = \dfrac{(308 \text{ m}/\text{s})^2 \left(1.20 \times 10^{-3} \text{ kg}\right)}{0.700 \text{ m}} = 163 \text{ N} \quad \Diamond$

(b) For the third harmonic, the tension, linear density, and speed are the same, but the string vibrates in three segments. Thus, that the wavelength is one third as long as in the fundamental.

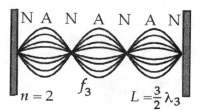

$$\lambda_3 = \lambda / 3$$

From the equation $\quad v = f\lambda$, we find the frequency is three times as high:

$$f_3 = \frac{v}{\lambda_3} = 3\frac{v}{\lambda} = 3f = 660 \text{ Hz} \qquad \Diamond$$

L: The tension seems reasonable, and the third harmonic is three times the fundamental frequency as expected. Related to part (b), some stringed instrument players use a technique to double the frequency of a note by "cutting" a vibrating string in half. When the string is lightly touched at its midpoint to form a node, the second harmonic is formed, and the resulting note is one octave higher (twice the original fundamental frequency).

31. Calculate the length of a pipe that has a fundamental frequency of 240 Hz if the pipe is (a) closed at one end and (b) open at both ends.

Solution   The relationship between the frequency and the wavelength of a sound wave is

$$v = f\lambda \qquad \text{or} \qquad \lambda = v/f$$

Next, we draw the pipes to help us visualize where the relationship between λ and L.

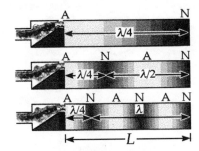

(a)   For the fundamental mode in a closed pipe, $\lambda = 4L$.

So $$L = \frac{\lambda}{4} = \frac{v/f}{4}$$

and $$L = \frac{(343 \text{ m / s})/(240 \text{ s}^{-1})}{4} = 0.357 \text{ m} \quad \lozenge$$

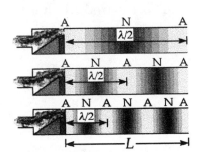

(b)   For the fundamental mode in an open pipe, $\lambda = 2L$

so $$L = \frac{\lambda}{2} = \frac{v/f}{2}$$

and $$L = \frac{(343 \text{ m / s})/(240 \text{ s}^{-1})}{2} = 0.715 \text{ m} \quad \lozenge$$

---

37.   A shower stall measures 86.0 cm × 86.0 cm × 210 cm.   If you were singing in this shower, which frequencies would sound the richest (because of resonance)?   Assume that the shower stall acts as a pipe closed at both ends, with nodes at opposite sides.   Assume that the voices of various singers range from 130 Hz to 2000 Hz.   Let the speed of sound in the hot shower stall be 355 m/s.

Solution   For a box closed at both ends, the pattern of standing waves will have the same properties as a string fixed at both ends. The resonant frequencies will have nodes at both sides, and we can use equation 18.7,

$$f_n = \frac{nv}{2L} \qquad\qquad \text{for} \qquad (n = 1, 2, 3, \ldots)$$

Therefore, with $L = 0.860$ m, the side-to-side resonant frequencies are

$$f_n = \frac{nv}{2L} = n\frac{355 \text{ m / s}}{2(0.860 \text{ m})} = n(206 \text{ Hz}), \qquad \text{for each } n \text{ from 1 to 9} \qquad\qquad \lozenge$$

With $L' = 2.10$ m, the top-to-bottom resonance frequencies are

$$f_n = \frac{nv}{2L'} = n\frac{355 \text{ m / s}}{2(2.10 \text{ m})} = n(84.5 \text{ Hz}), \qquad \text{for each } n \text{ from 1 to 23} \qquad\qquad \lozenge$$

---

43.   A glass tube is open at one end and closed at the other by a movable piston. The tube is filled with air warmer than that at room temperature, and a 384-Hz tuning fork is held at the open end. Resonance is heard when the piston is 22.8 cm from the open end and again when it is 68.3 cm from the open end. (a) What speed of sound is implied by these data? (b) How far from the open end will the piston be when the next resonance is heard?

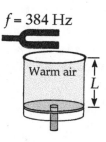

$f = 384$ Hz

Warm air

Solution   For an air column closed at one end, resonances will occur when the length of the column is equal to $\lambda/4$, $3\lambda/4$, $5\lambda/4$, and so on. Thus, the change in the length of the pipe from one resonance to the next is $\lambda/2$. In this case,

$$\lambda/2 = (0.683 \text{ m} - 0.228 \text{ m}) = 0.455 \text{ m} \qquad \text{and} \qquad \lambda = 0.910 \text{ m}$$

(a) $v = f\lambda = (384 \text{ Hz})(0.910 \text{ m}) = 349 \text{ m / s}$ $\qquad\qquad\qquad\qquad\qquad\qquad \lozenge$

(b) $L = 0.683 \text{ m} + 0.455 \text{ m} = 1.14 \text{ m}$ $\qquad\qquad\qquad\qquad\qquad\qquad\qquad \lozenge$

---

47. An aluminum rod 1.60 m long is held at its center. It is stroked with a resin-coated cloth to set up a longitudinal vibration. (a) What is the fundamental frequency of the waves established in the rod? (b) What harmonics are set up in the rod held in this manner? (c) What would be the fundamental frequency if the rod were copper?

Solution Either using $v = \sqrt{Y/\rho}$ or Table 17.1, we can find the speed of sound in aluminum and copper to be:

$$v_{Al} = 5100 \text{ m/s} \qquad \text{and} \qquad v_{Cu} = 3560 \text{ m/s}$$

(a) Since the central clamp establishes a node at the center, the fundamental mode of vibration will be ANA. Our first harmonic frequency, then, is

$$f_1 = \frac{v}{\lambda_1} = \frac{v}{2L} = \frac{5100 \text{ m/s}}{3.20 \text{ m}} = 1.59 \text{ kHz} \qquad \Diamond$$

(b) Since the rod is free at each end, the ends will be antinodes. The next vibration state will be ANANANA, as shown, with a wavelength and frequency of:

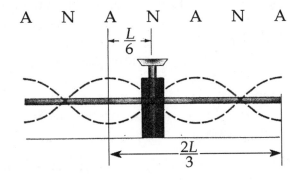

A N A N A N A

$$\lambda = \frac{2L}{3} \qquad \text{and} \qquad f = \frac{v}{\lambda} = \frac{3v}{2\lambda} = 3f_1$$

Since $f = 2f_1$ was bypassed, we know that we get odd harmonics.

That is, $\qquad f = \dfrac{nv}{2L} \qquad$ for $\qquad n = 1, 3, 5, \ldots \qquad \Diamond$

(c) For a copper rod, the speed of sound changes:

$$f_1 = \frac{v}{2L} = \frac{3560 \text{ m/s}}{3.20 \text{ m}} = 1.11 \text{ kHz} \qquad \Diamond$$

49. In certain ranges of a piano keyboard, more than one string is tuned to the same note to provide extra loudness. For example, the note at 110 Hz has two strings that vibrate at this frequency. If one string slips from its normal tension of 600 N to 540 N, what beat frequency is heard when the hammer strikes the two strings simultaneously?

Solution

G: Directly noticeable beat frequencies are usually only a few Hertz, so we should not expect a frequency much greater than this.

O: As in previous problems, the two wave speed equations can be used together to find the frequency of vibration that corresponds to a certain tension. The beat frequency is then just the difference in the two resulting frequencies from the two strings with different tensions.

A: Combining the velocity and the tension equations,

$$v = f\lambda \qquad \text{and} \qquad v = \sqrt{T/\mu}$$

we find that
$$f = \sqrt{\frac{T}{\mu \lambda^2}}$$

Since μ and λ are constant, we can apply that equation to both frequencies, and then divide the two equations to get

$$\frac{f_1}{f_2} = \sqrt{\frac{T_1}{T_2}}$$

With $f_1 = 110$ Hz, $T_1 = 600$ N, and $T_2 = 540$ N:

$$f_2 = (110 \text{ Hz})\sqrt{\frac{540 \text{ N}}{600 \text{ N}}} = 104.4 \text{ Hz}$$

The beat frequency is: $\qquad f_b = |f_1 - f_2| = 110 \text{ Hz} - 104.4 \text{ Hz} = 5.64 \text{ Hz}$ ◊

L: As expected, the beat frequency is only a few cycles per second. This result from the interference of the two sound waves with slightly different frequencies has a tone that varies in amplitude over time, similar to the sound made by saying "wa-wa-wa..."

Note: The beat frequency above is written with three significant figures on the assumption that the original data is known precisely enough to warrant them. This assumption implies that the original frequency is known somewhat more precisely than to the three significant digits quoted in "110 Hz." For example, if the original frequency of the strings were 109.6 Hz, the beat frequency would be 5.62 Hz.

---

51. A student holds a tuning fork oscillating at 256 Hz. He walks toward a wall at a constant speed of 1.33 m/s. (a) What beat frequency does he observe between the tuning fork and its echo? (b) How fast must he walk away from the wall to observe a beat frequency of 5.00 Hz?

Solution For an echo, $f' = \left(\dfrac{v + v_s}{v - v_s}\right)f$ and the beat frequency $f_b = |f - f_0|$

Solving for f_b gives $f_b = \left(\dfrac{2v_s}{v - v_s}\right)f$ when approaching the wall.

(a) $f_b = (256 \text{ Hz})\left(\dfrac{2(1.33 \text{ m / s})}{343 \text{ m / s} - 1.33 \text{ m / s}}\right) = 1.99 \text{ Hz}$ ◊

(b) When moving away from wall, the + and − signs are interchanged. Solving for v_s gives

$$v_s = f_b\left(\dfrac{v}{2f_0 - f_b}\right) = (5.00 \text{ Hz})\left(\dfrac{343 \text{ m / s}}{2(256 \text{ Hz}) - 5.00 \text{ Hz}}\right) = 3.38 \text{ m / s}$$ ◊

---

57. Two train whistles have identical frequencies of 180 Hz. When one train is at rest in the station and is sounding its whistle, a beat frequency of 2.00 Hz is heard from a train moving nearby. What are the two possible speeds and directions that the moving train can have?

Solution

We know from our beat equation $f_b = |f_1 - f_2|$ that the moving train can have apparent frequencies of $f' = 182$ Hz or $f' = 178$ Hz.

If we assume the train is moving away from the station, the apparent frequency of the moving train is lower (178 Hz):

$$f' = f \frac{v}{v + v_s}$$

and the train is **moving away** at
$$v_s = v\left(\frac{f}{f'} - 1\right) = (343 \text{ m / s})\left(\frac{180 \text{ Hz}}{178 \text{ Hz}} - 1\right)$$

$$v_s = 3.85 \text{ m / s} \qquad \lozenge$$

If the train is pulling into the station, then the apparent frequency is 182 Hz.

Again from the Doppler shift,
$$f' = f \frac{v}{v - v_s}$$

The train is **approaching** at:
$$v_s = v\left(1 - \frac{f}{f'}\right) = (343 \text{ m / s})\left(1 - \frac{180 \text{ Hz}}{182 \text{ Hz}}\right)$$

$$v_s = 3.77 \text{ m / s} \qquad \lozenge$$

63. Two wires are welded together. The wires are of the same material, but the diameter of one wire is twice that of the other. They are subjected to a tension of 4.60 N. The thin wire has a length of 40.0 cm and a linear mass density of 2.00 g/m. The combination is fixed at both ends and vibrated in such a way that two antinodes are present with the node between them being right at the weld. (a) What is the frequency of vibration? (b) How long is the thick wire?

Solution (a) Since the first node is at the weld, the wavelength in the thin wire is $2L$ or 80.0 cm. The frequency and tension are the same in both sections, so

$$f = \frac{1}{2L}\sqrt{\frac{T}{\mu}} = \frac{1}{2(0.400 \text{ m})}\sqrt{\frac{4.60 \text{ N}}{0.00200 \text{ kg / m}}} = 59.9 \text{ Hz} \quad \Diamond$$

(b) Since the thick wire is twice the diameter, it will have 4 times the cross-sectional area, and a linear density μ that is 4 times that of the thin wire.

$$\mu' = 4(2.00 \text{ g / m}) = 0.00800 \text{ kg / m}$$

L' varies accordingly: $\quad L' = \dfrac{1}{2f}\sqrt{\dfrac{T}{\mu'}} = \dfrac{1}{2(59.9 \text{ Hz})}\sqrt{\dfrac{4.60 \text{ N}}{0.00800 \text{ kg / m}}} = 20.0 \text{ cm} \quad \Diamond$

Note that the thick wire is half the length of the thin wire.

---

65. A standing wave is set up in a string of variable length and tension by a vibrator of variable frequency. When the vibrator has a frequency f, in a string of length L and under tension T, n antinodes are set up in the string. (a) If the length of the string is doubled, by what factor should the frequency be changed so that the same number of antinodes is produced? (b) If the frequency and length are held constant, what tension produces $n+1$ antinodes? (c) If the frequency is tripled and the length of the string is halved, by what factor should the tension be changed so that twice as many antinodes are produced?

Solution Combining the equations $v = f\lambda$ and $v = \sqrt{T/\mu}$, and noting that $\lambda_n = 2L/n$, we find that

$$f_n = \frac{n}{2L}\sqrt{\frac{T}{\mu}} \qquad (1)$$

(a) Keeping n, f, and μ constant,

$$f_n L = \frac{n}{2}\sqrt{\frac{T}{\mu}}$$

and

$$f_n' L' = \frac{n}{2}\sqrt{\frac{T}{\mu}}$$

Dividing the equations, we find $\qquad f_n/f_n' = L'/L$

If $L' = 2L$, then $\qquad f_n' = \tfrac{1}{2}f_n$

Therefore, in order to double the length but keep the same number of antinodes, the frequency should be halved. ◊

(b) Holding L and f_n constant in Eq. (1), $\qquad n'/n = \sqrt{T/T'}$

From this relation, we see that to produce $n+1$ antinodes, the tension must be decreased to $\qquad T' = Tn^2/(n+1)^2$ ◊

(c) Rearrange equation (1),

$$\frac{2f_n L}{n} = \sqrt{\frac{T}{\mu}}$$

and

$$\frac{2f_n' L'}{n'} = \sqrt{\frac{T'}{\mu}}$$

Dividing, we get $\qquad \dfrac{T'}{T} = \left(\dfrac{f_n'}{f_n}\cdot\dfrac{n}{n'}\cdot\dfrac{L'}{L}\right)^2 = \left(\dfrac{3f}{f_n}\right)^2\left(\dfrac{n}{2n}\right)^2\left(\dfrac{L/2}{L}\right)^2 = \dfrac{9}{16}$ ◊

67. If two adjacent natural frequencies of an organ pipe are determined to be 0.550 kHz and 0.650 kHz, calculate the fundamental frequency and length of the pipe. (Use $v = 343$ m / s.)

Solution

Because harmonic frequencies are given by $f_0 n$ for open pipes, and $f_0(2n-1)$ for closed pipes, the difference between all adjacent harmonics is constant. Therefore, we can find each harmonic below 650 Hz by subtracting

$$\Delta f_{\text{Harmonic}} = (650 \text{ Hz} - 550 \text{ Hz}) = 100 \text{ Hz}$$

from the previous value.

The harmonics are {650 Hz, 550 Hz, 450 Hz, 350 Hz, 250 Hz, 150 Hz, 50.0 Hz}

(a) The fundamental frequency, then, is 50.0 Hz. ◊

(b) The wavelength can be calculated from the velocity:

$$\lambda = \frac{v}{f} = \frac{340 \text{ m / s}}{50.0 \text{ Hz}} = 6.80 \text{ m}$$

Because the step size Δf is twice the fundamental frequency, we know the pipe is closed, with an antinode at the open end, and a node at the closed end. The wavelength in this situation is four times the pipe length, so

$$L = \frac{\lambda}{4} = 1.70 \text{ m}$$ ◊

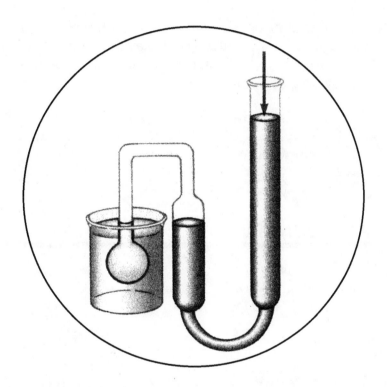

Temperature

TEMPERATURE

INTRODUCTION

A quantitative description of thermal phenomena requires a careful definition of the concepts of temperature, heat, and internal energy. The laws of thermodynamics provide us with a relationship among heat flow, work, and the internal energy of a system.

This chapter also includes a discussion of ideal gases on the macroscopic scale. This involves the relationships among such quantities as pressure, volume, and temperature.

EQUATIONS AND CONCEPTS

Pressure is force per unit area and has units of N/m^2. The SI unit of pressure is the pascal (Pa).

$$1\,Pa \equiv 1\,N/m^2$$

One atmosphere of pressure is equal to a standardized atmospheric pressure at sea level.

$$1\,atm = 1.013 \times 10^5\,Pa$$

The **Celsius temperature** T_C is related to the absolute temperature T (in kelvins) according to Equation 19.1, where $0°C$ corresponds to $273.15\,K$.

$$T_C = T - 273.15 \tag{19.1}$$

The **Fahrenheit temperature** T_F can be converted to degrees Celsius using Equation 19.2. Note that $0\,°C = 32\,°F$ and $100\,°C = 212\,°F$.

$$T_F = \frac{9}{5}T_C + 32\,°F \qquad (19.2)$$

If a body has a length L_i the **change in its length** ΔL due to a change in temperature is proportional to the change in temperature and the length. The proportionality constant α is called the **average coefficient of linear expansion.**

$$\Delta L = \alpha L_i \Delta T \qquad (19.4)$$

or

$$\alpha = \frac{1}{L_i}\frac{\Delta L}{\Delta T}$$

If the temperature of a body of volume V_i changes by an amount ΔT at constant pressure, **the change in its volume** is proportional to ΔT and the original volume. The constant of proportionality β is the **average coefficient of volume expansion**. For an isotropic solid, $\beta \cong 3\alpha$.

$$\Delta V = \beta V_i \Delta T \qquad (19.6)$$

A sample of any substance contains a number of moles which equals the ratio of the mass of the sample to the molar mass characteristic of that particular substance.

$$n = \frac{m}{M} \qquad (19.7)$$

If n moles of a dilute gas occupy a volume V, the **equation of state** which relates the variables P, V, and T at equilibrium is that of an **ideal gas**, Equation 19.8, where R is the **universal gas constant**.

$$PV = nRT \qquad (19.8)$$

$$R = 8.315 \text{ J} / \text{mol} \cdot \text{K} \qquad (19.9)$$

or

$$R = 0.0821 \text{ L} \cdot \text{atm} / \text{mol} \cdot \text{K}$$

The **equation of state of an ideal gas** can also be expressed in the form of Equation 19.10, where N is the total number of gas molecules and k_B is **Boltzmann's constant**.

$$PV = Nk_BT \qquad (19.10)$$

$$k_B = \frac{R}{N_A} = 1.38 \times 10^{-23} \text{ J} / \text{K} \qquad (19.11)$$

REVIEW CHECKLIST

▷ Understand the concepts of thermal equilibrium and thermal contact between two bodies, and state the zeroth law of thermodynamics.

▷ Discuss some physical properties of substances which change with temperature, and the manner in which these properties are used to construct thermometers.

▷ Describe the operation of the constant-volume gas thermometer and how it is used to define the ideal gas temperature scale.

▷ Convert between the various temperature scales, especially the conversion from degrees Celsius into kelvins, degrees Fahrenheit into kelvins, and degrees Celsius into degrees Fahrenheit.

▷ Provide a qualitative description of the origin of thermal expansion of solids and liquids; define the linear expansion coefficient and volume expansion coefficient for an isotropic solid, and learn how to deal with these coefficients in practical situations involving expansion or contraction.

ANSWERS TO SELECTED CONCEPTUAL QUESTIONS

2. A piece of copper is dropped into a beaker of water. If the water's temperature increases, what happens to the temperature of the copper? Under what conditions are the water and copper in thermal equilibrium?

Answer If the water's temperature increases, that means that heat is being transferred to the water. This can happen either if the copper changes phase from liquid to solid, or if the temperature of the copper is above that of the water, and falling.

In this case, the copper is referred to as a "piece" of copper, so it is already in the solid phase; therefore, its temperature must be falling. When the temperature of the copper reaches that of the water, the copper and water will reach equilibrium, and the subsequent net heat transfer between the two will be zero.

□   □   □   □

15. When the metal ring and metal sphere shown in Figure Q19.15 are both at room temperature, the sphere can just be passed through the ring. After the sphere is heated, it cannot be passed through the ring. Explain.

Answer The thermal expansion of the ring is not like the inflation of a blood-pressure cuff. Rather, it is like a photographic enlargement; every linear dimension, including the hole diameter, increases by the same factor. The reason for this is that the atoms everywhere, including those around the inner circumference, push away from each other. The only way that the atoms can accommodate the greater distances is for the circumference—and corresponding diameter— to grow. This property was once used to fit metal rims to wooden wagon and horse-buggy wheels.

Figure Q19.15

□   □   □   □

SOLUTIONS TO SELECTED END-OF-CHAPTER PROBLEMS

3. A constant-volume gas thermometer is calibrated in dry ice (that is, carbon dioxide in the solid state, which has a temperature of –80.0 °C) and in boiling ethyl alcohol (78.0 °C). The two pressures are 0.900 atm and 1.635 atm. (a) What Celsius value of absolute zero does the calibration yield? What is the pressure at (b) the freezing point of water and (c) the boiling point of water?

Solution   Since we have a linear graph, the pressure is related to the temperature as $P = A + BT$, where A and B are constants. To find A and B, we use the given data:

$$0.900 \text{ atm} = A + (-80.0 \text{ °C})B \tag{1}$$

$$1.635 \text{ atm} = A + (78.0 \text{ °C})B \tag{2}$$

Solving (1) and (2) simultaneously, we find

$$A = 1.272 \text{ atm} \quad \text{and} \quad A = 4.652 \text{ atm} \times 10^{-3} \text{ atm/°C}$$

Therefore, $P = 1.272 \text{ atm} + \left(4.652 \times 10^{-3} \text{ atm/°C}\right)T$

(a) At 0 K, $P = 0 = 1.272 \text{ atm} + \left(4.652 \times 10^{-3} \text{ atm/°C}\right)T$

so $T = -274 \text{ °C}$ ◊

(b)   At the freezing point of water,

$$P = 1.272 \text{ atm} + 0 = 1.27 \text{ atm}$$ ◊

(c)   and at the boiling point,

$$P = 1.272 \text{ atm} + \left(4.652 \times 10^{-3} \text{ atm/°C}\right)(100 \text{ °C}) = 1.74 \text{ atm}$$ ◊

5. Liquid nitrogen has a boiling point of –195.81 °C at atmospheric pressure. Express this temperature in (a) degrees Fahrenheit, and (b) kelvins.

Solution

(a) $$T_F = \tfrac{9}{5} T_C + 32.000 \text{ °F} = \tfrac{9}{5}(-195.81 \text{ °C}) + 32.000 \text{ °F} = -320 \text{ °F} \qquad \lozenge$$

(b) The ice point in kelvin is 273.15 K, so liquid N_2 boils at

$$273.15 \text{ K} - 195.81 \text{ K} = 77.3 \text{ K} \qquad \lozenge$$

Note that a kelvin is the same size change as a degree Celsius.

___

9. A copper telephone wire has essentially no sag between two poles 35.0 m apart on a winter day when the temperature is –20.0 °C. How much longer is the wire on a summer day when $T_C = 35.0$ °C?

Solution

G: Normally, we do not notice a change in the length of the telephone wires. Thus, we might expect the wire to expand by less than a meter.

O: The change in length can be found from the linear expansion of copper wire (we will assume that the insulation around the copper wire can stretch more easily than the wire itself). From Table 19.2, the coefficient of linear expansion for copper is 17.0×10^{-6} $(°C)^{-1}$.

A: The change in length between cold and hot conditions is

$$\Delta L = \alpha L_i \Delta T = \left[17.0 \times 10^{-6} \, (°C)^{-1} \right] (35.0 \text{ m})(35.0°C - (-20.0°C))$$
$$\Delta L = 3.27 \times 10^{-2} \text{ m} = 3.27 \text{ cm} \qquad \lozenge$$

L: This expansion agrees with our expectation that the change in length is less than a meter. From ΔL, we can find that if the wire sags, its midpoint can be displaced downward by 0.757 m on the hot summer day. This also seems reasonable based on everyday observations.

___

17. The active element of a certain laser is a glass rod 30.0 cm long by 1.50 cm in diameter. If the temperature of the rod increases by 65.0 °C, what is the increase in (a) its length, (b) its diameter, and (c) its volume. (Assume that $\alpha = 9.00 \times 10^{-6} (°C)^{-1}$)

Solution

(a) $\Delta L = \alpha L_i \, \Delta T = \left[9.00 \times 10^{-6} (°C)^{-1} \right] (0.300 \text{ m})(65.0 \text{ °C}) = 1.76 \times 10^{-4} \text{ m}$ ◊

(b) The diameter is a linear dimension, so the same equation applies:

$$\Delta D = \alpha D_i \, \Delta T = \left[9.00 \times 10^{-6} (°C)^{-1} \right] (0.0150 \text{ m})(65.0 \text{ °C}) = 8.78 \times 10^{-6} \text{ m} \quad ◊$$

(c) The original volume $V = \pi r^2 L = \dfrac{\pi}{4}(0.0150 \text{ m})^2 (0.300 \text{ m}) = 5.30 \times 10^{-5} \text{ m}^3$

Using the volumetric coefficient of expansion, β, $\quad \Delta V = \beta V_i \, \Delta T \cong 3\alpha V \, \Delta T$

$$\Delta V \cong 3\left(9.00 \times 10^{-6} (°C)^{-1} \right)\left(5.30 \times 10^{-5} \text{ m}^3 \right)(65.0 \text{ °C}) = 93.0 \times 10^{-9} \text{ m}^3 \quad ◊$$

Related Calculation The above calculation ignores ΔL^2 and ΔL^3 terms. Calculate the change in volume exactly, and compare your answer with the approximate solution above. The volume will increase by a factor of

$$\Delta V / V_i = \left(1 + \Delta D / D_i \right)^2 \left(1 + \Delta L / L_i \right) - 1$$

$$\frac{\Delta V}{V} = \left(1 + \frac{8.78 \times 10^{-6} \text{ m}}{0.0150 \text{ m}} \right)^2 \left(1 + \frac{1.76 \times 10^{-4} \text{ m}}{0.300 \text{ m}} \right) - 1 = 1.76 \times 10^{-3}$$

So, $\Delta V = \dfrac{\Delta V}{V_i} V_i = \left(1.76 \times 10^{-3} \right)\left(5.30 \times 10^{-5} \text{ m}^3 \right) = 93.0 \text{ mm}^3$ ◊

The answer is virtually identical, so the approximation $\beta \cong 3\alpha$ is a good one.

23. A hollow aluminum cylinder 20.0 cm deep has an internal capacity of 2.000 L at 20.0 °C. It is completely filled with turpentine, and then warmed to 80.0 °C. (a) How much turpentine overflows? (b) If the cylinder is then cooled back to 20.0 °C, how far below the surface of the cylinder's rim does the turpentine's surface recede?

Solution

(a) When the temperature is increased from 20.0 °C to 80.0 °C, both the cylinder and the turpentine increase in volume by $\Delta V = \beta V \Delta T$.

The overflow, $V_{over} = \Delta V_{Turp} - \Delta V_{Al}$

$$V_{over} = (\beta V_i\,\Delta T)_{Turp} - (\beta V_i\,\Delta T)_{Al} = V_i\,\Delta T(\beta_{Turp} - 3\alpha_{Al})$$

$$V_{over} = (2.000\ \text{L})(60.0\ ^\circ\text{C})(9.00\times10^{-4}\ ^\circ\text{C}^{-1} - 0.720\times10^{-4}\ ^\circ\text{C}^{-1})$$

$$V_{over} = 0.0994\ \text{L} \qquad \Diamond$$

(b) The whole new volume of turpentine is

$$V' = 2000\ \text{cm}^3 + \left(9.00\times10^{-4}\ ^\circ\text{C}^{-1}\right)\left(2000\ \text{cm}^3\right)(60.0\ ^\circ\text{C}) = 2108\ \text{cm}^3$$

The fraction lost is $\dfrac{99.4\ \text{cm}^3}{2108\ \text{cm}^3} = 4.71\times10^{-2}$

This also the fraction of the cylinder that will be empty after cooling:

$$\Delta h = \left(4.71\times10^{-2}\right)(20.0\ \text{cm}) = 0.943\ \text{cm} \qquad \Diamond$$

27. An auditorium has dimensions 10.0 m × 20.0 m × 30.0 m. How many molecules of air fill the auditorium at 20.0 °C and a pressure of 101 kPa?

Solution

G: The given room conditions are close to Standard Temperature and Pressure (STP is 0°C and 101.3 kPa), so we can use the estimate that one mole of an ideal gas at STP occupies a volume of about 22 L. The volume of the auditorium is 6×10^3 m^3 and 1 m^3 = 1000 L , so we can estimate the number of molecules to be:

$$N \approx \left(6000 \text{ m}^3\right)\left(\frac{10^3 \text{ L}}{1 \text{ m}^3}\right)\left(\frac{1 \text{ mol}}{22 \text{ L}}\right)\left(\frac{6 \times 10^{23} \text{ molecules}}{1 \text{ mol}}\right)$$

$$N \approx 1.6 \times 10^{29} \text{ molecules of air}$$

O: The number of molecules can be found more precisely by applying $PV = nRT$.

A: The equation of state of an ideal gas is $PV = nRT$. We need to solve for the number of moles to find N.

$$n = \frac{PV}{RT} = \frac{(1.01 \times 10^5 \text{ N} / \text{m}^2)[(10.0 \text{ m})(20.0 \text{ m})(30.0 \text{ m})]}{(8.314 \text{ J} / \text{mol} \cdot \text{K})(293 \text{ K})} = 2.49 \times 10^5 \text{ mol}$$

$$N = n(N_A) = (2.49 \times 10^5 \text{ mol})\left(6.022 \times 10^{23} \frac{\text{molecules}}{\text{mol}}\right)$$

$$N = 1.50 \times 10^{29} \text{ molecules} \qquad \Diamond$$

L: This result agrees quite well with our initial estimate. The numbers would match even better if the temperature of the auditorium were 0 °C.

29. The mass of a hot-air balloon and its cargo (not including the air inside) is 200 kg. The air outside is at 10.0 °C and 101 kPa. The volume of the balloon is 400 m³. To what temperature must the air in the balloon be heated before the balloon will lift off? (Air density at 10.0 °C is 1.25 kg/m³.)

Solution

G: The air inside the balloon must be significantly hotter than the outside air in order for the balloon to have a net upward force, but the temperature must also be less than the melting point of the nylon used for the balloon's envelope. (Rip-stop nylon melts around 200 °C.) Otherwise the results could be disastrous!

O: The density of the air inside the balloon must be sufficiently low so that the buoyant force is equal to the weight of the balloon, its cargo, and the air inside. The temperature of the air required to achieve this density can be found from the equation of state of an ideal gas.

A: The buoyant force equals the weight of the air at 10.0 °C displaced by the balloon: $B = m_{air}g = \rho_{air}Vg$

$$B = \left(1.25 \text{ kg} / \text{m}^3\right)\left(400 \text{ m}^3\right)\left(9.80 \text{ m} / \text{s}^2\right) = 4900 \text{ N}$$

The weight of the balloon and its cargo is

$$F_g = m_b g = \left(200 \text{ kg}\right)\left(9.80 \text{ m} / \text{s}^2\right) = 1960 \text{ N}$$

Since $B > F_g$, the balloon has a chance of lifting off as long as the weight of the air inside the balloon is less than the difference in these forces:

$$F_{air} < B - F_g = 4900 \text{ N} - 1960 \text{ N} = 2940 \text{ N}$$

The mass of this air is $\qquad m_{air} = \dfrac{F_{air}}{g} = \dfrac{2940 \text{ N}}{9.80 \text{ m} / \text{s}^2} = 300 \text{ kg}$

To find the required temperature of this air from $PV = nRT$, we must find the corresponding number of moles of air. Dry air is approximately 20% O_2, and 80% N_2. Using data from a periodic table, we can calculate the molar mass of the air to be approximately

$$M = (0.80)(28.0 \text{ g} / \text{mol}) + (0.20)(32.0 \text{ g} / \text{mol}) = 28.8 \text{ g} / \text{mol}$$

so the number of moles, $\quad n = \dfrac{m}{M} = (300 \text{ kg}) \left(\dfrac{1000 \text{ g} / \text{kg}}{28.8 \text{ g} / \text{mol}} \right) = 1.04 \times 10^4 \text{ mol}$

The pressure of this air is the ambient pressure; from $PV = nRT$, we can now find the minimum temperature required for lift off:

$$T = \dfrac{PV}{nR} = \dfrac{\left(1.01 \times 10^3 \text{ N} / \text{m}^2 \right)\left(400 \text{ m}^3 \right)}{\left(1.04 \times 10^4 \text{ mol} \right)(8.315 \text{ J} / \text{mol} \cdot \text{K})} = 466 \text{ K} = 193 \text{ °C} \qquad \Diamond$$

L: The average temperature of the air inside the balloon required for lift off appears to be close to the melting point of the nylon fabric, so this seems like a dangerous situation! A larger balloon would be better suited for the given weight of the balloon. (A quick check on the internet reveals that this balloon is only about 1/10 the size of most sport balloons, which have a volume of about 3000 m^3).

If the buoyant force were less than the weight of the balloon and its cargo, the balloon would not lift off no matter how hot the air inside might be! If this were the case, then either the weight would have to be reduced or a bigger balloon would be required.

33. An automobile tire is inflated with air originally at 10.0 °C and normal atmospheric pressure. During the process, the air is compressed to 28.0% of its original volume and its temperature is increased to 40.0 °C. (a) What is the tire pressure? (b) After the car is driven at high speed, the tire air temperature rises to 85.0 °C and the interior volume of the tire increases by 2.00%. What is the new tire pressure (absolute) in pascals?

Solution

(a) Taking $PV = nRT$ in the initial (i) and final (f) states, and dividing, we have

$$P_iV_i = nRT_i \qquad \text{and} \qquad P_fV_f = nRT_f \qquad \text{yield} \qquad \frac{P_fV_f}{P_iV_i} = \frac{T_f}{T_i}$$

So $P_f = P_i\dfrac{V_iT_f}{V_fT_i} = (1.013\times10^5\ \text{Pa})\left(\dfrac{V_i}{0.280V_i}\right)\left(\dfrac{273\ \text{K}+40.0\ \text{K}}{273\ \text{K}+10.0\ \text{K}}\right) = 4.00\times10^5\ \text{Pa}$ ◊

(b) Introducing the hot (h) state, $\qquad \dfrac{P_hV_h}{P_fV_f} = \dfrac{T_h}{T_f}$

So $P_h = P_f\left(\dfrac{V_f}{V_h}\right)\left(\dfrac{T_h}{T_f}\right) = (4.00\times10^5\ \text{Pa})\left(\dfrac{V_f(358\ \text{K})}{1.02V_f(313\ \text{K})}\right) = 4.49\times10^5\ \text{Pa}$ ◊

39. The pressure gauge on a tank registers the gauge pressure, which is the difference between the interior and exterior pressures. When the tank is full of oxygen (O_2), it contains 12.0 kg of oxygen at a gauge pressure of 40.0 atm. Determine the mass of oxygen that has been withdrawn from the tank when the pressure reading is 25.0 atm. Assume the temperature of the tank remains constant.

Solution   The ideal gas law gives us $PV = nRT$. At constant volume and temperature,

$$\frac{P}{n} = \text{constant} \quad \text{or} \quad \frac{P_1}{n_1} = \frac{P_2}{n_2} ; \quad \text{and} \quad n_2 = \left(\frac{P_2}{P_1}\right) n_1$$

However, n is proportional to m, so

$$m_2 = \left(\frac{P_2}{P_1}\right) m_1 = \frac{26.0 \text{ atm}}{41.0 \text{ atm}} (12.0 \text{ kg}) = 7.61 \text{ kg}$$

The mass removed is $\Delta m = (12.0 - 7.61) \text{ kg} = 4.39 \text{ kg}$ ◊

---

47. A mercury thermometer is constructed as shown in Figure P19.47. The capillary tube has a diameter of 0.00400 cm, and the bulb has a diameter of 0.250 cm. Neglecting the expansion of the glass, find the change in height of the mercury column with a temperature change of 30.0 °C.

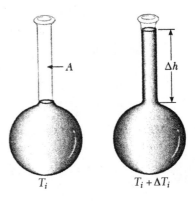

Figure P19.47

Solution   Neglecting the expansion of the glass, the volume of liquid in the capillary will be $\Delta V = A \, \Delta h$ where A is the cross-sectional area of the capillary and V is the volume of the bulb.

$$\Delta V = V_i \beta \, \Delta T$$

$$\Delta h = \left(\frac{V_i}{A}\right) \beta \, \Delta T = \left[\frac{\frac{4}{3} \pi R_{\text{bulb}}^3}{\pi R_{\text{cap}}^2}\right] \beta \, \Delta T$$

$$\Delta h = \frac{4}{3} \frac{(0.125 \text{ cm})^3}{(0.00200 \text{ cm})^2} \left(1.82 \times 10^{-4} \, (\text{°C})^{-1}\right)(30.0 \text{ °C}) = 3.55 \text{ cm} \quad ◊$$

---

49. A liquid has a density ρ. (a) Show that the fractional change in density for a change in temperature ΔT is $\Delta\rho / \rho = -\beta\,\Delta T$. What does the negative sign signify? (b) Fresh water has a maximum density of 1.0000 g/cm³ at 4.00 °C. At 10.0 °C, its density is 0.9997 g/cm³. What is β for water over this temperature interval?

Solution   Since the mass is invariant and the density is $\rho = m / V$,

(a) $\dfrac{\Delta\rho}{\rho} = \dfrac{\rho - \rho_i}{\rho} = 1 - \dfrac{\rho_i}{\rho} = 1 - \dfrac{m / V_i}{m / V} = 1 - \dfrac{V}{V_i} = 1 - \dfrac{V_i + \beta V_i \Delta T}{V_i} = -\beta\,\Delta T$ ◊

The negative sign means that any increase in temperature causes the density to decrease and vice versa.

(b)   For water we have

$$\beta = \left| \frac{\Delta\rho}{\rho\,\Delta T} \right| = \left| \frac{(1.0000\ \text{g}/\text{cm}^3 - 0.9997\ \text{g}/\text{cm}^3)}{(1.0000\ \text{g}/\text{cm}^3)(10.0\ ^\circ\text{C} - 4.00\ ^\circ\text{C})} \right| = 5.00 \times 10^{-5}\ (^\circ\text{C})^{-1}$$ ◊

51. A vertical cylinder of cross-sectional area A is fitted with a tight-fitting, frictionless piston of mass m (Fig. P19.51). (a) If n mol of an ideal gas are in the cylinder at a temperature T, what is the height h at which the piston is in equilibrium under its own weight? (b) What is the value for h if $n = 0.200$ mol, $T = 400$ K, $A = 0.00800$ m² and $m = 20.0$ kg?

Solution

(a)   We suppose that air at atmospheric pressure P_0 is above the piston. Taking P to be the pressure exerted by the gas in the piston, and applying $\Sigma F_y = ma_y$ to the piston's equilibrium,

$$-P_0 A - mg + PA = 0$$

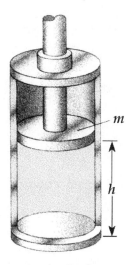

Figure P19.51

417

Noting that $V = Ah$, and that n, T, m, g, A, and P_0 are given,

$PV = nRT$ becomes $PA = nRT/h$, and $\qquad P_0 A - mg + \dfrac{nRT}{Ah} A = 0$

Rearranging terms to isolate h $\qquad\qquad h = \dfrac{nRT}{P_0 A + mg}$ ◊

(b) $\quad h = \dfrac{(0.200 \text{ mol})(8.314 \text{ J} / \text{mol} \cdot \text{K})(400 \text{ K})}{\left(1.013 \times 10^5 \text{ N} / \text{m}^2\right)\left(0.00800 \text{ m}^2\right) + (20.0 \text{ kg})\left(9.80 \text{ m} / \text{s}^2\right)} = 0.661 \text{ m}$ ◊

---

53. The rectangular plate shown in Figure P19.53 has an area A equal to ℓw. If the temperature increases by ΔT, show that the increase in area is $\Delta A = 2\alpha A_i \Delta T$, where α is the average coefficient of linear expansion. What approximation does this expression assume? (**Hint:** Note that each dimension increases according to $\Delta L = \alpha L_i \Delta T$.)

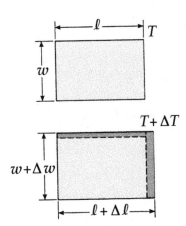

Figure P19.53

Solution From the diagram in Figure 19.53, we see that the **change** in area is

$$\Delta A = \ell \Delta w + w \Delta \ell + \Delta w \Delta \ell$$

Since $\Delta \ell$ and Δw are each small quantities, the product $\Delta w \Delta \ell$ will be very small. Therefore, we assume

$$\Delta w \Delta \ell \approx 0.$$

Since $\Delta w = w\alpha \Delta T$ and $\Delta \ell = \ell \alpha \Delta T$, we then have

$$\Delta A = \ell w \alpha \Delta T + w \ell \alpha \Delta T$$

Finally, since $A = \ell w$, $\qquad \Delta A = 2\alpha A \Delta T$ ◊

59. Starting with Equation 19.10, show that the total pressure P in a container filled with a mixture of several ideal gases is $P = P_1 + P_2 + P_3 + \ldots$, where $P_1, P_2, \ldots$, are the pressures that each gas would exert if it alone filled the container (these individual pressures are called the **partial pressures** of the respective gases). This is known as **Dalton's law of partial pressures.**

Solution For each gas alone, $P_1 = \dfrac{N_1 k_B T}{V_1}$, $P_2 = \dfrac{N_2 k_B T}{V_2}$, $P_3 = \dfrac{N_3 k_B T}{V_3}$ $\ldots$

For the gases combined, $\qquad N_1 + N_2 + N_3 + \ldots = N = \dfrac{PV}{k_B T}$

Therefore, $\qquad \dfrac{P_1 V_1}{k_B T} + \dfrac{P_2 V_2}{k_B T} + \dfrac{P_3 V_3}{k_B T} + \ldots = \dfrac{PV}{k_B T}$

But $\qquad V_1 = V_2 = V_3 = \ldots = V,$ so $\qquad P_1 + P_2 + P_3 + \ldots = P$ $\qquad \lozenge$

69. A steel guitar string with a diameter of 1.00 mm is stretched between supports 80.0 cm apart. The temperature is 0.0 °C. (a) Find the mass per unit length of this string. (Use the value 7.86×10^3 kg / m^3 as the mass density.) (b) The fundamental frequency of transverse oscillations of the string is 200 Hz. What is the tension in the string? (c) If the temperature is raised to 30.0 °C, find the resulting values of the tension and the fundamental frequency. (Assume that both the Young's modulus [Table 12.1] and the average coefficient of expansion [Table 19.2] have constant values between 0.0 °C and 30.0 °C.)

Solution (a) $\mu = \frac{1}{4} \rho (\pi d^2) = \frac{1}{4} \pi (1.00 \times 10^{-3} \text{ m})^2 (7.86 \times 10^{-3} \text{ kg / m}^3)$

$\mu = 6.17 \times 10^{-3}$ kg / m $\qquad\qquad\qquad \lozenge$

(b) Since $f_1 = \dfrac{v}{2L}$, $\qquad v = \sqrt{F/\mu}$ and $\qquad f_1 = \dfrac{1}{2L}\sqrt{F/\mu}$

$F = \mu(2Lf_1)^2 = (6.17 \times 10^{-3} \text{ kg / m})[(2)(0.800 \text{ m})(200 \text{ s}^{-1})]^2 = 632 \text{ N}$ $\quad \lozenge$

(c)   At 0 °C, the length of the guitar string will be $L_{0°C} = L_{natural}\left(1 + \dfrac{F}{AY}\right)$

We know the string's cross section $A = \left(\dfrac{\pi}{4}\right)\left(1.00 \times 10^{-3} \text{ m}\right)^2 = 7.85 \times 10^{-7} \text{ m}^2$

and modulus $Y = 20.0 \times 10^{10} \text{ N / m}^2$

Therefore, $\dfrac{F}{AY} = \dfrac{632 \text{ N}}{\left(7.85 \times 10^{-7} \text{ m}^2\right)\left(20.0 \times 10^{10} \text{ N / m}^2\right)} = 4.02 \times 10^{-3}$

$$L_{0°C} = \dfrac{0.800 \text{ m}}{1 + 4.02 \times 10^{-3}} = 0.7968 \text{ m}$$

Then at 30 °C, $L_{30°C} = (0.799 \text{ m})\left(1 + (30.0 \text{ °C})\left(11.0 \times 10^{-6} \text{ °C}^{-1}\right)\right) = 0.7971 \text{ m}$

Since $0.800 \text{ m} = (0.7971 \text{ m})\left[1 + \dfrac{F'}{A'Y}\right]$

$\dfrac{F'}{A'Y} = \dfrac{0.8000}{0.7971} - 1 = 3.689 \times 10^{-3}$ and $F' = A'Y\left(3.689 \times 10^{-3}\right)$

$F' = \left(7.85 \times 10^{-7} \text{ m}^2\right)\left(20.0 \times 10^{10} \text{ N / m}^2\right)\left(3.689 \times 10^{-3}\right)(1 + \alpha \, \Delta T)^2$

$F' = (579.1 \text{ N})\left(1 + 3.30 \times 10^{-4}\right)^2 = 579 \text{ N}$ ◊

Also, $\dfrac{f_1'}{f_1} = \sqrt{\dfrac{F'}{F}}$ so $f_1' = (200 \text{ Hz})\sqrt{\dfrac{579 \text{ N}}{632 \text{ N}}} = 191 \text{ Hz}$ ◊

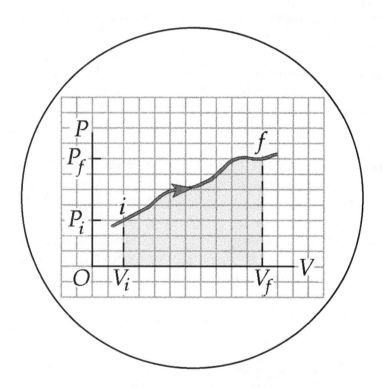

Heat and the First Law of Thermodynamics

HEAT AND THE FIRST LAW
OF THERMODYNAMICS

INTRODUCTION

This chapter focuses on the concept of heat, the first law of thermodynamics, processes by which thermal energy is transferred, and some important applications. The first law of thermodynamics is merely the law of conservation of energy. It tells us only that an increase in one form of energy must be accompanied by a decrease in some other form of energy. The first law places no restrictions on the types of energy conversions that can occur. Furthermore, it makes no distinction between heat and work. According to the first law, a system's internal energy can be increased either by transfer of thermal energy to the system or by work done on the system. An important difference between thermal energy and mechanical energy is not evident from the first law; it is possible to convert work completely to thermal energy but impossible to convert thermal energy completely to work.

EQUATIONS AND CONCEPTS

Every substance requires a unique quantity of heat to change the temperature of 1 kg of the substance by 1.00 °C. This quantity is a measure of the **specific heat** of the substance.

$$c \equiv \frac{Q}{m \, \Delta T} \tag{20.3}$$

The **heat energy** Q that must be transferred between a system of mass m and its surroundings to produce a temperature change ΔT varies with the substance.

$$Q = mc\,\Delta T \qquad (20.4)$$

A substance may undergo a phase change when heat is transferred between the substance and its surroundings. The heat Q required to change the phase of a mass m depends on a property of the material called the **latent heat**, L. The phase change process occurs at constant temperature. The value of L depends on the nature of the phase change and the thermal properties of the substance.

$$Q = mL \qquad (20.6)$$

The **work done by a gas** which undergoes an expansion or compression from volume V_i to volume V_f depends on the path between the initial and final states. The pressure is generally not constant, so you must exercise care in evaluating W using this equation. In general, the work done is the area under the PV curve bounded by V_i and V_f, and the function P.

$$W = \int_{V_i}^{V_f} P\,dV \qquad (20.8)$$

The **first law of thermodynamics** is a generalization of the law of conservation of energy that includes possible changes in internal energy. It states that the **change** in internal energy of a system, ΔE_{int}, equals the quantity $Q - W$, where Q is the heat added to the system. The quantity $(Q - W)$ is independent of the path taken between the initial and final states.

$$\Delta E_{int} = Q - W \qquad (20.9)$$

In an **adiabatic process**, $Q = 0$, and the change in internal energy equals the negative of the work done **by** the gas.

$$\Delta E_{int} = -W \qquad (20.10)$$

In an isovolumetric (constant volume) process, zero work is done and all heat added to the system increases the internal energy.

$$\Delta E_{int} = Q \qquad (20.12)$$

An **isothermal process** is one which occurs at constant temperature. During such a process, the change in internal energy results from both heat transfer and work done.

$$W = nRT \ln\left(\frac{V_f}{V_i}\right) \qquad (20.13)$$

SUGGESTIONS, SKILLS, AND STRATEGIES

Many applications of the first law of thermodynamics deal with the work done by (or on) a system which undergoes a change in state. As a gas is taken from a state whose initial pressure and volume are P_i, V_i, and whose final pressure and volume are P_f, V_f, the work can be calculated if the process can be drawn on a PV diagram as in the figure to the right. The work done during the expansion is

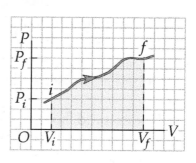

$$W = \int_{V_i}^{V_f} P \, dV$$

which numerically represents the **area** under the PV curve (the shaded region) shown in the figure. It is important to recognize that the work **depends on the path taken as the gas goes from i to f**. That is, W depends on the specific manner in which the pressure P changes during the process.

PROBLEM SOLVING STRATEGY: CALORIMETRY PROBLEMS

If you are having difficulty with calorimetry problems, consider the following factors:

- Be sure your units are consistent throughout. For instance, if you are using specific heats in cal/g·°C, be sure that masses are in grams and temperatures are Celsius throughout.

- Heat loss and gains are found by using $Q = mc\Delta T$ only for those intervals in which no phase changes are occurring. The equations $Q = mL_f$ and $Q = mL_v$ are to be used only when phase changes **are** taking place.

- Often sign errors occur in calorimetry problems. Remember that *heat gained = −heat lost*.

REVIEW CHECKLIST

▷ Understand the concepts of heat, internal energy, and thermodynamic processes.

▷ Define and discuss the calorie, heat capacity, specific heat, and latent heat.

▷ Understand how work is defined when a system undergoes a change in state, and the fact that work (like heat) depends on the path taken by the system. You should also know how to sketch processes on a PV diagram, and calculate work using these diagrams.

▷ State the first law of thermodynamics ($\Delta E_{int} = Q - W$), and explain the meaning of the three forms of energy contained in this statement.

▷ Discuss the implications of the first law of thermodynamics as applied to (a) an isolated system, (b) a cyclic process, and (c) an adiabatic process.

ANSWERS TO SELECTED CONCEPTUAL QUESTIONS

7. What is wrong with the statement, "Given any two bodies, the one with the higher temperature contains more heat."

Answer The statement shows a misunderstanding of the concept of heat. Heat is a form of energy that is being transferred, not a form of energy that is held or contained. If you wish to speak of energy that is "contained," you speak of **internal energy**, not **heat**.

Further, even if the statement used the term "internal energy," it would still be incorrect, since the effects of specific heat and mass are both ignored. A 1-kg mass of water at 20 °C has more internal energy than a 1-kg mass of air at

30 °C. Similarly, the earth has far more internal energy than a drop of molten titanium metal.

Correct statements would be: (1) "given any two bodies in thermal contact, the one with the higher temperature will transfer heat to the other." (2) "given any two bodies of equal mass, the one with the higher product of absolute temperature and specific heat contains more internal energy."

☐　　☐　　☐　　☐

29. Using the first law of thermodynamics, explain why the **total** energy of an isolated system is always constant.

Answer The first law of thermodynamics says that the net change in energy of a system is equal to the heat added, minus the work done by the system.

$$\Delta U = Q - W$$

However, an isolated system is defined as an object or set of objects that exchanges neither work nor heat with its surroundings. Specifically, it says that $Q = W = 0$, so the change in total energy of the system at all times must be zero.

It should be noted that although $\Delta U = 0$, the energy may change from one form to another or move from one object to another within the system. For example, "bomb calorimeters" are closed systems that consist of a sturdy steel container, a water bath, an item of food, and oxygen. The food is burned in the presence of an excess of oxygen, and chemical energy is converted to heat energy. In the process of oxidation, heat energy is absorbed by the water bath, raising its temperature. The change in temperature of the water bath is used to determine the caloric content of the food. The process works specifically **because** the total energy remains unchanged in a closed system.

☐　　☐　　☐　　☐

SOLUTIONS TO SELECTED END-OF-CHAPTER PROBLEMS

1. Water at the top of Niagara Falls has a temperature of 10.0 °C. It falls through a distance of 50.0 m. Assuming that all of its potential energy goes into warming of the water, calculate the temperature of the water at the bottom of the Falls.

Solution

G: Water has a high specific heat, so the difference in water temperature between the top and bottom of the falls is probably less than 1 °C. (Besides, if the difference was significantly large, we might have heard about this phenomenon at some point.)

O: The temperature change can be found from the potential energy that is converted to internal energy. The final temperature is this change added to the initial temperature of the water.

A: The change in potential energy is $\Delta E = mgy$. It will produce the same temperature change as the same amount of heat entering the water from a stove, as described by $Q = mc\,\Delta T$. Thus, $mgy = mc\,\Delta T$

Isolating ΔT, $$\Delta T = \frac{gy}{c} = \frac{(9.80 \text{ m}/\text{s}^2)(50.0 \text{ m})}{4.186 \times 10^3 \text{ J}/\text{kg}\cdot°\text{C}} = 0.117 \text{ °C}.$$

$$T_g = T_i + \Delta T = 10.0 °C + 0.117 °C = 10.1 °C \qquad \Diamond$$

L: The water temperature rose less than 1 °C as expected; however, the final temperature might be less than we calculated since this solution does not account for cooling of the water due to evaporation as it falls. It is interesting to note that the change in temperature is independent of the amount of water. James Joule took a large thermometer with him on his honeymoon in Switzerland, where there are nice waterfalls. He tried but was unable to make consistent measurements of the expected temperature increase.

3. The temperature of a silver bar rises by 10.0 °C when it absorbs 1.23 kJ of energy by heat. The mass of the bar is 525 g. Determine the specific heat of silver.

Solution $\Delta Q = mc_{silver}\Delta T$ or $c_{silver} = \dfrac{Q}{m\,\Delta T}$

$$c_{silver} = \frac{1.23 \times 10^3 \text{ J}}{(0.525 \text{ kg})(10.0°\text{C})} = 234 \text{ J} / \text{kg·°C} \qquad \Diamond$$

5. A 1.50-kg iron horseshoe initially at 600 °C is dropped into a bucket containing 20.0 kg of water at 25.0 °C. What is the final temperature? (Neglect the heat capacity of the container and assume that a negligible amount of water boils away.)

Solution

G: Even though the horseshoe is much hotter than the water, the mass of the water is significantly greater, so we might expect the water temperature to rise less than 10°C.

O: The heat lost by the iron will be gained by the water, and from this heat difference, the change in water temperature can be found.

A: $\Delta Q_{iron} = -\Delta Q_{water}$ or $(mc\,\Delta T)_{iron} = -(mc\,\Delta T)_{water}$

$$m_{Fe}c_{Fe}\,(T - 600\ °\text{C}) = -m_w c_w\,(T - 25.0\ °\text{C})$$

$$T = \frac{m_w c_w (25.0\ °\text{C}) + m_{Fe} c_{Fe}\,(600\ °\text{C})}{m_{Fe} c_{Fe} + m_w c_w}$$

$$T = \frac{(20.0 \text{ kg})(4186 \text{ J} / \text{kg·°C})(25.0\ °\text{C}) + (1.50 \text{ kg})(448 \text{ J} / \text{kg·°C})(600\ °\text{C})}{(1.50 \text{ kg})(448 \text{ J} / \text{kg·°C}) + (20.0 \text{ kg})(4186 \text{ J} / \text{kg·°C})}$$

$$T = 29.6\ °\text{C} \qquad \Diamond$$

L: The temperature only rose about 5 °C, so our answer seems reasonable. The specific heat of the water is about 10 times greater than the iron, so this effect also reduces the change in water temperature. In this problem, we assumed that a negligible amount of water boiled away, but in reality, the final temperature of the water would be somewhat less than we calculated, since some of the heat energy would be used to vaporize a bit of water.

---

13. A 3.00-g lead bullet at 30.0 °C is fired at a speed of 240 m/s into a large block of ice at 0 °C, in which it becomes embedded. What quantity of ice melts?

Solution

G: The amount of ice that melts is probably small, maybe only a few grams based on the size, speed, and initial temperature of the bullet.

O: We will assume that all of the initial kinetic and excess internal energy of the bullet goes into internal energy to melt the ice, the mass of which can be found from the latent heat of fusion.

A: At thermal equilibrium, the energy lost by the bullet equals the energy gained by the ice: $\Delta K_b + |\Delta Q_b| = \Delta Q_{ice}$ gives

$$\frac{1}{2}m_b v_b^2 + m_b c_{Pb}|\Delta T| = m_{ice}L_f \quad \text{or} \quad m_{ice} = m_b\left(\frac{\frac{1}{2}v_b^2 + c_{Pb}|\Delta T|}{L_f}\right)$$

$$m_{ice} = \left(3.00\times10^{-3}\text{ kg}\right)\left(\frac{\frac{1}{2}(240\text{ m/s})^2 + (128\text{ J/kg}\cdot°\text{C})(30.0\,°\text{C})}{3.33\times10^3\text{ J/kg}}\right)$$

$$m_{ice} = \frac{86.4\text{ J} + 11.5\text{ J}}{3.33\times10^3\text{ J/kg}} = 2.94\times10^{-4}\text{ kg} = 0.294\text{ g} \qquad \lozenge$$

L: The amount of ice that melted is less than a gram, which agrees with our prediction. It appears that most of the energy used to melt the ice comes from the kinetic energy of the bullet (88%), while the excess internal energy of the bullet only contributes 12% to melt the ice. Small chips of ice probably fly off when the bullet makes impact. Some of the energy is transferred to their kinetic energy, so in reality, the amount of ice that would melt should be less than what we calculated. If the block of ice were colder than 0 °C (as is usually the case), then the melted ice would refreeze.

---

17. In an insulated vessel, 250 g of ice at 0 °C is added to 600 g of water at 18.0 °C. (a) What is the final temperature of the system? (b) How much ice remains when the system reaches equilibrium?

Solution

(a) When 250 g of ice is melted,

$$\Delta Q_f = mL_f = (0.250 \text{ kg})(3.33 \times 10^5 \text{ J / kg}) = 83.3 \text{ kJ}$$

The heat energy released when 600 g of water cools from 18.0 °C to 0 °C is

$$\Delta Q - mc\,\Delta T = (0.600 \text{ kg})(4186 \text{ J / kg} \cdot {}^\circ\text{C})(18.0\ {}^\circ\text{C}) = 45.2 \text{ kJ}$$

Since the heat required to melt 250 g of ice at 0 °C **exceeds** the heat released by cooling 600 g of water from 18.0 °C to 0 °C, the final temperature of the system (water + ice) must be 0 °C. ◊

(b) The thermal energy released by the water (45.2 kJ) will melt a mass of ice m, where $Q = mL_f$.

Solving for the mass, $$m = \frac{Q}{L_f} = \frac{45.2 \times 10^3 \text{ J}}{3.33 \times 10^5 \text{ J / kg}} = 0.136 \text{ kg}$$

Thus, the ice remaining is $m' = 0.250 \text{ kg} - 0.136 \text{ kg} = 0.114 \text{ kg}$ ◊

---

21. A sample of ideal gas is expanded to twice its original volume of 1.00 m³ in a quasi-static process for which $P = \alpha V^2$, with $\alpha = 5.00$ atm / m⁶, as shown in Figure P20.21. How much work was done by the expanding gas?

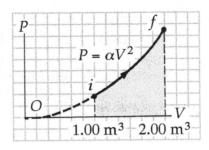

Solution $W = \int_{V_i}^{V_f} P\, dV$

Figure P20.21

$$V_f = 2V_i = 2(1.00 \text{ m}^3) = 2.00 \text{ m}^3$$

The work done by the gas is the area under the curve $P = \alpha V^2$, from V_i to V_f.

$$W = \int_{V_i}^{V_f} \alpha\, V^2 dV = \frac{1}{3}\alpha\left(V_f{}^3 - V_i{}^3\right)$$

$$W = \frac{1}{3}\left(5.00\, \frac{\text{atm}}{\text{m}^6}\right)\left(1.013 \times 10^5\, \frac{\text{Pa}}{\text{atm}}\right)\left[\left(2.00 \text{ m}^3\right)^3 - \left(1.00 \text{ m}^3\right)^3\right]$$

$$W = 1.18 \times 10^6 \text{ J} \qquad \lozenge$$

---

25. An ideal gas is enclosed in a cylinder with a movable piston on top. The piston has a mass of 8000 g and an area of 5.00 cm² and is free to slide up and down, keeping the pressure of the gas constant. How much work is done as the temperature of 0.200 mol of the gas is raised from 20.0 °C to 300 °C?

Solution For constant pressure, $W = P\,\Delta V = P\left(\dfrac{nR}{P}\right)(T_h - T_c)$

Therefore, $W = nR\,\Delta T = (0.200 \text{ mol})(8.314 \text{ J / mol} \cdot \text{K})(280 \text{ K}) = 466 \text{ J} \qquad \lozenge$

---

29. A thermodynamic system undergoes a process in which its internal energy decreases by 500 J. If, at the same time, 220 J of work is done on the system, what is the thermal energy transferred to or from it by heat?

Solution $\Delta E = Q - W$, where W is negative, because work is done **on** the system:

$$Q = \Delta E + W = -500 \text{ J} - 220 \text{ J} = -720 \text{ J}$$

+ 720 J of heat is transferred **from** the system. ◊

---

33. An ideal gas initially at 300 K undergoes an isobaric expansion at 2.50 kPa. If the volume increases from 1.00 m³ to 3.00 m³ and 12.5 kJ of energy is transferred to the gas by heat, what are (a) the change in its internal energy and (b) its final temperature?

Solution

(a) $\Delta E = Q - W$ where $W = P\Delta V$ so that $\Delta E = Q - P\Delta V$

$$\Delta E = 1.25 \times 10^4 \text{ J} \pm \left(2.50 \times 10^3 \text{ N} / \text{m}^2\right)\left(3.00 \text{ m}^3 - 1.00 \text{ m}^3\right) = 7500 \text{ J}$$ ◊

(b)   Since $\dfrac{V_1}{T_1} = \dfrac{V_2}{T_2}$,

$$T_2 = \left(\frac{V_2}{V_1}\right)T_1 = \left(\frac{3.00 \text{ m}^3}{1.00 \text{ m}^3}\right)(300 \text{ K}) = 900 \text{ K}$$ ◊

---

35. How much work is done by the steam when 1.00 mol of water at 100 °C boils and becomes 1.00 mol of steam at 100 °C and at 1.00 atm pressure? Assuming the steam to be an ideal gas, determine the change in internal energy of the steam as it vaporizes.

Solution

$$W = P\,\Delta V = P(V_s - V_w)$$

Substituting, $\qquad PV_s = nRT \qquad$ and $\qquad PV_w = nM\left(\dfrac{P}{\rho}\right)$

Solving each work term, $\quad PV_s = (1.00 \text{ mol})\left(8.31 \dfrac{J}{K \cdot mol}\right)(373 \text{ K}) = 3101 \text{ J}$

$$PV_w = (1.00 \text{ mol})(18.0 \text{ g / mol})\left(\dfrac{1.01 \times 10^5 \text{ N/m}^2}{1.00 \times 10^6 \text{ g / m}^3}\right)$$

$$PV_w = 1.82 \text{ J}$$

Thus the work done is $\quad W = 3.10 \text{ kJ}$ ◊

$$Q = mL_V = (18.0 \text{ g})\left(2.26 \times 10^6 \text{ J / kg}\right) = 40.7 \text{ kJ}$$

$$\Delta E = Q - W = 37.6 \text{ kJ}$$ ◊

37. A 2.00-mole sample of helium gas initially at 300 K and 0.400 atm is compressed isothermally to 1.20 atm. Find (a) the final volume of the gas, (b) the work done by the gas, and (c) the energy transferred by heat.

Solution   Rearranging $PV = nRT$, we get

(a) $V_i = \dfrac{nRT}{P_i} = \dfrac{(2.00 \text{ mol})(8.31 \text{ J / mol} \cdot \text{K})(300 \text{ K})}{(0.400 \text{ atm})\left(1.01 \times 10^5 \text{ N / m}^2 \text{ / atm}\right)} = 0.123 \text{ m}^3$

For isothermal compression, PV = constant, so $P_i V_i = P_f V_f$

$$V_f = V_i\left(\frac{P_i}{P_f}\right) = V_i\left(\frac{0.400 \text{ atm}}{1.20 \text{ atm}}\right) = \frac{0.123 \text{ m}^3}{3} = 0.0410 \text{ m}^3 \qquad \lozenge$$

(b) $W = \int P\,dV = \int \dfrac{nRT}{V}\,dV = nRT\ln\left(\dfrac{V_f}{V_i}\right) = (4986 \text{ J})\ln\left(\dfrac{1}{3}\right) = -5.48 \text{ kJ} \qquad \lozenge$

(c) $\Delta E_{\text{int}} = 0 = Q - W$ and $Q = -5.48 \text{ kJ}$ $\qquad\qquad\qquad\lozenge$

---

45. A bar of gold is in thermal contact with a bar of silver of the same length and area (Fig. P20.45). One end of the compound bar is maintained at 80.0 °C while the opposite end is at 30.0 °C. When the rate of energy transfer by conduction reaches steady state, what is the temperature at the junction?

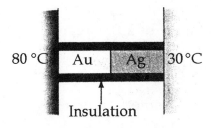

Figure P20.45

Solution

Call the gold bar Object 1 and the silver bar Object 2. When heat flow reaches a steady state, the flow rate through each will be the same:

$$\mathscr{P}_1 = \mathscr{P}_2 \qquad\qquad \text{or} \qquad\qquad \frac{k_1 A_1\,\Delta T_1}{L_1} = \frac{k_2 A_2\,\Delta T_2}{L_2}$$

In this case, $L_1 = L_2$ and $A_1 = A_2$ so $\quad k_1 \, \Delta T_1 = k_2 \, \Delta T_2$

Let T_3 = temperature at the junction; $\quad k_1 \left(80.0 \, °C - T_3 \right) = k_2 \left(T_3 - 30.0 \, °C \right)$

Solving, $\qquad T_3 = \dfrac{\left(80.0 \, °C \right) k_1 + \left(30.0 \, °C \right) k_2}{k_1 + k_2}$

$$T_3 = \frac{\left(80.0 \, °C \right)\left(314 \text{ W/ m·°C} \right) + \left(30.0 \, °C \right)\left(427 \text{ W/ m·°C} \right)}{\left(314 \text{ W/ m·°C} + 427 \text{ W/ m·°C} \right)} = 51.2 \, °C \qquad \lozenge$$

55. An aluminum rod 0.500 m in length and with a cross-sectional area 2.50 cm^2 is inserted into a thermally insulated vessel containing liquid helium at 4.20 K. The rod is initially at 300 K. (a) If one half of the rod is inserted into the helium, how many liters of helium boil off by the time the inserted half cools to 4.20 K? (Assume that the upper half does not yet cool.) (b) If the upper end of the rod is maintained at 300 K, what is the approximate boil-off rate of liquid helium after the lower half has reached 4.20 K? (Aluminum has thermal conductivity of 31.0 J/s·cm·K at 4.2 K; ignore its temperature variation. Aluminum has a specific heat of 0.210 cal/g·°C and density of 2.70 g/cm^3. The density of liquid helium is 0.125 g/cm^3..)

Solution

G: Demonstrations with liquid nitrogen give us some indication of the phenomenon described. Since the rod is much hotter than the liquid helium and of significant size (almost 2 cm in diameter), a substantial volume (maybe as much as a liter) of helium will boil off before thermal equilibrium is reached. Likewise, since aluminum conducts rather well, a significant amount of helium will continue to boil off as long as the upper end of the rod is maintained at 300 K.

O: Until thermal equilibrium is reached, the excess internal energy of the rod will be used to vaporize the liquid helium, which is already at its boiling point (so there is no change in the temperature of the helium).

A: As you solve this problem, be careful not to confuse L (the **conduction length** of the rod) with L_v (the **heat of vaporization** of the helium).

(a) Before heat conduction has time to become important, we suppose the heat energy lost by half the rod equals the heat energy gained by the helium. Therefore, $(mL_v)_{He} = (mc\,\Delta T)_{Al}$ or $(\rho VL_v)_{He} = (\rho Vc\,\Delta T)_{Al}$

So $V_{He} = \dfrac{(\rho Vc\,\Delta T)_{Al}}{(\rho L_v)_{He}} = \dfrac{(2.70\text{ g / cm}^3)(62.5\text{ cm}^3)(0.210\text{ cal / g·°C})(295.8°\text{C})}{(0.125\text{ g / cm}^3)(4.99\text{ cal / g})}$

$$V_{He} = 1.68 \times 10^4 \text{ cm}^3 = 16.8 \text{ liters} \qquad \Diamond$$

(b) Heat will be conducted along the rod at
$$\frac{dQ}{dt} = \mathcal{P} = \frac{kA\,\Delta T}{L} \qquad (1)$$

and will boil off helium according to $Q = mL_v$:
$$\frac{dQ}{dt} = \left(\frac{dm}{dt}\right)L_v \qquad (2)$$

Combining (1) and (2) gives us the "boil-off" rate:
$$\frac{dm}{dt} = \frac{kA\,\Delta T}{L \cdot L_v}$$

Set the conduction length $L = 25$ cm, and use $k = 7.41$ cal / s·cm·K

$$\frac{dm}{dt} = \frac{(7.41\text{ cal / s·cm·K})(2.50\text{ cm}^2)(295.8\text{ K})}{(25.0\text{ cm})(4.99\text{ cal / g})} = 43.9 \text{ g / s}$$

or $\dfrac{dV}{dt} = \dfrac{43.9\text{ g / s}}{0.125\text{ g / cm}^3} = 351\text{ cm}^3\text{ / s} = 0.351 \text{ liters / s} \qquad \Diamond$

L: The volume of helium boiled off initially is much more than expected. If our calculations are correct, that sure is a lot of liquid helium that is wasted! Since liquid helium is much more expensive than liquid nitrogen, most low-temperature equipment is designed to avoid unnecessary loss of liquid helium by surrounding the liquid with a container of liquid nitrogen.

63. A "solar cooker" consists of a curved reflecting mirror that focuses sunlight onto the object to be warmed (Fig. P20.63). The solar power per unit area reaching the Earth at the location is 600 W/m², and the cooker has a diameter of 0.600 m. Assuming that 40% of the incident energy is transferred to the water, how long does it take to completely boil off 0.500 L of water initially at 20.0 °C? (Neglect the heat capacity of the container.)

Figure P20.63

Solution

The power incident on the solar collector is

$$\mathcal{P}_i = IA = \left(600 \text{ W / m}^2\right)\pi(0.300 \text{ m})^2 = 170 \text{ W}$$

For a 40% reflector, the collected power is

$$\mathcal{P}_c = 67.9 \text{ W} = 67.9 \text{ J / s}$$

The total energy required to increase the temperature of the water to the boiling point and to evaporate it is

$$Q = mc\,\Delta T + mL_v$$

$$Q = (0.500 \text{ kg})(4186 \text{ J / kg·°C})(80.0 \text{ °C}) + (0.500 \text{ kg})(2.26 \times 10^6 \text{ J / kg})$$

$$Q = 1.30 \times 10^6 \text{ J}$$

The time required is

$$\Delta t = \frac{Q}{\mathcal{P}_c} = \frac{1.30 \times 10^6 \text{ J}}{67.9 \text{ J / s}} = 1.91 \times 10^4 \text{ s} = 5.31 \text{ h} \qquad \Diamond$$

71. The passenger section of a jet airliner is in the shape of a cylindrical tube with a length of 35.0 m and inner radius of 2.50 m. Its walls are lined with an insulating material 6.00 cm in thickness and having a thermal conductivity of 4.00×10^{-5} cal/s·cm·°C. A heater must maintain the interior temperature at 25.0 °C while the outside temperature is at –35.0 °C. What power must be supplied to the heater if this temperature difference is to be maintained? (Use the result you obtained in Problem 70.)

Solution

From problem 70, the rate of heat flow through the wall is:

$$\frac{dQ}{dt} = \frac{2\pi kL(T_a - T_b)}{\ln(b/a)}$$

$$\frac{dQ}{dt} = \frac{2\pi(3500 \text{ cm})(4.00 \times 10^{-5} \text{ cal}/\text{s}\cdot\text{cm}\cdot°\text{C})(60.0\ °\text{C})}{\ln(2.56/2.50)}$$

and $\dfrac{dQ}{dt} = 2.23 \times 10^3$ cal/s = 9.32 kW $\lozenge$

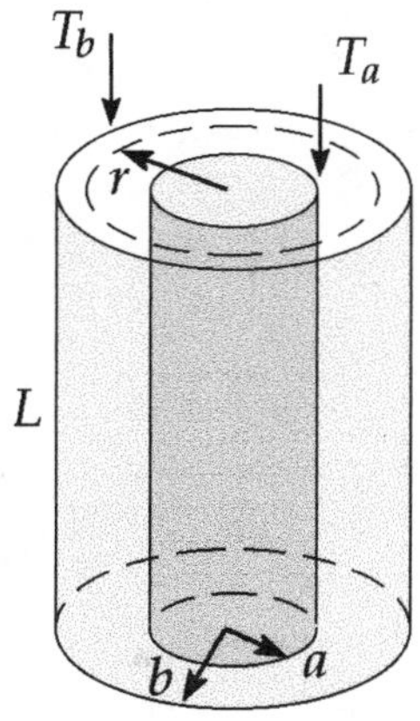

Figure P20.70

This is the rate of heat loss from the plane, and consequently the rate at which energy must be supplied in order to maintain an equilibrium temperature.

Chapter 21

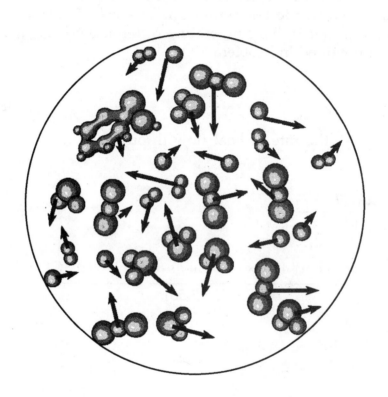

The Kinetic Theory of Gases

THE KINETIC THEORY OF GASES

INTRODUCTION

In Chapter 19, we discussed the properties of an ideal gas using such macroscopic variables as pressure, volume, and temperature. We shall now show that such large-scale properties can be described on a microscopic scale, where matter is treated as a collection of molecules. Newton's laws of motion applied in a statistical manner to a collection of particles provide a reasonable description of thermodynamic processes. In order to keep the mathematics relatively simple, we shall consider only the molecular behavior of gases, where the interactions between molecules are much weaker than in liquids or solids. In the current view of gas behavior, called the **kinetic theory**, gas molecules move about in a random fashion, colliding with the walls of their container and with each other. Perhaps the most important consequence of this theory is that it shows the equivalence between the kinetic energy of molecular motion and the internal energy of the system. Furthermore, the kinetic theory provides us with a physical basis upon which the concept of temperature can be understood.

EQUATIONS AND CONCEPTS

In the kinetic theory of an ideal gas, the pressure of the gas is proportional to the number of molecules per unit volume and the average translational kinetic energy per molecule.

$$P = \tfrac{2}{3}\left(\frac{N}{V}\right)\left(\tfrac{1}{2}m\overline{v^2}\right) \qquad (21.2)$$

From Equation 21.2, and the equation of state for an ideal gas, $PV = Nk_BT$, we find that **the absolute temperature of an ideal gas is a direct measure of the average molecular kinetic energy.**

$$T = \frac{2}{3k_B}\left(\tfrac{1}{2}m\overline{v^2}\right) \qquad (21.3)$$

The **total internal energy** E_{int} of N molecules (or n mols) of a monatomic ideal gas is proportional to the absolute temperature.

$$E_{int} = \tfrac{3}{2}Nk_BT = \tfrac{3}{2}nRT \qquad (21.10)$$

or

$$E_{int} = nC_VT \qquad (21.12)$$

If we apply the first law of thermodynamics to a monatomic **ideal gas** in which heat is transferred at **constant volume,** we find that the **specific heat at constant volume** is equal to $\tfrac{3}{2}R$.

$$C_V = \tfrac{3}{2}R \quad \text{(monatomic only)} \qquad (21.14)$$

If heat is transferred to an ideal gas at **constant pressure,** the first law of thermodynamics shows that the **specific heat at constant pressure** is greater than the specific heat at constant volume by an amount R.

$$C_P - C_V = R \quad \text{(any ideal gas)} \qquad (21.16)$$

or

$$C_P = \tfrac{5}{2}R \quad \text{(monatomic)}$$

The **ratio of specific heats** is a dimensionless quantity γ which, for a **monatomic ideal gas,** is equal to 1.67.

$$\gamma = \frac{C_P}{C_V} \qquad \text{(always)}$$

$$\gamma = \frac{C_P}{C_V} = 1.67 \quad \text{(monatomic)} \qquad (21.17)$$

If an **ideal gas** undergoes a **quasi-static, adiabatic expansion,** and we assume $PV = nRT$ is valid at any time, then the pressure and volume of the gas obey Equation 21.18.

$$PV^\gamma = \text{constant} \qquad (21.18)$$

The law of atmospheres predicts that the density of the atmosphere (molecules per unit volume, n_V) decreases exponentially with altitude, y. Atmospheric pressure also decreases exponentially.

$$n_V(y) = n_0 e^{-mgy/k_BT} \qquad (21.23)$$

$$n_0 = 2.69 \times 10^{25} \text{ molecules} / \text{m}^3$$

$$P = P_0 e^{-mgy/k_BT} \qquad (21.24)$$

The Boltzmann distribution law (which is valid for a system of a large number of particles) states that the probability of finding the particles in a particular arrangement (distribution of energy values) varies exponentially as the negative of the reciprocal of the absolute temperature.

$$n_V(E) = n_0 e^{-E/k_BT} \qquad (21.25)$$

The Maxwell-Boltzmann speed distribution function describes the most probable distribution of speeds of N gas molecules at temperature T (where k_B is the Boltzmann constant).

$$N_v = 4\pi N \left(\frac{m}{2\pi k_B T}\right)^{3/2} v^2 e^{-mv^2/2k_B T} \quad (21.26)$$

Specific expressions can be derived for the rms, average, and most probable speeds.

$$v_{rms} = 1.73\sqrt{k_B T / m} \quad (21.27)$$

$$\bar{v} = 1.60\sqrt{k_B T / m} \quad (21.28)$$

$$v_{mp} = 1.41\sqrt{k_B T / m} \quad (21.29)$$

The average distance ℓ between collisions is called the **mean free path** (where d is the molecular "diameter" and n_V is the density of molecules).

$$\ell = \frac{1}{\sqrt{2}\,\pi d^2 n_V} \quad (21.30)$$

SUGGESTIONS, SKILLS, AND STRATEGIES

CALCULATING SPECIFIC HEATS

It is important to recognize that not all ideal gases are equal. Depending on the type of molecule and the number of degrees of freedom, the ratio of specific heats for a gas (γ) might be 1.67, 1.40, or some other value. However, in any ideal gas, two equations are always valid and can be used to derive the specific heats C_P and C_V from a known value of γ:

$$\gamma = C_P/C_V \qquad \text{and} \qquad C_P - C_V = R$$

Substituting C_P in the first equation for C_P in the second equation, we find

$$C_V = \frac{R}{\gamma - 1} \qquad \text{and} \qquad C_P = \frac{\gamma R}{\gamma - 1}$$

This may be used for ideal gases that are not monatomic.

REVIEW CHECKLIST

▷ State and understand the assumptions made in developing the molecular model of an ideal gas.

▷ Recognize that the temperature of an ideal gas is proportional to the average molecular kinetic energy. State the theorem of equipartition of energy, noting that each degree of freedom of a molecule contributes an equal amount of energy, of magnitude $\frac{1}{2}k_B T$.

▷ Recognize that the internal energy of an ideal gas is proportional to the absolute temperature, and be able to derive the specific heat of an ideal gas at constant volume from the first law of thermodynamics.

▷ Define an adiabatic process, and be able to derive the expression $PV^\gamma = \text{constant}$, which applies to a quasi-static, adiabatic process.

▷ Understand the meaning of the Maxwell-Boltzmann speed distribution function, and recognize the differences among rms speed, average speed, and most probable speed.

ANSWERS TO SELECTED CONCEPTUAL QUESTIONS

5. When alcohol is rubbed on your body, your body temperature decreases. Explain this effect.

Answer

As the alcohol evaporates, high speed molecules leave the liquid. This reduces the average speed of the remaining molecules. Since the average speed is lowered, the temperature of the alcohol is reduced. This process helps to carry energy away from the skin of the patient, resulting in cooling of the skin. The alcohol plays the same role of evaporative cooling as does perspiration, but alcohol evaporates much more quickly than perspiration.

□   □   □   □

9. If a helium-filled balloon initially at room temperature is placed in a freezer, will its volume increase, decrease, or remain the same?

Answer

Helium is a fairly ideal gas. Therefore, we think in terms of the ideal gas law.

$$PV = nRT$$

In this case, the pressure stays nearly constant, being equal to 1 atm. Since the temperature may decrease by 10% (from 293 K to 263 K), the volume should also **decrease** by 10%. This process is called "isobaric cooling," or "isobaric contraction."

□   □   □   □

10. What happens to a helium-filled balloon released into the air? Will it expand or contract? Will it stop rising at some height?

Answer   Imagine the balloon rising into the air. The air cannot be uniform in pressure, because the lower layers support the weight of all the air above. Therefore, as the balloon rises, the pressure will decrease. At the same time, the temperature decreases slightly, but not enough to overcome the effects of the pressure.

By the ideal gas law, $PV = nRT$. Since the decrease in temperature has little effect, the decreasing pressure results in an increasing volume; thus the balloon expands quite dramatically. By the time the balloon reaches an altitude of 10 miles, its volume will be 90 times larger.

   Long before that happens, one of two events will occur. The first possibility, of course, is that the balloon will break, and the pieces will fall to the earth. The other possibility is that the balloon will come to a spot where the average density of the balloon and its payload is equal to the average density of the air. At that point, buoyancy will cause the balloon to "float" at that altitude, until it loses helium and descends.

   People who remember releasing balloons with pen-pal notes will perhaps also remember that replies most often came back when the balloons were low on helium, and were barely able to take off. Such balloons found a fairly low floatation altitude, and were the most likely to be found intact at the end of their journey.

□   □   □   □

SOLUTIONS TO SELECTED END-OF-CHAPTER PROBLEMS

7. A spherical balloon with a volume of 4 000 cm^3 contains helium at an (inside) pressure of 1.20×10^5 Pa. How many moles of helium are in the balloon if each helium atom has an average kinetic energy of 3.60×10^{-22} J?

Solution

G: The balloon has a volume of 4.00 L and a diameter of about 20 cm, which seems like a reasonable size for a typical toy helium balloon. The pressure of the balloon is only slightly more than 1 atm, and if the temperature is anywhere close to room temperature, we can use the estimate of 22 L/mol for an ideal gas at STP conditions. If this is valid, the balloon should contain about 0.2 moles of helium.

O: The average kinetic energy can be used to find the temperature of the gas, which can be used with $PV = nRT$ to find the number of moles.

A: The gas temperature must be that implied by $\frac{1}{2}m\overline{v^2} = \frac{3}{2}k_BT$ for a monatomic gas like Helium.

$$T = \frac{2}{3}\left(\frac{\frac{1}{2}m\overline{v^2}}{k_B}\right) = \frac{2}{3}\left(\frac{3.60 \times 10^{-22}\text{ J}}{1.38 \times 10^{-23}\text{ J / K}}\right) = 17.4\text{ K}$$

Now $PV = nRT$ gives

$$n = \frac{PV}{RT} = \frac{(1.20 \times 10^5\text{ N / m}^2)(4.00 \times 10^{-3}\text{ m}^3)}{(8.314\text{ J / mol} \cdot \text{K})(17.4\text{ K})} = 3.32\text{ mol} \qquad \Diamond$$

L: This result is more than ten times the number of moles we predicted, primarily because the temperature of the helium is **much colder** than room temperature. In fact, T is only slightly above the temperature at which the helium would liquefy (4.2 K at 1 atm). We should hope this balloon is not being held by a child; not only would the balloon sink in the air, it is cold enough to cause frostbite!

9. (a) How many atoms of helium gas fill a balloon to diameter 30.0 cm at 20.0 °C and 1.00 atm? (b) What is the average kinetic energy of the helium atoms? (c) What is the root-mean-square speed of each helium atom?

Solution

(a) The volume is

$$V = \tfrac{4}{3}\pi r^3 = \tfrac{4}{3}\pi(0.150 \text{ m})^3 = 1.41 \times 10^{-2} \text{ m}^3$$

Now $PV = nRT$

$$n = \frac{PV}{RT} = \frac{(1.013 \times 10^5 \text{ N} / \text{m}^2)(1.41 \times 10^{-2} \text{ m}^3)}{(8.314 \text{ N} \cdot \text{m} / \text{mol} \cdot \text{K})(293 \text{ K})}$$

$$n = 0.588 \text{ mol}$$

$$N = nN_A = (0.588 \text{ mol})\left(6.02 \times 10^{23} \text{ molecule} / \text{mol}\right)$$

$$N = 3.54 \times 10^{23} \text{ helium atoms} \qquad \Diamond$$

(b) The kinetic energy is $\overline{K} = \tfrac{1}{2}m\overline{v^2} = \tfrac{3}{2}k_B T$

$$\overline{K} = 1.5(1.38 \times 10^{-23} \text{ J} / \text{K})(293 \text{ K}) = 6.07 \times 10^{-21} \text{ J} \quad \Diamond$$

(c) An He atom has mass $m = \dfrac{M}{N_A} = \dfrac{4.0026 \text{ g} / \text{mol}}{6.02 \times 10^{23} \text{ molecule} / \text{mol}}$

$$m = 6.65 \times 10^{-24} \text{ g} = 6.65 \times 10^{-27} \text{ kg}$$

So the kinetic energy is $\tfrac{1}{2}\left(6.65 \times 10^{-27} \text{ kg}\right)\overline{v^2} = 6.07 \times 10^{-21} \text{ J}$

and $v_{rms} = \sqrt{\overline{v^2}} = 1.35 \text{ km} / \text{s} \qquad \Diamond$

11. A cylinder contains a mixture of helium and argon gas in equilibrium at 150 °C. (a) What is the average kinetic energy of each type of gas molecule? (b) what is the root-mean-square speed for each type of molecule?

Solution

(a) Both kinds of molecules have the same average kinetic energy, sharing it in collisions. Also, since this is a monatomic gas, all of the kinetic energy is translational.

$$\overline{K} = \tfrac{3}{2}k_BT = \tfrac{3}{2}(1.38\times10^{-23}\text{ J / K})(273+150)\text{K} = 8.76\times10^{-21}\text{ J} \qquad \lozenge$$

(b) Since $\overline{K} = \tfrac{3}{2}k_BT = \tfrac{1}{2}m\overline{v^2}$, the RMS velocity depends upon the mass of each atom.

Thus,
$$v_{rms} = \sqrt{\overline{v^2}} = \sqrt{\frac{m\overline{v^2}}{m_{atom}}} = \sqrt{\frac{2\overline{K}}{m_{atom}}}$$

For Helium,
$$m_{He} = \frac{M}{N_A} = \frac{(4.0026\text{ g / mol})(10^{-3}\text{ kg / g})}{6.02\times10^{23}\text{ molecule / mol}} = 6.65\times10^{-27}\text{ kg}$$

and
$$v_{rms,\ He} = \sqrt{\frac{2(8.76\times10^{-21}\text{ J})}{6.65\times10^{-27}\text{ kg}}} = 1620\text{ m / s} = 1.62\text{ km / s} \qquad \lozenge$$

For Argon,
$$m_{He} = \frac{M}{N_A} = \frac{(39.962\text{ g / mol})(10^{-3}\text{ kg / g})}{6.02\times10^{23}\text{ molecule / mol}} = 6.64\times10^{-26}\text{ kg}$$

and
$$v_{rms,\ Ar} = \sqrt{\frac{2(8.76\times10^{-21}\text{ J})}{6.65\times10^{-27}\text{ kg}}} = 514\text{ m / s} = 0.514\text{ km / s} \qquad \lozenge$$

15. One mole of hydrogen gas is heated at constant pressure from 300 K to 420 K. Calculate (a) the energy transferred by heat to the gas, (b) the increase in its internal energy, and (c) the work done by the gas.

Solution Since this is a constant-pressure process, $Q = nC_P\Delta T$

(a)
$$Q = (1.00 \text{ mol})(28.8 \text{ J / mol} \cdot \text{K})(120 \text{ K}) = 3.46 \text{ kJ} \qquad \lozenge$$

(b) For any system, $\Delta E_{int} = nC_V\Delta T$

so
$$\Delta E_{int} = (1.00 \text{ mol})(20.4 \text{ J / mol} \cdot \text{K})(120 \text{ K}) = 2.45 \text{ kJ} \qquad \lozenge$$

(c) $\Delta E_{int} = Q - W$ so $W = Q - \Delta E_{int} = 3.46 \text{ kJ} - 2.45 \text{ kJ} = 1.01 \text{ kJ} \qquad \lozenge$

---

25. Two moles of an ideal gas ($\gamma = 1.40$) expands slowly and adiabatically from a pressure of 5.00 atm and a volume of 12.0 L to a final volume of 30.0 L. (a) What is the final pressure of the gas? (b) What are the initial and final temperatures? (c) Find Q, W, and ΔE_{int}.

Solution

(a) $P_i V_i^\gamma = P_f V_f^\gamma$

$$P_f = \left(\frac{V_i}{V_f}\right)^\gamma P_i = \left(\frac{12.0 \text{ L}}{30.0 \text{ L}}\right)^{1.40} (5.00 \text{ atm}) = 1.39 \text{ atm} \qquad \lozenge$$

(b) $T_i = \dfrac{P_i V_i}{nR} = \dfrac{(5.00 \text{ atm})(1.01 \times 10^5 \text{ Pa / atm})(12.0 \times 10^{-3} \text{ m}^3)}{(2.00 \text{ mol})(8.31 \text{ N} \cdot \text{m / mol} \cdot \text{K})} = 366 \text{ K} \qquad \lozenge$

$$T_f = \frac{P_f V_f}{nR} = 254 \text{ K} \qquad \lozenge$$

451

(c) This is an adiabatic process, so by definition $Q = 0$ ◊

For any process, $\Delta E_{int} = n C_V \Delta T$, and for this gas, $C_V = \dfrac{R}{\gamma - 1} = \dfrac{5}{2} R$

Thus, $\Delta E_{int} = \dfrac{5}{2}(2.00 \text{ mol})(8.31 \text{ J / mol} \cdot \text{K})254(366 \text{ K} - 366 \text{ K}) = -4650 \text{ J}$ ◊

Now, $W = Q - \Delta E_{int} = 0 - (-4650 \text{ J}) = 4650 \text{ J}$ ◊

Note that in this case, the work is positive, so work is done by the gas.

---

27. Air in a thundercloud expands as it rises. If its initial temperature was 300 K, and if no energy is lost by thermal conduction on expansion, what is its temperature when the initial volume has doubled?

Solution

G: The air should cool as it expands, so we should expect $T_f < 300 \text{ K}$.

O: The air expands adiabatically, losing no heat but dropping in temperature as it does work on the air around it, so we assume that $PV^\gamma = \text{constant}$ (where $\gamma = 1.40$ for an ideal diatomic gas).

A: By Equation 21.20, $T_1 V_1^{\gamma-1} = T_2 V_2^{\gamma-1}$

$$T_2 = T_1 \left(\frac{V_1}{V_2}\right)^{\gamma-1} = 300 \text{ K} \left(\frac{1}{2}\right)^{(1.40-1)} = 227 \text{ K} \quad ◊$$

L: The air does cool, but the temperature is not inversely proportional to the volume. The temperature drops 24% while the volume doubles.

---

33. Consider 2.00 mol of an ideal diatomic gas. Find the total heat capacity at constant volume and at constant pressure (a) if the molecules rotate but do not vibrate and (b) if the molecules both rotate and vibrate.

Solution　(a) Count degrees of freedom. A diatomic molecule oriented along the y axis can possess energy by moving in x, y, and z directions and by rotating around the x and z axes. Outside forces can get no lever arms to make it spin around the y axis. The molecule will have average energy $\frac{1}{2}k_BT$ for each of these five degrees of freedom.

The gas will have internal energy $\qquad E_{int} = N\left(\frac{5}{2}\right)k_BT = nN_A\left(\frac{5}{2}\right)\left(\frac{R}{N_A}\right)T = \left(\frac{5}{2}\right)nRT$

so the constant-volume heat capacity of the whole sample is

$$\frac{\Delta E_{int}}{\Delta T} = \left(\frac{5}{2}\right)nR = \left(\frac{5}{2}\right)(2.00 \text{ mol})(8.314 \text{ J / mol} \cdot \text{K}) = 41.6 \text{ J / K} \quad \lozenge$$

For one mole, $\qquad C_P = C_V + R;$　　For the sample, $\quad nC_P = nC_V + nR$

With P constant, $\quad nC_P = 41.6 \text{ J / K} + (2.00 \text{ mol})(8.314 \text{ J / mol} \cdot \text{K}) = 58.2 \text{ J / K} \quad \lozenge$

(b) Vibration adds a degree of freedom for kinetic energy and a degree of freedom for elastic energy. Now the molecule's average energy is $\left(\frac{7}{2}\right)k_BT$,

the sample's internal energy is $\qquad\qquad N\left(\frac{7}{2}\right)k_BT = \left(\frac{7}{2}\right)nRT$

and the sample's constant-volume heat capacity is

$$\left(\frac{7}{2}\right)nR = \left(\frac{7}{2}\right)(2.00 \text{ mol})(8.314 \text{ J / mol} \cdot \text{K}) = 58.2 \text{ J / K} \quad\qquad \lozenge$$

At constant pressure, its heat capacity is

$$nC_V + nR = 58.2 \text{ J / K} + (2.00 \text{ mol})(8.314 \text{ J / mol} \cdot \text{K}) = 74.8 \text{ J / K} \quad \lozenge$$

39. Fifteen identical particles have the following speeds: one has a speed of 2.00 m/s; two have a speed of 3.00 m/s; three have a speed of 5.00 m/s; four have a speed of 7.00 m/s; three have a speed of 9.00 m/s; and two have a speed of 12.0 m/s. Find (a) the average speed, (b) the rms speed, and (c) the most probable speed of these particles.

Solution

(a) $\bar{v} = \dfrac{\Sigma n_i v_i}{\Sigma n_i} = \dfrac{1(2.00) + 2(3.00) + 3(5.00) + 4(7.00) + 3(9.00) + 2(12.0)}{1 + 2 + 3 + 4 + 3 + 2}$ m/s

$\bar{v} = 6.80$ m/s ◊

(b) $\overline{v^2} = \dfrac{\Sigma n_i v_i^2}{\Sigma n_i}$

$\overline{v^2} = \dfrac{1(2.00^2) + 2(3.00^2) + 3(5.00^2) + 4(7.00^2) + 3(9.00^2) + 2(12.0^2)}{15}$ m^2/s^2

$\overline{v^2} = 54.9$ m^2/s^2

$v_{rms} = \sqrt{\overline{v^2}} = \sqrt{54.9 \text{ m}^2/\text{s}^2} = 7.41$ m/s ◊

(c) More particles have $v_{mp} = 7.00$ m/s than any other speed. ◊

41. From the Maxwell—Boltzmann speed distribution, show that the most probable speed of a gas molecule is given by Equation 21.29. Note that the most probable speed corresponds to the point at which the slope of the speed distribution curve, dN_v / dv, is zero.

Solution   From the Maxwell speed distribution function,

$$N_v = 4\pi N\left(\frac{m}{2\pi k_B T}\right)^{3/2} v^2\, e^{\left(-mv^2/2k_B T\right)}$$

we locate the peak in the graph of N_v versus v by evaluating dN_v/dv and setting it equal to zero, to solve for the most probable speed:

$$\frac{dN_v}{dv} = 4\pi N\left(\frac{m}{2\pi k_B T}\right)^{3/2}\left(\frac{-2mv}{2k_B T}\right)v^2\, e^{\left(-mv^2/2k_B T\right)}$$

$$+4\pi N\left(\frac{m}{2\pi k_B T}\right)^{3/2} 2v e^{\left(-mv^2/2k_B T\right)} = 0$$

This equation is solved by $v = 0$ and $v \to \infty$, but those correspond to minimum-probability speeds, so we divide by v and by the exponential function.

$$v_{mp}\left(-\frac{m2v_{mp}}{2k_B T}\right) + 2 = 0$$

$$v_{mp} = \left(\frac{2k_B T}{m}\right)^{1/2} \quad \text{as in Equation 21.29.} \qquad \lozenge$$

---

45. In an ultrahigh vacuum system, the pressure is measured to be 1.00×10^{-10} torr (where 1 torr = 133 Pa). Assume that the gas molecules have a molecular diameter of 3.00×10^{-10} m and that the temperature is 300 K. Find (a) the number of molecules in a volume of 1.00 m³, (b) the mean free path of the molecules, and (c) the collision frequency, assuming an average speed of 500 m/s.

Solution

G: Since high vacuum means low pressure as a result of a low molecular density, we should expect a relatively low number of molecules, a long free path, and a low collision frequency compared with the values found in Example 21.7 for normal air. Since the ultrahigh vacuum absolute pressure is 13 orders of magnitude lower than atmospheric pressure, we might expect corresponding values of $N \sim 10^{12}$ molecules/m^3, $\ell \sim 10^6$ m, and $f \sim 0.0001$/s.

O: The equation of state for an ideal gas can be used with the given information to find the number of molecules in a specific volume. The mean free path can be found directly from equation 21.30, and this result can be used with the average speed to find the collision frequency.

A: (a) $PV = \left(\dfrac{N}{N_A}\right)RT$ means $N = \dfrac{PVN_A}{RT}$, so that

$$N = \frac{(1.00 \times 10^{-10})(133)(6.02 \times 10^{23})}{(8.31)(300)} = 3.21 \times 10^{12} \text{ molecules} \qquad \Diamond$$

(b) $\ell = \dfrac{1}{n_V \pi d^2 \sqrt{2}} = \dfrac{V}{N \pi d^2 \sqrt{2}} = \dfrac{1.00 \text{ m}^3}{(3.21 \times 10^{12} \text{ molecules})\pi(3.00 \times 10^{-10} \text{ m})^2 \sqrt{2}}$

$\ell = 7.78 \times 10^5 \text{ m} = 778 \text{ km}$ $\qquad \Diamond$

(c) $f = v/\ell = \dfrac{500 \text{ m}/\text{s}}{7.78 \times 10^5 \text{ m}} = 6.42 \times 10^{-4} \text{ s}^{-1}$ $\qquad \Diamond$

L: The calculated results differ from the results in Example 21.7 by about 13 orders of magnitude, as we expected. This ultrahigh vacuum is an environment very different from the ocean floor to which we are accustomed, under the Earth's atmosphere. These conditions can provide a "clean" environment for a variety of experiments and manufacturing processes that would otherwise be impossible.

53. A cylinder contains n mol of an ideal gas that undergoes an adiabatic process. (a) Starting with the expression $W = \int P \, dV$ and using $PV^\gamma = $ constant, show that the work done is

$$W = \left(\frac{1}{\gamma-1}\right)(P_iV_i - P_fV_f)$$

(b) Starting with the first-law equation in differential form, prove that the work done also is equal to $nC_V(T_i - T_f)$. Show that this result is consistent with the equation given in part (a).

Solution (a) $PV^\gamma = k$ so $W = \int_i^f P \, dV = k \int_{V_i}^{V_f} \frac{dV}{V^\gamma} = \left.\frac{kV^{1-\gamma}}{1-\gamma}\right|_{V_i}^{V_f}$

$$W = \frac{P_fV_f^\gamma V_f^{1-\gamma} - P_iV_i^\gamma V_i^{1-\gamma}}{1-\gamma} = \frac{P_iV_i - P_fV_f}{\gamma-1} \quad \Diamond$$

(b) $dE_{int} = dQ - dW$ and $dQ = 0$ for an adiabatic process. Therefore,

$$W = -\Delta E_{int} = -nC_V\Delta T = nC_V(T_i - T_f) \quad \Diamond$$

To show consistency between these two equations, consider that

$$\gamma = \frac{C_P}{C_V} \quad \text{and} \quad C_P - C_V = R, \quad \text{so that} \quad \frac{1}{\gamma-1} = \frac{C_V}{R}$$

Thus part (a) becomes $\quad W = (P_iV_i - P_fV_f)\dfrac{C_V}{R}$

Then, $\quad \dfrac{PV}{R} = nT \quad$ so that $\quad W = nC_V(T_i - T_f) \quad \Diamond$

57. The compressibility κ of a substance is defined as the fractional change in volume of that substance for a given change in pressure:

$$\kappa = -\frac{1}{V}\frac{dV}{dP}$$

(a) Explain why the negative sign in this expression ensures that κ is always positive. (b) Show that if an ideal gas is compressed isothermally, its compressibility is given by $\kappa_1 = 1/P$. (c) Show that if an ideal gas is compressed adiabatically, its compressibility is given by $\kappa_2 = 1/\gamma P$. (d) Determine values for κ_1 and κ_2 for a monatomic ideal gas at a pressure of 2.00 atm.

Solution   Since pressure increases as volume decreases (and vice versa), and the volume is always positive,

(a) $$\frac{dV}{dP} < 0 \qquad \text{and} \qquad -\left(\frac{1}{V}\right)\frac{dV}{dP} > 0 \qquad \Diamond$$

(b)   For an ideal gas, $$V = \frac{nRT}{P} \qquad \text{and} \qquad \kappa_1 = -\frac{1}{V}\frac{d}{dP}\left(\frac{nRT}{P}\right)$$

For isothermal compression, T is constant : $$\kappa_1 = -\frac{nRT}{V}\left(\frac{-1}{P^2}\right) = \frac{1}{P} \qquad \Diamond$$

(c)   For an adiabatic compression, $PV^\gamma = C$ (where C is a constant) and

$$\kappa_2 = -\left(\frac{1}{V}\right)\frac{d}{dP}\left(\frac{C}{P}\right)^{1/\gamma} = \left(\frac{1}{V\gamma}\right)\frac{C^{1/\gamma}}{P\left(P^{1/\gamma+1}\right)} = \frac{V}{V\gamma P} = \frac{1}{\gamma P} \qquad \Diamond$$

(d)   For a monatomic ideal gas, $\gamma = C_P/C_V = 5/3$

$$\kappa_1 = \frac{1}{P} = \frac{1}{2.00 \text{ atm}} = 0.500 \text{ atm}^{-1}; \quad \kappa_2 = \frac{1}{\gamma P} = \frac{1}{(5/3)(2.00 \text{ atm})} = 0.300 \text{ atm}^{-1} \; \Diamond$$

59. For a Maxwellian gas, use a computer or programmable calculator to find the numerical value of the ratio $\{N_v(v)/N_v(v_{mp})\}$ for the following values of v: $\quad v = (v_{mp}/50)$, $(v_{mp}/10)$, $(v_{mp}/2)$, v_{mp}, $2v_{mp}$, $10v_{mp}$, $50v_{mp}$. Give your results to three significant figures.

Solution Substitute $v_{mp} = \left(\dfrac{2k_BT}{m}\right)^{1/2}$

into $\qquad\qquad N_v(v) = 4\pi N\left(\dfrac{m}{2\pi k_BT}\right)^{3/2} v^2\, e^{-mv^2/2k_BT}$

and evaluate $\qquad\qquad N_v(v_{mp}) = 4\pi N\left(\dfrac{m}{2\pi k_BT}\right)^{3/2} \dfrac{2k_BT}{m}\, e^{-1}$

$$N_v(v_{mp}) = 4\pi^{-1/2}\, e^{-1}\, N\left(\dfrac{m}{2k_BT}\right)^{1/2} = 4\pi^{-1/2}\, e^{-1}\, Nv_{mp}^{-1}$$

Then $\qquad\quad \dfrac{N_v(v)}{N_v(v_{mp})} = \dfrac{4\pi^{-1/2}\, Nv_{mp}^{-3}\, v^2 e^{-\left(v\,/v_{mp}\right)^2}}{4\pi^{-1/2}\, Nv_{mp}^{-1}\, e^{-1}} = \left(\dfrac{v^2}{v_{mp}^2}\right)e^{1-v^2/v_{mp}^2}$

For $\quad v = \dfrac{v_{mp}}{50}$, $\qquad \dfrac{N_v(v)}{N_v(v_{mp})} = \left(\dfrac{1}{50}\right)^2 e^{1-(1/50)^2} = 1.09\times10^{-3}$ $\qquad\qquad$ ◊

For $\quad v = \dfrac{v_{mp}}{10}$, $\qquad \dfrac{N_v(v)}{N_v(v_{mp})} = \left(\dfrac{1}{10}\right)^2 e^{1-1/100} = 2.69\times10^{-2}$ $\qquad\qquad$ ◊

For $\quad v = \dfrac{v_{mp}}{2}$, $\qquad \dfrac{N_v(v)}{N_v(v_{mp})} = \left(\dfrac{1}{2}\right)^2 e^{1-1/4} = 0.529$ $\qquad\qquad$ ◊

For $\quad v = v_{mp}$, $\qquad \dfrac{N_v(v)}{N_v(v_{mp})} = (1)^2\, e^{1-1} = 1.00$ $\qquad\qquad$ ◊

Similarly,

$\dfrac{v}{v_{mp}}$	$\dfrac{N_v(v)}{N_v(v_{mp})}$
1/50	1.09×10^{-3}
1/10	2.69×10^{-2}
1/2	0.529
1	1.00
2	0.199
10	1.01×10^{-41}
50	1.25×10^{-1082}

To find the last,

$$\left(50^2\right)e^{1-2500} = \left(50^2\right)e^{-2499}$$

$$= 10^{\log 2500}\, e^{\ln 10(-2499/\ln 10)}$$

$$= 10^{(\log 2500 - 2499/\ln 10)} = 10^{-1080} \qquad \diamond$$

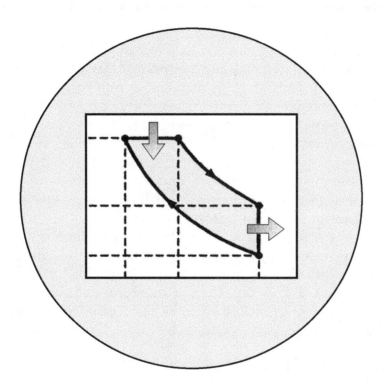

Heat Engines, Entropy, and the Second Law of Thermodynamics

HEAT ENGINES, ENTROPY, AND THE SECOND LAW OF THERMODYNAMICS

INTRODUCTION

The first law of thermodynamics, studied in Chapter 20, is a statement of conservation of energy, generalized to include heat as a form of energy transfer. This law tells us only that an increase in one form of energy must be accompanied by a decrease in some other form of energy. It places no restrictions on the types of energy conversions that can occur. Furthermore, it makes no distinction between heat and work. However, such a distinction exists. One manifestation of this difference is the fact that it is impossible to convert thermal energy to mechanical energy entirely and continuously. The second law of thermodynamics establishes which energy conversion processes can and cannot occur in nature.

From an engineering viewpoint, perhaps the most important application of the second law of thermodynamics is the limited efficiency of heat engines. The second law says that a machine capable of continuously converting thermal energy completely to other forms of energy in a cyclic process cannot be constructed.

EQUATIONS AND CONCEPTS

The net **work done** by a heat engine during one cycle equals the net heat flowing into the engine.

$$W = Q_h - Q_c \qquad (22.1)$$

The **thermal efficiency**, e, of a heat engine is defined as the ratio of the net work done to the heat absorbed during one cycle of the process.

$$e = \frac{W}{Q_h} = 1 - \frac{Q_c}{Q_h} \qquad (22.2)$$

The **coefficient of performance** of a refrigerator or a heat pump used in the cooling cycle is defined by the ratio of the heat absorbed, Q_c, to the work done. A good refrigerator has a high coefficient of performance.

$$\text{COP (cooling mode)} = \frac{Q_c}{W} \qquad (22.7)$$

No real engine operating between the temperatures T_c and T_h can be more efficient than an engine operating reversibly in a Carnot cycle between the same two temperatures.

$$e_c = 1 - \frac{T_c}{T_h} \qquad (22.4)$$

The efficiency of the Otto cycle (gasoline engine) depends on the compression ratio (V_1 / V_2) and the ratio of molar specific heats.

$$e = 1 - \frac{1}{(V_1 / V_2)^{\gamma - 1}} \qquad (22.5)$$

The effectiveness of a heat pump is described in terms of the coefficient of performance, COP.

$$\text{COP (heating mode)} \equiv$$
$$\frac{\text{energy transferred at high temp}}{\text{work done by pump}} = \frac{Q_h}{W} \qquad (22.6)$$

If a system changes from one equilibrium state to another, under a reversible, quasi-static process, and a quantity of heat dQ_r is added (or removed) at the absolute temperature T, the **change in entropy** is defined by the ratio dQ_r / T.

$$dS = \frac{dQ_r}{T} \qquad (22.8)$$

The **change in entropy** of a system which undergoes a reversible process between the states i and f depends only on the properties of the initial and final equilibrium states.

$$\Delta S = \int_i^f \frac{dQ_r}{T} \text{ (reversible path)} \qquad (22.9)$$

The change in entropy of a system for any arbitrary **reversible cycle** is identically **zero.**

$$\oint \frac{dQ_r}{T} = 0 \qquad (22.10)$$

The change in entropy of an ideal gas which undergoes a quasi-static and reversible process depends only on the initial and final states.

$$\Delta S = nC_V \ln \frac{T_f}{T_i} + nR \ln \frac{V_f}{V_i} \qquad (22.12)$$

During a **free expansion** process, the entropy of a gas **increases.**

$$\Delta S = nR \ln \frac{V_f}{V_i} \qquad (22.13)$$

During the process of irreversible heat transfer between two masses, the entropy of the system will increase.

$$\Delta S = m_1 c_1 \ln \frac{T_f}{T_1} + m_2 c_2 \ln \frac{T_f}{T_2} \qquad (22.15)$$

All systems tend toward disorder, and the entropy of the system is related to the number of possible microstates of the system, W.

$$S \equiv k_B \ln W \qquad (22.18)$$

REVIEW CHECKLIST

▷ Understand the basic principle of the operation of a heat engine, and be able to define and discuss the **thermal efficiency** of a heat engine.

▷ State the second law of thermodynamics, and discuss the difference between reversible and irreversible processes. Discuss the importance of the first and second laws of thermodynamics as they apply to various forms of energy conversion and thermal pollution.

▷ Describe the processes which take place in an ideal heat engine taken through a **Carnot cycle**.

▷ Calculate the efficiency of a Carnot engine, and note that the efficiency of real heat engines is always less than the Carnot efficiency.

▷ Calculate entropy changes for reversible processes (such as one involving an ideal gas).

▷ Calculate entropy changes for irreversible processes, recognizing that the entropy change for an irreversible process is equivalent to that of a reversible process between the same two equilibrium states.

ANSWERS TO SELECTED CONCEPTUAL QUESTIONS

2. What are some factors that affect the efficiency of automobile engines?

Answer The most basic limit of efficiency for automobile engines is the Carnot cycle; the efficiency cannot **ever** exceed this ideal model. The efficiency is therefore affected by the maximum temperature that the engine block can stand, and the intake and exhaust pressures. However, the Carnot cycle assumes that the exhaust gases can be released over an infinite period of time. If you want your engine to act more quickly than "never," you must take additional losses. Other limits on efficiency are imposed by friction, the mass and acceleration of the engine parts, and the timing of the ignition.

□   □   □   □

4. A steam-driven turbine is one major component of an electric power plant. Why is it advantageous to have the temperature of the steam as high as possible?

Answer The most optimistic limit of efficiency in a steam engine is the ideal (Carnot) efficiency. This can be calculated from the high and low temperatures of the steam, as it expands:

$$e_c = \frac{T_h - T_c}{T_h} = 1 - \frac{T_c}{T_h}$$

The engine will be most efficient when the low temperature is extremely low, and the high temperature is extremely high. However, since the electric power plant is typically placed on the surface of the earth, there is a limit to how much the steam can expand, and how low the lower temperature can be. The only way to further increase the efficiency, then, is to raise the temperature of the hot steam, as high as possible.

□   □   □   □

466

20. The device shown in Figure Q22.20, which is called a thermoelectric converter, uses a series of semiconductor cells to convert thermal energy to electrical energy. In the photograph on the left, both legs of the device are at the same temperature and no electrical energy is produced. However, when one leg of is at a higher temperature than the other, as shown in the photograph on the right, electrical energy is produced as the device drives a small electric motor. (a) Why does the temperature differential produce electrical energy in this demonstration? (b) In what sense does this intriguing experiment demonstrate the second law of thermodynamics?

Figure Q22.20

Answer (a) The semiconductor converter operates essentially like a thermocouple, which is a pair of wires of different metals, with a junction at each end. When the junctions are at different temperatures, a small voltage appears around the loop, so that the device can be used to measure temperature or (here) to drive a small motor.

(b) The second law's statement is that it is impossible for a cycling engine simply to extract heat from one reservoir, and convert it into work. This exactly describes the first situation, where both legs are in contact with a single reservoir, and the thermocouple fails to produce work. To convert heat to work, the device must transfer heat from a hot reservoir to a cold reservoir, as in the second situation.

SOLUTIONS TO SELECTED END-OF-CHAPTER PROBLEMS

3. A particular engine has a power output of 5.00 kW and an efficiency of 25.0%. Assuming that the engine expels 8000 J of energy in each cycle, find (a) the energy absorbed in each cycle and (b) the time for each cycle.

Solution We are given that $Q_c = 8000$ J

(a) We have
$$e = \frac{W}{Q_h} = \frac{Q_h - Q_c}{Q_h} = 1 - \frac{Q_c}{Q_h} = 0.250$$

Isolating Q_h,
$$Q_h = \frac{Q_c}{1-e} = \frac{8000 \text{ J}}{1 - 0.250} = 10.7 \text{ kJ} \qquad \lozenge$$

(b) The work per cycle is
$$W = Q_h - Q_c = 2667 \text{ J}$$

From $\mathcal{P} = \dfrac{W}{t}$,
$$t = \frac{W}{\mathcal{P}} = \frac{2667 \text{ J}}{5000 \text{ J/s}} = 0.533 \text{ s} \qquad \lozenge$$

7. One of the most efficient engines ever built (actual efficiency 42.0%) operates between 430 °C and 1870 °C. (a) What is its maximum theoretical efficiency? (b) How much power does the engine deliver if it absorbs 1.40×10^5 J of energy each second from the hot reservoir?

Solution $T_c = 430 \,°\text{C} = 703$ K and $T_h = 1870 \,°\text{C} = 2143$ K

(a) $e_c = \dfrac{\Delta T}{T_h} = \dfrac{1440 \text{ K}}{2143 \text{ K}} = 0.672$ or 67.2% $\qquad \lozenge$

(b) $Q_h = 1.40 \times 10^5$ J $\qquad W = 0.420 Q_h = 5.88 \times 10^4$ J

$$\mathcal{P} = \frac{W}{t} = \frac{5.88 \times 10^4 \text{ J}}{1 \text{ s}} = 58.8 \text{ kW} \qquad \lozenge$$

11. An ideal gas is taken through a Carnot cycle. The isothermal expansion occurs at 250 °C, and the isothermal compression takes place at 50.0 °C. Assuming that the gas absorbs 1200 J of energy from the hot reservoir during the isothermal expansion, find (a) the energy expelled to the cold reservoir in each cycle and (b) the net work done by the gas in each cycle.

Solution   (a)   For a Carnot cycle, $e_c = 1 - T_c/T_h$

For any engine, $e = \dfrac{W}{Q_h} = 1 - \dfrac{Q_c}{Q_h}$

Therefore, for a Carnot engine, $1 - \dfrac{T_c}{T_h} = 1 - \dfrac{Q_c}{Q_h}$

Then, since $Q_c = Q_h\left(\dfrac{T_c}{T_h}\right)$, $Q_c = (1200 \text{ J})\left(\dfrac{323 \text{ K}}{523 \text{ K}}\right) = 741 \text{ J}$ ◊

(b) The work is calculated as $W = Q_h - Q_c = 1200 \text{ J} - 741 \text{ J} = 459 \text{ J}$ ◊

---

19. In a cylinder of an automobile engine just after combustion, the gas is confined to a volume of 50.0 cm³ and has an initial pressure of 3.00×10^6 Pa. The piston moves outward to a final volume of 300 cm³, and the gas expands without energy loss by heat. (a) If $\gamma = 1.40$ for the gas, what is the final pressure? (b) How much work is done by the gas in expanding?

Solution

G: The pressure will decrease as the volume increases. If the gas were to expand at constant temperature, the final pressure would be $\frac{50}{300}(3 \times 10^6 \text{ Pa}) = 5 \times 10^5$ Pa. But the temperature must drop a great deal as the gas spends internal energy in doing work. Thus the final pressure must be lower than 500 kPa. The order of magnitude of the work can be estimated from the average pressure and change in volume:

$$W \sim (10^6 \text{ N}/\text{m}^2)(100 \text{ cm}^3)(10^{-6} \text{ m}^3/\text{cm}^3) = 100 \text{ J}$$

O: The gas expands adiabatically (there is not enough time for significant heat transfer), so equation 21.18 can be applied to find the final pressure. With $Q = 0$, the amount of work can be found from the change in internal energy.

A: (a) For adiabatic expansion, $\qquad P_i V_i^{\gamma} = P_f V_f^{\gamma}$

Therefore, $P_f = P_i \left(\dfrac{V_i}{V_f}\right)^{\gamma} = \left(3.00 \times 10^6 \text{ Pa}\right)\left(\dfrac{50.0 \text{ cm}^3}{300 \text{ cm}^3}\right)^{1.40} = 2.44 \times 10^5 \text{ Pa}$ ◊

(b) Since $Q = 0$, we have $W = Q - \Delta E = -\Delta E = -nC_V \Delta T = -nC_V\left(T_f - T_i\right)$

From $\gamma = \dfrac{C_P}{C_V} = \dfrac{C_V + R}{C_V}$, we get $(\gamma - 1)C_V = R$

So that $C_V = \dfrac{R}{1.40 - 1} = 2.50R$

$$W = n(2.50R)\left(T_i - T_f\right) = 2.50 P_i V_i - 2.50 P_f V_f$$

$$W = 2.50\left[\left(3.00 \times 10^6 \text{ Pa}\right)\left(50.0 \times 10^{-6} \text{ m}^2\right) - \left(2.44 \times 10^5 \text{ Pa}\right)\left(300 \times 10^{-6} \text{ m}^2\right)\right]$$

$$W = 192 \text{ J} \qquad\qquad\qquad ◊$$

L: The final pressure is about half of the 500 kPa, in agreement with our qualitative prediction. The order of magnitude of the work is as we predicted.

From the work done by the gas in part (a), the average power of the engine could be calculated if the time for one cycle was known. Adiabatic expansion is the power stroke of our industrial civilization.

25. An ideal refrigerator or ideal heat pump is equivalent to a Carnot engine running in reverse. That is, energy Q_c is absorbed from a cold reservoir, and energy Q_h is rejected to a hot reservoir. (a) Show that the work that must be supplied to run the refrigerator or heat pump is

$$W = \frac{T_h - T_c}{T_c} Q_c$$

(b) Show that the coefficient of performance (COP) of the ideal refrigerator is

$$COP = \frac{T_c}{T_h - T_c}$$

Solution

(a) For a complete cycle $\Delta E_{int} = 0$, and $\qquad W = Q_h - Q_c = Q_c\left(\frac{Q_h}{Q_c} - 1\right)$

Since, for a Carnot cycle (and only a Carnot cycle) $\quad \dfrac{Q_h}{Q_c} = \dfrac{T_h}{T_c}$

Then $\qquad\qquad\qquad\qquad\qquad\qquad\qquad\qquad W = \dfrac{T_h - T_c}{T_c} Q_c \qquad\qquad \Diamond$

(b) From Equation 22.7, $\quad COP = \dfrac{Q_c}{W}$, so $\qquad COP = \dfrac{T_c}{T_h - T_c} \qquad\qquad \Diamond$

---

27. How much work does an ideal Carnot refrigerator require to remove 1.00 J of energy from helium at 4.00 K and reject this thermal energy to a room-temperature (293-K) environment?

Solution $\qquad (COP)_{He,\ refrig} = \dfrac{T_c}{\Delta T} = \dfrac{4.00\ K}{293\ K - 4.00\ K} = 0.0138$

$$W = \frac{Q_c}{COP} = \frac{1.00\ J}{0.0138} = 72.3\ J \qquad\qquad \Diamond$$

---

33. Calculate the change in entropy of 250 g of water heated slowly from 20.0 °C to 80.0 °C. (**Hint:** Note that $dQ = mc\,dT$.)

Solution

To do the heating reversibly, put the water pot successively into contact with reservoirs at temperatures 20.0 °C + δ, 20.0 °C + 2δ, ... 80.0 °C, where δ is some small increment. Then

$$\Delta S = \int_i^f \frac{dQ}{T} = \int_{T_i}^{T_f} mc\frac{dT}{T}$$

Here T means the absolute temperature. We would ordinarily think of dT as the change in the Celsius temperature, but one Celsius degree of temperature change is the same size as one Kelvin of change, so dT is also the change in absolute T. Then

$$\Delta S = mc\ln T\Big|_{T_i}^{T_f} = mc\ln\left(\frac{T_f}{T_i}\right)$$

$$\Delta S = (0.250\text{ kg})(4186\text{ J / kg}\cdot\text{K})\ln\left(\frac{353\text{ K}}{293\text{ K}}\right) = 195\text{ J / K} \qquad \lozenge$$

---

37. A 1500-kg car is moving at 20.0 m/s. The driver brakes to a stop. The brakes cool off to the temperature of the surrounding air, which is nearly constant at 20.0 °C. What is the total entropy change?

Solution   The original kinetic energy of the car,

$$K = \tfrac{1}{2}mv^2 = \tfrac{1}{2}(1500\text{ kg})(20.0\text{ m / s})^2 = 300\text{ kJ}$$

becomes irreversibly 300 kJ of extra internal energy in the brakes, the car, and its surroundings. Since their total heat capacity is so large, their temperature ends up very nearly still at 20.0° C. To carry them reversibly to this same final state, imagine putting 300 kJ from a heater into the car and its environment. Then

$$\Delta S = \int_i^f \frac{dQ_r}{T} = \frac{1}{T}\int dQ_r = \frac{Q}{T} = \frac{300 \text{ kJ}}{293 \text{ K}}$$

$$\Delta S = 1.02 \text{ kJ / K} \qquad \Diamond$$

39. One mole of H_2 gas is contained in the left-hand side of the container shown in Figure P22.39, which has equal volumes left and right. The right-hand side is evacuated. When the valve is opened, the gas streams into the right side. What is the final entropy change of the gas? Does the temperature of the gas change?

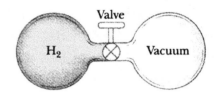

Figure P22.39

Solution

(a) This is an example of free expansion; from Equation 22.13 we have

$$\Delta S = nR \ln\left(\frac{V_f}{V_i}\right) \qquad \Delta S = (1.00 \text{ mole})\left(8.31\frac{\text{J}}{\text{mole} \cdot \text{K}}\right)\ln\left(\frac{2}{1}\right)$$

$$\Delta S = 5.76 \text{ J/K} \qquad \text{or} \qquad \Delta S = 1.38 \text{ cal/K} \qquad \Diamond$$

(b) The gas is expanding into an evacuated region. Therefore, $W = 0$; and it expands so fast that heat has no time to flow; $Q = 0$. But $\Delta E_{int} = Q - W$, so in this case $\Delta E_{int} = 0$. For an ideal gas, the internal energy is a function of the temperature, so if $\Delta E_{int} = 0$, the temperature remains constant. $\qquad \Diamond$

47. Repeat the procedure used to construct Table 22.1 (a) for the case in which you draw three marbles from your bag rather than four and (b) for the case in which you draw five rather than four.

Solution   The combinations are as follows:

(a)

Result	Possible combinations	Total
All Red	RRR	1
2R, 1G	RRG, RGR, GRR	3
1R, 2G	RGG, GRG, GGR	3
All Green	GGG	1

(b)

Result	Possible combinations	Total
All Red	RRRRR	1
4R, 1G	RRRRG, RRRGR, RRGRR, RGRRR, GRRRR	5
3R, 2G	RRRGG, RRGRG, RGRRG, GRRRG, RRGGR, RGRGR, GRRGR, RGGRR, GRGRR, GGRRR	10
2R, 3G	GGGRR, GGRGR, GRGGR, RGGGR, GGRRG, GRGRG, RGGRG, GRRGG, RGRGG, RRGGG	10
1R, 4G	GGGGR, GGGRG, GGRGG, GRGGG, RGGGG	5
All Green	GGGGG	1 ◊

---

51. A house loses energy through the exterior walls and roof at a rate of 5 000 J/s = 5.00 kW when the interior temperature is 22.0 °C and the outside temperature is –5.00 °C. Calculate the electric power required to maintain the interior temperature at 22.0 °C for the following two cases: (a) The electric power is used in electric resistance heaters (which convert all of the electricity supplied to internal energy). (b) The electric power is used to drive an electric motor that operates the compressor of a heat pump (which has a coefficient of performance [COP] equal to 60.0% of the Carnot-cycle value).

Solution

G: The electric heater should be 100% efficient, so P = 5 kW in part (a). It sounds as if the heat pump is only 60% efficient, so we might expect P ≅ 9 kW in (b).

O: Power is the change of energy per unit of time, so we can find the power for both cases by examining the change in heat energy.

A: (a) We know that $P_{electric} = \Delta E_{int}/\Delta t$, so if all of the electricity is converted into thermal energy, $\Delta E_{int} = \Delta Q$.

Therefore,
$$P_{electric} = \frac{\Delta Q}{\Delta t} = 5000 \text{ W} \qquad \Diamond$$

(b) For a heat pump,
$$(COP)_{Carnot} = \frac{T_h}{\Delta T} = \frac{295 \text{ K}}{27.0 \text{ K}} = 10.92$$

$$\text{Actual COP} = (0.600)(10.92) = \frac{Q_h}{W} = \frac{Q_h/t}{W/t}$$

Therefore, to bring 5000 W of heat into the house only requires input power

$$P_{heat\ pump} = \frac{W}{t} = \frac{Q_h/t}{COP} = \frac{5000 \text{ W}}{6.56} = 763 \text{ W} \qquad \Diamond$$

L: The result for the electric heater's power is consistent with our prediction, but the heat pump actually requires far **less** power than we expected. Since both types of heaters use electricity to operate, we can now see why it is more cost effective to use a heat pump even though it is less than 100% efficient!

53. Figure P22.53 represents n mol of an ideal monatomic gas being taken through a cycle that consists of two isothermal processes at temperatures $3T_i$ and T_i and two constant-volume processes. For each cycle, determine, in terms of n, R, and T_i, (a) the net energy transferred by heat to the gas and (b) the efficiency of an engine operating in this cycle.

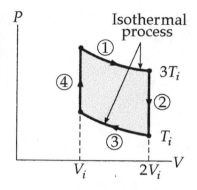

Figure P22.53

Solution

(a) For an isothermal process, $Q = nRT \ln\left(\dfrac{V_f}{V_i}\right)$

Thus, $Q_1 = nR(3T_i)\ln 2$ and $Q_3 = nR(T_i)\ln\left(\frac{1}{2}\right)$

The internal energy of a monatomic ideal gas is, from equation 16.17, $E_{int} = \frac{3}{2}nRT$. In the constant-volume processes,

$$Q_2 = \Delta E_{int,\, 2} = \frac{3}{2}nR(T_i - 3T_i) \quad \text{and} \quad Q_4 = \Delta E_{int,\, 4} = \frac{3}{2}nR(3T_i - T_i)$$

The net heat transferred is then

$$Q = Q_1 + Q_2 + Q_3 + Q_4 \qquad \text{or} \qquad Q = 2nRT_i \ln 2 \qquad \Diamond$$

(b) Q_h is the sum of the positive contributions to Q.

$$Q_h = Q_1 + Q_4 = 3nRT_i(1 + \ln 2)$$

Since the change in temperature for the complete cycle is zero, $\Delta E_{int} = 0$ and $W = Q$.

Therefore, the efficiency is $e = \dfrac{W}{Q_h} = \dfrac{Q}{Q_h} = \dfrac{2\ln 2}{3(1 + \ln 2)} = 0.273 = 27.3\% \qquad \Diamond$

63. One mole of a monatomic ideal gas is taken through the cycle shown in Figure P22.63. At point A, the pressure, volume, and temperature are P_i, V_i, and T_i, respectively. In terms of R and T_i, find (a) the total energy entering the system by heat per cycle, (b) the total energy leaving the system by heat per cycle, (c) the efficiency of an engine operating in this cycle, and (d) the efficiency of an engine operating in a Carnot cycle between the same temperature extremes.

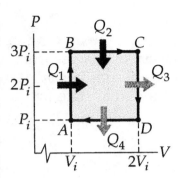

Figure P22.63

Solution At point A, $P_iV_i = nRT_i$,   with $n = 1$

At point B, $3P_iV_i = RT_B$,   and $T_B = 3T_i$

At point C, $(3P_i)(2V_i) = RT_C$ and $T_C = 6T_i$

At point D, $P_i(2V_i) = RT_i$ and $T_D = 2T_i$

The heat transfer for each step in the cycle is found using

$$C_V = \frac{3R}{2} \qquad \text{and} \qquad C_P = \frac{5R}{2}$$

$$Q_{AB} = C_V(3T_i - T_i) = 3RT_i \qquad Q_{BC} = C_P(6T_i - 3T_i) = 7.5RT_i$$

$$Q_{CD} = C_V(2T_i - 6T_i) = -6RT_i \qquad Q_{DA} = C_P(T_i - 2T_i) = -2.5RT_i$$

Therefore,   (a) $Q_{in} = Q_h = Q_{AB} + Q_{DA} = 10.5RT_i$ ◊

(b) $Q_{out} = Q_c = |Q_{CD} + Q_{DA}| = 8.5RT_i$ ◊

(c) $e = \dfrac{Q_h - Q_c}{Q_h} = 0.190 = 19\%$ ◊

(d)   Carnot Efficiency, $e_C = 1 - \dfrac{T_C}{T_h} = 1 - \dfrac{T_i}{6T_i} = 0.833 = 83.3\%$ ◊

65. A system consisting of n mol of an ideal gas undergoes a reversible, **isobaric** process from a volume V_i to a volume $3V_i$. Calculate the change in entropy of the gas. (**Hint:** Imagine that the system goes from the initial state to the final state first along an isotherm and then along an adiabatic path—no change in entropy occurs along the adiabatic path.)

Solution

The isobaric process (AB) is shown along with an isotherm (AC) and an adiabat (CB) in the PV diagram.

Since the change in entropy is path independent,

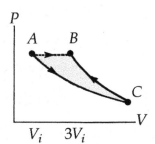

$$\Delta S_{AB} = \Delta S_{AC} + \Delta S_{CB}$$

But because (CB) is adiabatic, $\Delta S_{CB} = 0$.

For isotherm (AC), $\qquad P_A V_A = P_C V_C$

For adiabat (CB), $\qquad P_C V_C{}^{\gamma} = P_B V_B{}^{\gamma}$

Combining these gives

$$V_C = \left(\frac{P_B V_B{}^{\gamma}}{P_A V_A}\right)^{1/(\gamma-1)} = \left[\left(\frac{P_i}{P_i}\right)\frac{(3V_i)^{\gamma}}{V_i}\right]^{1/(\gamma-1)} = 3^{\gamma/\gamma-1}V_i$$

Therefore, $\quad \Delta S_{AC} = (nR)\ln\left(\frac{V_C}{V_A}\right) = nR\ln\left[3^{(\gamma/\gamma-1)}\right] = \frac{nR\gamma \ln 3}{\gamma - 1}$ $\qquad\qquad \Diamond$

67. An idealized Diesel engine operates in a cycle known as the **air-standard Diesel cycle,** shown in Figure 22.13. Fuel is sprayed into the cylinder at the point of maximum compression, B. Combustion occurs during the expansion $B \rightarrow C$, which is approximated as an isobaric process. Show that the efficiency of an engine operating in this idealized Diesel cycle is

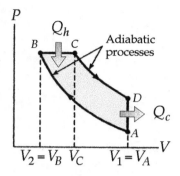

Figure 22.13

$$e = 1 - \frac{1}{\gamma}\left(\frac{T_D - T_A}{T_C - T_B}\right)$$

Solution

The heat transfer over the paths CD and BA are zero since they are adiabats.

Over path BC: $\qquad Q_{BC} = nC_P(T_C - T_B) > 0$

Over path DA: $\qquad Q_{DA} = nC_V(T_A - T_D) < 0$

Therefore, $\qquad Q_c = |Q_{DA}| \qquad$ and $\qquad Q_h = Q_{BC}$

Hence, the efficiency is

$$e = 1 - \frac{Q_c}{Q_h} = 1 - \left(\frac{T_D - T_A}{T_C - T_B}\right)\frac{C_V}{C_P}$$

$$e = 1 - \frac{1}{\gamma}\left(\frac{T_D - T_A}{T_C - T_B}\right) \qquad\qquad \Diamond$$